ADVANCES IN AUTOMATION AND ROBOTICS

Theory and Applications

Volume 1 • 1985

to Youla

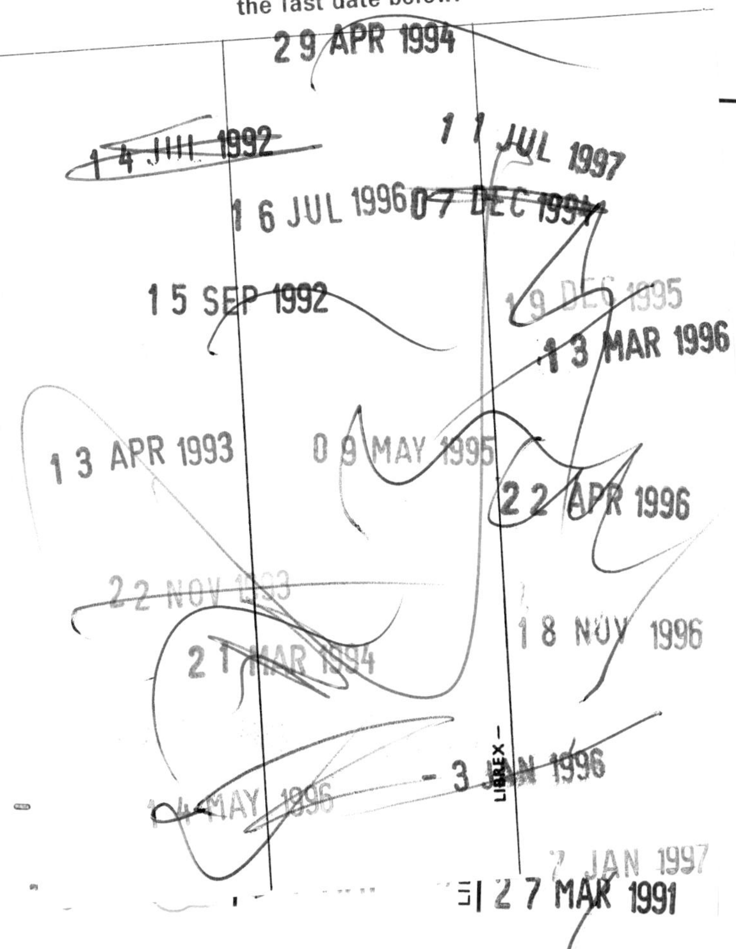
29 APR 1994
14 JUL 1992
11 JUL 1997
16 JUL 1996
07 DEC 1994
15 SEP 1992
19 DEC 1995
13 MAR 1996
13 APR 1993
09 MAY 1995
22 APR 1996
22 NOV 1993
18 NOV 1996
21 MAR 1994
14 MAY 1996
- 3 JAN 1996
LIBREX
27 MAR 1991

ADVANCES IN AUTOMATION AND ROBOTICS

Theory and Applications

Editor: G.N. SARIDIS
Director, Robotics and Automation Laboratory
Department of Electrical, Computer and Systems Engineering
Rensselaer Polytechnic Institute

VOLUME 1 • 1985

Greenwich, Connecticut *London, England*

36 Sherwood Place
Greenwich, Connecticut 06836

JAI PRESS LTD.
3 Henrietta Street
London WC2E 8LU
England

ISBN: 0-89232-399-X

Manufactured in the United States of America

CONTENTS

LIST OF CONTRIBUTORS vii

PREFACE
G.N. Saridis ix

Chapter 1
INTELLIGENT CONTROLS FOR ROBOTICS AND ADVANCED AUTOMATION
G.N. Saridis 1

Chapter 2
ROBOT ARM KINEMATICS AND DYNAMICS
C.S.G. Lee 21

Chapter 3
ON THE ROBOTIC MANIPULATOR CONTROL
C.S.G. Lee 65

Chapter 4
THE EVOLUTION OF ROBOT MANIPULATOR CONTROL
R.P. Paul 117

Chapter 5
ROBOTIC VISION
J. Mundy 141

Chapter 6
SHAPE FROM TOUCH
R. Bajcsy 209

Chapter 7
VEHICULAR LEGGED LOCOMOTION
R.B. McGhee 259

INDEX 285

LIST OF CONTRIBUTORS

R. Bajcsy — Department of Computer and Information Science, University of Pennsylvania

C.S.G. Lee — Department of Electrical and Computer Engineering, University of Michigan

R.B. McGhee — Department of Electrical Engineering, Ohio State University

J. Mundy — Corporate Research and Development, General Electric Company

R.P. Paul — Department of Computer and Information Science, University of Pennsylvania

G.N. Saridis — Department of Electrical, Computer and Systems Engineering, Rensselaer Polytechnic Institute

PREFACE

This is the first volume of a series of research annuals on *Advances in Automation and Robotics* published by JAI Press, Inc. The purpose of this series is to provide state-of-the-art information about a field that has been only recently recognized and has been experiencing exponential growth since. Such a situation should create severe problems in the selection of the proper material to be reported in a specific publication of high caliber. However, it is fortunate that such a problem never arose in the compilation of this volume. This is mainly due to the knowledge and expertise of my coauthors that made the task of editing this book a scientific experience.

Robotics and automation was not always a scientific discipline. In fact, industrial automation was introduced during World War II when many manufacturers were forced to accommodate wartime mass production. At first there were machines programmed to perform, repetitively, several tasks with minimal intervention of a human operator. Most of these machines could be found in machine shops or were parts of a semiautomated assembly line in automobile and appliance factories during the years after the war.

Industrial robots were developed during the 1950s, and their first units were produced in the early 1960s by companies like Unimation, Inc. and AMF Versatran. But what is a robot? The word, derived from the Czech *robota* (work), was first used by the Czech writer Karel Čapek in 1921 in his play *R.U.R.* (Rossum's Universal Robots) to describe machines created to replace humans. The official definition given by the Robot Institute of America is "a programmable, multifunction manipulator designed to move material, parts, tools, or specific devices through variable programmed motions for the performance of a variety of tasks." In this sense, there is a wide range of devices created to replace the human worker in hazardous and tedious tasks. As such, all automated industrial machines qualify as primitive robots.

The real revolution in robots and industrial automation took place in the late 1960s when fixed mechanical programming was replaced by a flexible digital or numerical "computer program." Numerical control (NC) machines replaced most semiautomatic machining and milling systems, and robots were developed for such tedious tasks as spot and arc welding, materials handling, painting, and so on. These machines, even though originally developed in the United States, were adapted and coordinated into the industrial mainstream by Japan during their 10-year automatization program which started in 1970. This program turned out to be so successful that in the 1980s most industrial countries in the world are modifying their manufacturing procedures around a fully automated factory to create the "Robotic Revolution." Of course, the center of such a revolution is the digital computer and the continuously evolving integrated electronics technology which made all this possible. What really made the modern robots and automated manufacturing systems different from their brothers and sisters is their flexibility and versatility, due to their reprogrammability. This is the direction of current robotic research: to develop machines to handle various human tasks in hazardous, unpleasant, or unfriendly environments.

At the present time, the problem faced by all researchers in the area of robotics and automation is the lack of a scientifically systematic approach to design robots. Such machines, endowed with one or two arms for manipulation, gripping, and tool handling and possible locomotion, have been manufactured by the heuristic principles of their ancestors, the remote manipulator and the NC machine. They may fill present industrial needs, but they definitely do not represent the prototypes of the future in pace with modern computer, integrated electronics, and control systems technologies.

This series will attempt to fill this gap and thus provide service to this growing robotics technology. This first volume presents a survey of the fundamental problems facing modern robotic systems. Such systems are cast in the intelligent machine framework which requires, in addition to high-level basic computational capability, advanced sensing, control, and locomotion systems. The chapters of this book attempt to meet the theoretical needs of such an advanced system. Chapter 1, written by me, presents hierarchically intelligent control theory as an integrating approach to the design of modern robotic systems. Chapter 2, by C. S. G. Lee, discusses the most recent formulation of kinematics and dynamics for robotic manipulators. Chapter 3, also by C. S. G. Lee, covers thoroughly the various control methods used in modern robotics. Chapter 4, by R. P. Paul, discusses the evolution of several programming languages that are associated with robots. Chapter 5, by J. Mundy, introduces the important subject of vision for sensing and feedback in robotic manipulators. Chapter 6, by R. Bajcsy, presents touch sensing as another device for understanding the environment for a robotic system. Finally, Chapter 7, by R. B. McGhee, deals with legged locomotion as an advanced technique for propelling modern robots.

The whole book covers but does not exhaust the basics of modern robotic manipulators. Future volumes will specialize in other aspects of this highly exciting discipline. From this position, though, I would like to express my gratitude to all my coauthors for their prompt response in producing such excellent contributions to this volume, Mrs. Sharon Sorell for typing parts of the manuscript, and finally the National Science Foundation for supporting my work and the work of some of my coauthors for several years, the results of which made this book possible.

G. N. Saridis
Series Editor

Chapter 1

INTELLIGENT CONTROL FOR ROBOTICS AND ADVANCED AUTOMATION

G. N. Saridis

1. INTRODUCTION

In the past few years, we have been experiencing the Robotic Revolution which has been affecting the ways of industrial production. Since then, the technological image of industrial and other commercial robots has been drastically changed. The reason for this change is that robots are and will play the key role in the new industrial and research environments.

The new robots must have the ability to operate in a human-made environment, sense its details, and provide information back for processing in order to execute various tasks with minimum interaction with a human operator [40]. For this purpose, it must be endowed with arms of flexible mechanical structure in order to manipulate beyond physical obstacles, a powerful vision system for object recognition and tracking, tactile and other sensing for precise handling, adaptable hardware controls for flexible task execution, and

Advances in Automation and Robotics, Volume 1, pages 1–20.

ISBN: 0-89232-399-X

legged locomotion for movement in rough terrains. But above all, it must have the machine intelligence to organize, coordinate, and execute complex diversified tasks without continuous supervision by an operator. With our modern technology and especially the evolution of the digital computer, such a robot is not a storywriter's dream but an intelligent machine of the near future. While the next chapters will discuss the details of the components of such a structure, this one will elaborate on the design and implementation of the machine intelligence for the robots of the future.

Various methodologies such as control systems theory, operations research, and artificial intelligence have been recently dealing with different aspects of

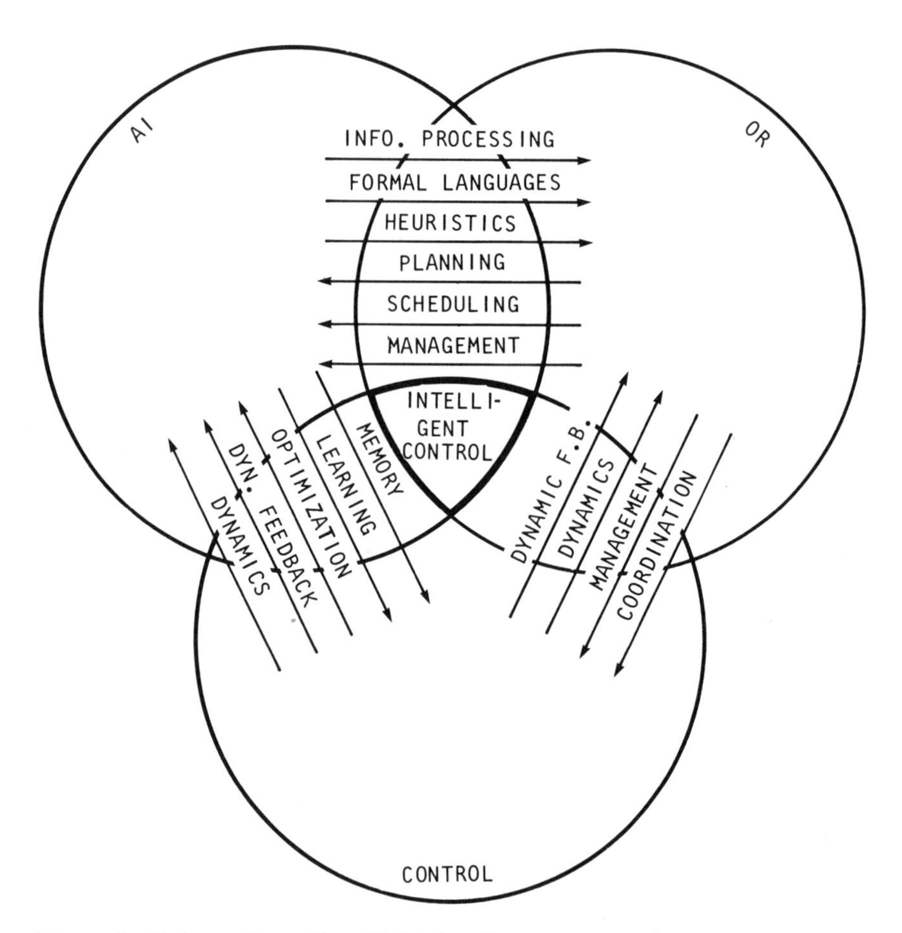

Figure 1. Intersection of artificial intelligence, operations research, and control theory and the resulting intelligent control.

machine intelligence and they may lead to the creation of intelligent control theory, as in Figure 1.

Such a discipline requires a rigorous mathematical modeling and subsequent analysis of the associated physical process and a systematic synthesis of precise controls resulting in the design and effective operation of industrial, economic, urban, and even space exploration systems that have become essential parts of the socioeconomic environment of the modern societies as in systems theory and operations research [28,38]. As the world economy is reaching a turning point due to the depletion of certain popular energy resources while higher demands are imposed from space exploration, work in hazardous environments, modernization of industrial plans, and efficient transportation of large groups of people, new methodologies are developed suitable for computer utilization and demonstrating advanced machine intelligence and decision [40]. For over 20 years, scientists have been developing the cognitive field of artificial intelligence, more or less in the image of a human brain. Significant results have been accomplished in speech recognition, image analysis and perception, data base analysis and decision making, learning, theorem proving and gains, autonomous robots, and so forth [1,8,9,13,14,16,20,23,29,30,34,35,37,49,50,51,54,55]. The discipline that couples these advanced methodologies with the system theoretic approached necessary for the solution of the current technological problems of our societies is called *intelligent controls* [12,40].

Intelligent control studies utilize the powerful high-level decision making of the digital computer with advanced mathematical modeling and synthesis techniques of system theory to produce a unified approach suitable for the engineering needs of the future.

One of the most important applications of intelligent control theory is in manipulative systems. These systems may involve the control of a general-purpose manipulator for space exploration, like the Mars-rover, or a hazardous environment robot for operation in a nuclear containment, or a hospital aid manipulator, an electrically driven prosthetic limb to replace an amputated arm or even an orthotic brace to assist paralyzed people [6,10,11,22,24,41,47,52]. Such devices impose special considerations and constraints in terms of small weight, small physical dimensions, real-time performance, human limb appearance and functionality, and most restrictive, a small number of noninteracting command sources, that is, of a command vocabulary and a small number of sensors. The above constraints exclude computationally complex algorithms or long computation time. Also, training of the operator to generate combinatorial command codes must be very limited. Hence, such systems must maximize flexibility of performance subject to a minimal input dictionary and minimal computational complexity.

This chapter will consider the general theory of intelligent controls first and then its application to general-purpose robotic manipulators.

2. COGNITIVE SYSTEMS ENGINEERING AND ARTIFICIAL INTELLIGENCE

Cognitive systems have been traditionally developed as part of the field of artificial intelligence to implement, on a computer, functions similar to the ones encountered in human behavior. Such functions as speech recognition and analysis, image and scene analysis, data base organization and dissemination, learning and high-level decision making have been based on methodologies emanating from simple logic operations to pattern recognition, linguistic, and fuzzy set theory approaches [56]. The results have been well documented in the literature [1,12,23,33,37,54,55].

In order to solve the modern technological problems that require control systems with intelligent functions such as simultaneous utilization of a memory, learning, or multilevel decision making in response to "fuzzy" or qualitative commands, a new generation of control systems have been developed. They are termed *intelligent controls* and utilize the results of cognitive systems' research effectively with various mathematical programming control techniques [28,38]. Each cognitive system associated with the specific process under consideration may be considered a subtask of the process requested by an original general qualitative command, programmed by a special high-level symbolic computer language, and sequentially executable along with decision making and control of the hardware part of the process.

Many systems have been designed to perform in the above manner. In the area of manipulators and robotics, many such systems have been developed for object handling in an industrial assembly line, remote manipulation in hazardous environments, the planet-exploration Mars vehicle, hospital aids for the disabled, and autonomous robots [2–4,11,27,33,36,42,53].

In most cases where the control process is remotely performed from the operator, its function is semiautonomous and the system must utilize some cognitive systems to understand the task requested to execute, identify the environment, and then choose the best plan to execute the task.

Various pattern recognition, linguistic, or even heuristic methods have been used to analyze and classify speech, images, or other information coming in through sensory devices as part of the cognitive system [1,5,6,23,24, 37]. Decision making and motion control have been performed by a dedicated digital computer using either kinematic methods, like trajectory tracking, or dynamic methods based on compliance, dynamic programming, or even approximately optimal and adaptive control [45,46].

Artificial intelligence has definitely provided significant contributions to the development of cognitive engineering. It always utilizes large-size mainframe computers to provide solutions to intellectual problems related to human intelligence. Vision, and other sensory systems as well as speech recognition and understanding are essential for intelligent techniques. Furthermore, manipulative and autonomous robotic systems have also been created with

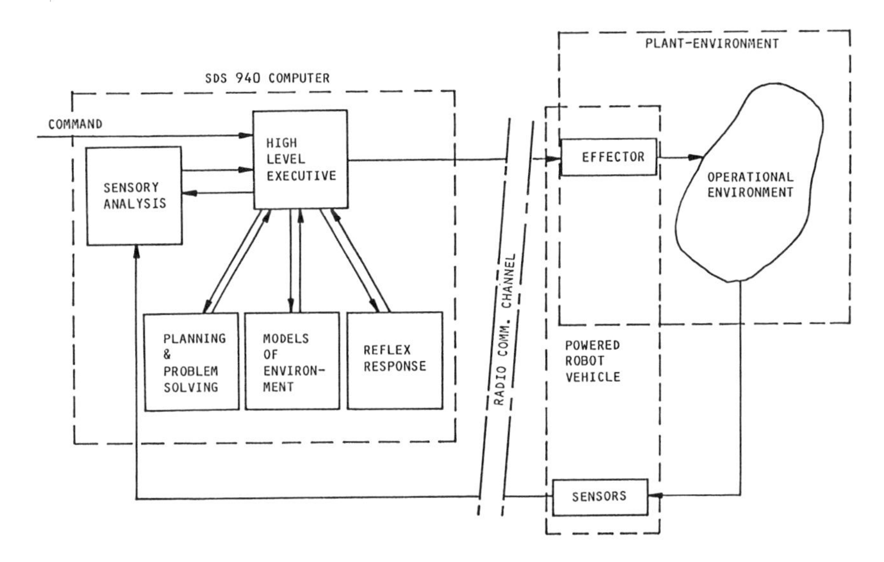

Figure 2. Artificial intelligent implementation of a robotic arm control system. This SRI's "Shakey Robot."

the use of artificial intelligence techniques. An example of robotic controls based on artificial intelligence concepts is given in Figure 2 [40].

However, modern intelligent robots require a more modest computer implementation in order to be practically feasible [12]. Machine intelligence,

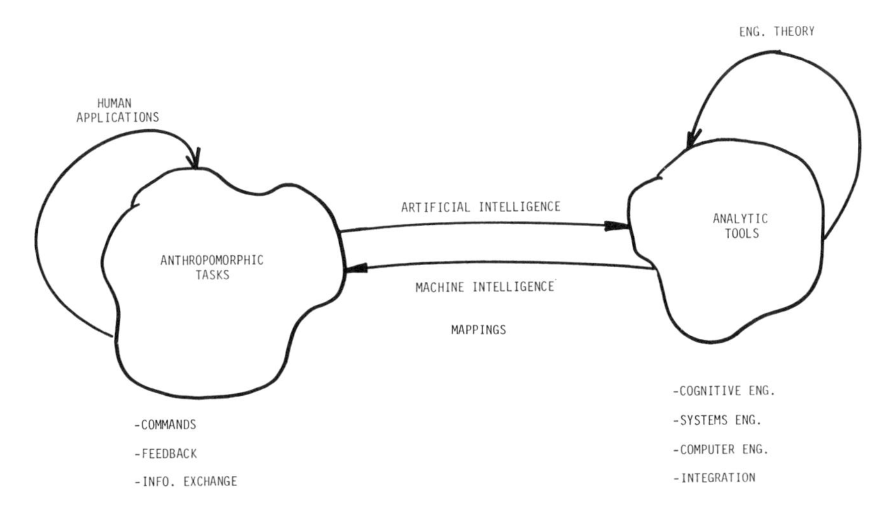

Figure 3. The problem of designing machines to execute anthropomorphic functions.

which may be thought of as the inverse mapping of artificial intelligence,—for example, the mapping from the theoretical space of engineering science to the space of human understanding and applications, shown in Figure 3—was assumed more suitable for the design of self-supported robots. The reason is that by using machine-generated intelligence and communications in a minimally man–machine interactive system, one may obtain more efficient, faster, and smaller-size computer systems to control an intelligent robot.

3. HIERARCHICALLY INTELLIGENT CONTROL THEORY

A hierarchically intelligent control approach has been proposed by Saridis as a unified theoretic approach of cognitive and control systems methodologies. The control intelligence is hierarchically distributed according to the principle of *decreasing precision with increasing intelligence* evident in all hierarchically management systems [40,41]. Such systems are composed of three basic levels of controls, although each level may contain more than one layer of tree-structured functions:

- The organization level
- The coordination level
- The hardware control level

The *organization level* is the mastermind of such a system. It accepts and interprets the input commands and related feedback from the system, defines the task to be executed, and segments it into subtasks in their appropriate order of execution. An appropriate subtask library and a learning scheme for continuous improvement provide additional intelligence to the organizer. Since the organization level takes place on a medium- to large-size computer, appropriate *translation and decision schemata* linguistically implement the desirable functions [21,41].

The *coordination level* receives instructions from the organizer and feedback information from the process for each subtask to be executed and coordinates the execution at the lowest level. The coordinator, composed usually of a decision-making automaton representing a context-free language, may assign both the performance index and end condition as well as possible penalty functions designed to avoid inaccesible areas in the space of the motion. The decisions of the coordinator are obtained with the aid of a performance library and a learning decision schema, recursively updated to minimize the cost of operation.

A *lowest-level control* process usually involves the execution of a certain motion and requires besides the knowledge of the mathematical model of the process the assignment of end conditions and a performance criterion or cost function defined by the coordinator. Optimal or approximately optimal control system theory may be used for the design of the lower-level controls of decentralized subprocesses of the overall process to be controlled [45].

The method has been successfully applied to control a general-purpose manipulator with visual feedback and voice inputs for effective end-point control tasks [31] at Purdue University's Advanced Automation Research Laboratory. Figure 4 depicts the above system. Other implications may be found in Refs. [40,46,47].

The success of implementation of such a control system depends greatly on the development and optimization of the linguistic decision schemata for high-level decision making and the effective application of modern control techniques for the optimal operation of the lowest-level control.

Hence, we have a linguistic decision schema:

$$D = \{N_i, N_o, \Sigma, \Delta, R, P, S\} \tag{1}$$

where N_i, N_o are the input and output nonterminal alphabets;
Σ, Δ are the input and output terminal alphabets;
R is the set of decision rules;
P is the set of associative probabilities;
S is the root of the trees.

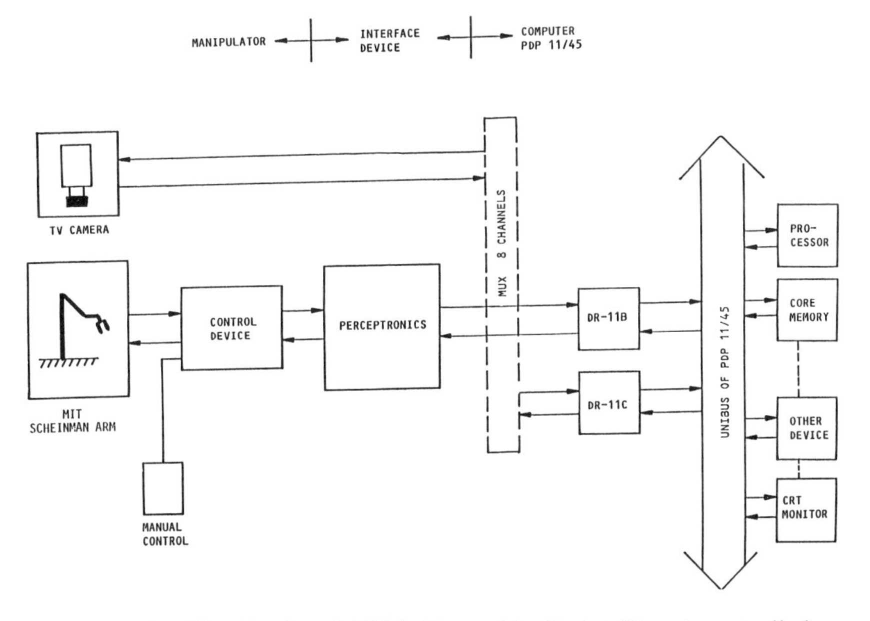

Figure 4. The Purdue AARL's hierarchically intelligent controlled robotic arm.

$$L(G_i) = \begin{Bmatrix} x_1 \\ \cdot \\ \cdot \\ \cdot \\ x_n \end{Bmatrix} \xrightarrow{P} \bigcup_{j=1}^{\ell} L(G_{oj}) = \begin{Bmatrix} y_{11} \\ \cdot \\ \cdot \\ \cdot \\ y_{1m_j} \end{Bmatrix} \cdots \begin{Bmatrix} y_{\ell 1} \\ \cdot \\ \cdot \\ \cdot \\ y_{\ell m_\ell} \end{Bmatrix}$$

Figure 5. The decision schema $D = \{N_i, N_o, \Sigma, \Delta, R, P, S\}$

This is a software device that maps a string from an input language $L(G_i)$ to each possible string belonging to one or more output languages $L(G_{oj})$ as depicted in Figure 5. To each mapping there is associated an index p_{ijk} which may be used to select the proper output string related to the particular input string. A rule for a unique association is obtained for a given input by defining p_i^* to be the highest value among p_{ijk} corresponding to the highest probability to minimize a cost function related to the process under consideration:

$$P^* = \{p_i^* = p_{ijk}, x_i \xrightarrow{P_i^*} y_{jk}/p_{ijk} = \max_{q,r} p_{iqr}, \qquad i = 1, \ldots, n\} \tag{2}$$

where $x_i \in L(G_i)$ is an input string and $y_{jk} \in L(G_{oj})$ is an output string. The sequential updating of the p_{ijk} can be obtained through two stochastic approximation algorithms [30]:

$$\begin{aligned} p_{ijk}(t+1) &= p_{ijk}(t) \pm \gamma_{ijk}(t+1)[\xi_{ijk}(t) - p_{ijk}(t)] \\ J_{ijk}(n_{ijk}+1) &= J_{ijk}(n_{ijk}) + \beta_{ijk}(n_{ijk}+1)[C(n_{ijk}+1) - J_{ijk}(n_{ijk})] \\ \xi_{ijk}(t) &= \begin{cases} 1 & \text{if } J_{ijk} = \min_{q,r} J_{iqr} \\ 0 & \text{otherwise} \end{cases} \end{aligned} \tag{3}$$

where J_{ijk} is the performance estimate, C is the observed cost; n_{ijk} is the number of occurrence of the event (x_i, y_{jk}); and $\gamma_{ijk}(t+1)$ and $\beta_{ijk}(n_{ijk}+1)$ are sequences satisfying Dvoketcky's condition for convergence of the algorithms.

Several special decision schemata and associated decision codes have been proposed in Ref. [21]. A translation schema is a special case of the above shown in Figure 6. The earliest implemented decision schema is the so-called *vocabulary optimal decision schema* in which the input and output languages retain the same sytax—for examples, structural form—but there are several

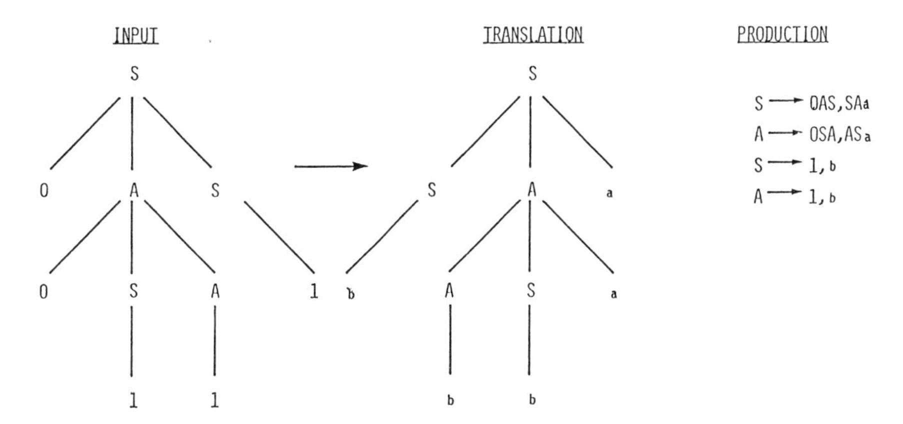

Figure 6. Production trees of the syntax-directed translation 00111 → bbbaa.

output terminal vocabularies to be selected for different tasks (Figure 7). This decision schema was thought most suitable to implement the coordination of the general-purpose manipulator.

Finally, decision schemata may be organized in a multilevel tree structure in order to implement a high-level decision maker. Information theoretic methods [7] are applicable in order to analyze the flow of information through the structure with decision schemata at the nodes using *entropy* as the cost function because of its accumulative property at each level. A block

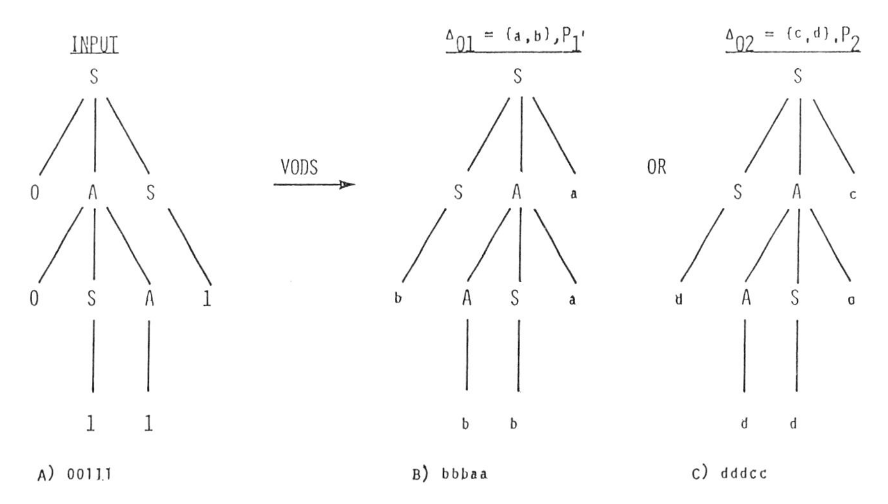

Figure 7. Example of vocabulary optimal decision schema where 00111 → bbbaa if $P_1 > P_2$ or 00111 → dddcc if $P_2 > P_1$.

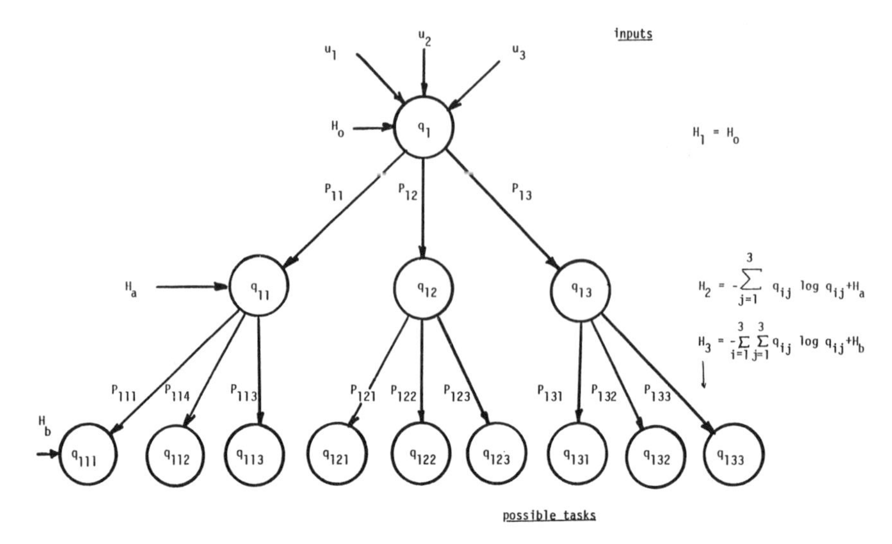

Figure 8. Organization of a linguistic decision tree using decision schemata at the nodes and entropies max H_k.

diagram of such a decision structure is given in Figure 8. Minimization of the entropy would yield the optimum path along the tree and therefore the optimal decision for executing a task on-line. Sequential minimization at the nodes provides a learning capability to this system.

4. A HIERARCHICAL INTELLIGENT CONTROL FOR A GENERAL-PURPOSE MANIPULATOR

A computer-controlled manipulator system consists of a general-purpose digital computer with appropriate peripheral devices, a seven-degrees-of-freedom Unimation PUMA 600 arm, as shown in Figure 9 (three for positioning the wrist, three for the orientation of the hand, and one for opening and closing the "fingers"), a CID digitizing camera, external tactile sensors for the arm, and an interfacing device between the arm and the computer [46].

In order to accomplish a complex manipulative task, a computer-controlled manipulator system and its control algorithm must show the following capabilities:

a. Man–manipulator communications—recognizes the linguistic commands from the operator and interacts with him
b. Coordinated motion control—possesses some levels of autonomous coordinated position and rate control without the assistance of the operator; that is, the operator is taken out of the control loop

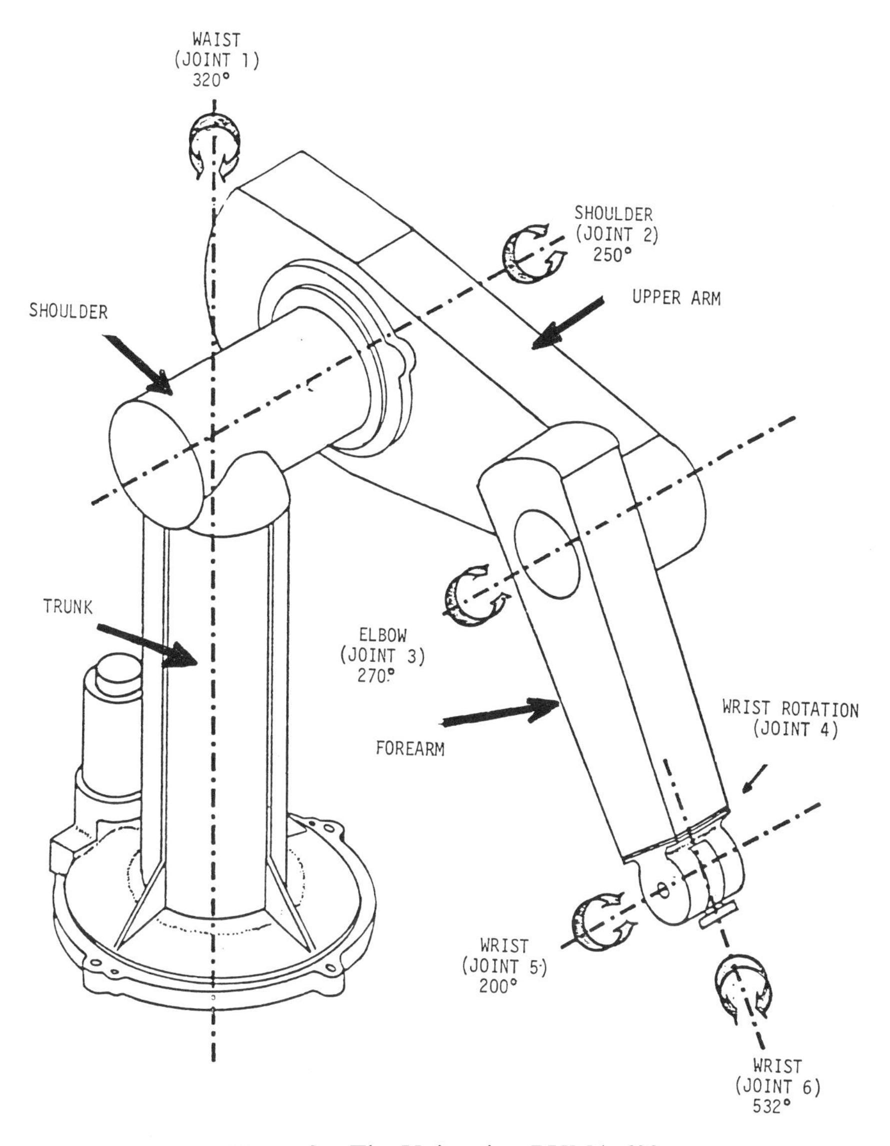

Figure 9. The Unimation PUMA 600.

c. Interaction with the environment—the ability to integrate the feedback signals from television camera and other external sensors into the system and update its strategies or sequences of control actions to accomplish the task

Hierarchically intelligent control is mostly suitable to control such a general-purpose manipulator. A diagram of a hierarchically intelligent control system for the PUMA arm developed at the RPI Robotics and Automation Laboratory is depicted in Figure 10. According to the principle of *increasing intelligence with decreasing precision* [46,47], the lowest level in the hierarchy, the run time control level, must execute a local task with high precision by satisfying certain local performance criteria, thus requiring a rather sophisticated and precise model. These models of the arm subsystems are formulated from the state space approach. The next higher level in the hierarchy is that of the coordinator, where the individual subtasks are put to work together by appropriately selecting their performance evaluation and modes of operations requiring less precision and higher-level decision making to improve the overall performance of the system. In addition to coordinating and supervising the decision-making units in the lower level, the coordinator should be endowed with learning capability to improve the system's performance under reappearing control situations. An automaton capable of executing fuzzy or stochastic inputs can be implemented to perform such a learning function. Finally, the organization level which serves as a linguistic organizer at the top of the hierarchy also possesses certain learning capabilities and

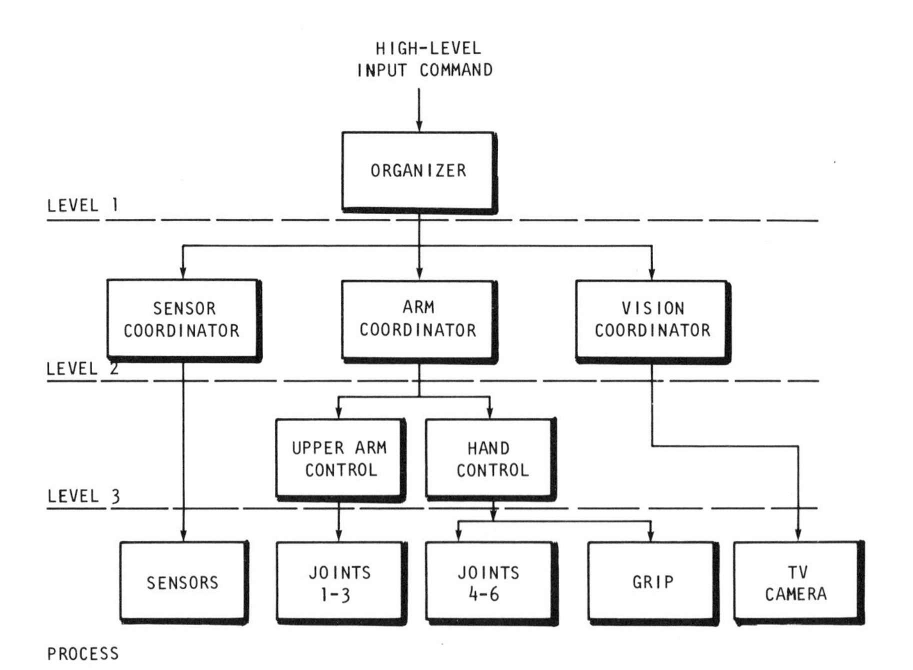

Figure 10. Hierarchically intelligent control system for the Unimation PUMA 600.

decision making. The highest-level decision making involves parsing the stochastic linguistic input strings, organizing the task, identifying the control situation, and assigning the appropriate control pattern without much knowledge of the detailed execution of the task.

4.1 The Approximately Optimal Control of the Mechanical Arm

The decision-making units in the lowest level of the hierarchy require precise models of the process in order for them to execute the task properly [46]. These models are obtained from modeling the subsystems of the arm. The arm is subdivided into two main subsystems, namely, the *wrist subsystem* and the *hand subsystem.* The motion of the hand is modeled by three independent separate motions which are described by a set of differential equations. A suboptimal controller is used for the coordination of the motion of the upper three joints of the arm $(\theta_1, \theta_2, \theta_3)$ which constitute the positioning system of the wrist. This subsystem transports the arm from any initial position in the work space to a prespecified terminal position. The lower three joints of the arm $(\theta_4, \theta_5, \theta_6)$ constitute the orientation system of the hand which aligns the grip position according to the specific orientation of the object.

The state space approach is used throughout to formulate the control problem for the wrist subsystem. The dynamic equations of the motion of the wrist are derived from the generalization Euler–Lagrange equations [3]. They are expressed in angular position and angular velocity coordinates

$x^T = [x_1, x_2, x_3, x_4, x_5, x_6] = [\theta_1, \theta_2, \theta_3, \dot{\theta}_1, \dot{\theta}_2, \dot{\theta}_3]$:

$$x(t) = Fx(t) + \begin{bmatrix} 0 \\ J^{-1}(x)N(x) \end{bmatrix} + \begin{bmatrix} 0 \\ J^{-1}(x) \end{bmatrix} u(t) \tag{4}$$

where $J(x)$ is the inertia matrix and $N(x)$ is the vector representing the nonlinear centripetal and coriolis force and friction of the arm.

The original control design [44] was based on implementing a feedback controller using the approximation theory of optimal control developed in Ref. [45]. The advantages of such an approach over the kinematic approach followed by other investigators [36] is that it provides a coordinated control action for all three degrees of freedom in the relative angular position–velocity space without repetitive reference to the real Cartesian coordinate space. In such a formulation, the associated performance index is given as

$$J_0(u) = \int_0^T [(x(t) - x^d)^T Q(x(t) - x^d) + u^T(t)u(t)]\, dt \tag{5}$$

The performance index physically represents some functions of the energy expenditure of the physical mechanical arm system. The gain of weighting matrix Q indicates how the errors of each joint angle and joint velocity are weighted or penalized in a motion. It is suggested that for each different motion of the arm, one associates a different weighting matrix Q; x^d defines the desired end position and velocity of the end point. Object avoidance may be obtained by defining enveloping surfaces and including them as penalty functions with the performance index.

The resulting suboptimal feedback control has the form

$$u(x) = -J(x)[N(x) + B^T S(x(t) - x^d)] \tag{6}$$

where x is a positive definite matrix which is selected to stabilize the systems and is adapted using the iterative procedure defined by the approximation theory.

Experimental investigation has conclusively established that the approximately optimal controller was very sensitive to parameter variations of moments of inertia and especially friction. Therefore, the model (4) of the system that assumes constant coefficients is not suitable for such a solution.

Instead of improving and thus complicating the system's model, the following hierarchical design has been proposed. An analog minor loop with a compensator built in to increase the robustness of the system is designed around each of the dc motors, obtaining measurements through torque sensors. Such a compensator may be designed using frequency domain methods [25,57]. The resulting compensated system, which will be insensitive to nonlinearity and noise variations, now could be globally controlled using the approximation optimal design procedure to retain the advantages of feedback and single-command control mentioned earlier [45]. A block diagram in Figure 11 explains the procedure. Work toward the evaluation of this approach is presently in progress at the Robotics and Automation Laboratory at RPI.

4.2 Task Coordination Level

The function of an intelligently controlled mechanical arm may be subdivided into three major tasks [47]:

a. Sensory
b. Vision
c. Mechanical motion

The first task deals with the collection of information from proximity sensors, pressure gauges, and other sensory devices [5]. The selection of the proper device and the processing of data to provide needed feedback infor-

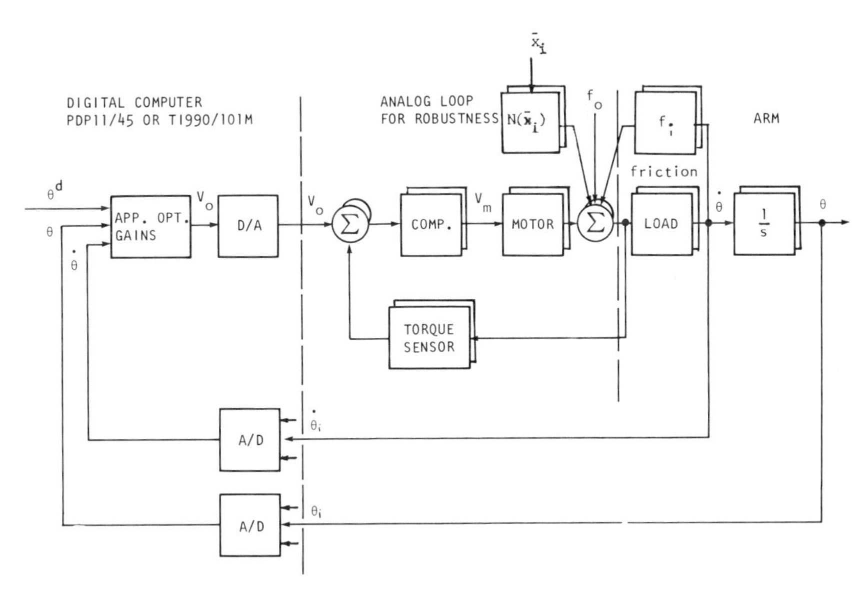

Figure 11. Hierarchically optimal and frequency domain design of arm control for servoing and robustness.

mation should be performed by an appropriate coordinator at the second level of the hierarchically intelligent control system.

The second task deals with processing visual information provided from one or more TV cameras fixed or moving with the arm [1,6]. This involves object recognition and classification for appropriate selection and end-point coordinate evaluation, object tracking using three-dimensional vision, object avoidance, and vision feedback for motion control purposes. A coordinating device may generate all close subtasks in the proper order at the second level of the hierarchically intelligent control system.

Finally, the third task deals with the selection of the proper control gains based on information about the end points and the type of motion of the upper three joints requested (for example, fast, slow, and so on), the orientation of the hand for object handling, and finally the coordination of the wrist motion with the hand orientation during special motions—for example, the transportation of a cup full of water.

A fuzzy or stochastic automaton was originally suggested to implement the coordinator. This is a finite-state machine designed to select one particular subtask from a library, using a learning (optimization) procedure to avoid external supervision in an unfamiliar environment [44,47]. Since the equivalence of such an automaton to a formal language has been established in Ref. [15], the coordinator was implemented by linguistic decision schemata

[21] in the form of software discussed in the previous section. Similar linguistic decision schemata are generated for the other tasks of the process.

Since the purpose of the coordinator is to move the upper three joints in an approximately optimal mode, orient the three degrees of freedom of the hand, and operate the gripper, it must generate an input command from position and velocity feedback and input from the organizer. This is done through a 14-string vector 10 bits long each. This vector is then parsed through a grammer G_D, where the part of the motion is specified. The output of the schema associated with the subtasks of selection of the approximate motion of the joints—for example, motion coordination or performance index for approximately optimal control—is given by

$$D = \{N, N, \Sigma, \Delta, R, P, S\} \tag{7}$$

where N = set of nonterminals = $\{M$ = motion, L = lower arm, C = code, U = upper arm, Q = qualifier, V = code, W = qualifier, J = code, S = start$\}$

Σ = input alphabet = $\{m$ = move, l = level, f = fast, o = optimal; 1, 2, 3$\}$

Δ = output alphabet = $\{$1 = level a, l = level b, p = plan a, p = plan b, m = move, 1, 2, 3$\}$;

R = decision rules;

P = probabilitie's associated with output vocabularies $\{p_1$ level a + plan a, $p_2 \sim$ level a + plan b, $p_3 \sim$ level b + plan a, $p_4 \sim$ level b + plan b$\}$.

All commands are of the form ⟨MOVE⟩⟨QUALIFIER⟩⟨CODE⟩, where qualifiers are level, fast, optimal, and code is 1, 2, 3. The probabilities p are obtained through on-line framing to minimize an overall cost and thus provide learning capabilities to the coordinator.

Similar linguistic decision schemata are generated for the other tasks of the process

4.3 The Task Organization Level

The organization level serves two major purposes; to interface with the operator and to organize the various tasks for different control and environmental situations [46].

External supervision of the manipulator has been minimized and replaced by the hierarchically intelligent controller, but it should be capable of accepting commands by a user to execute a certain job involving all the tasks pertinent to that process. Therefore, the organizer is designed to accept voice inputs from the user, decode them, and then organize the sequence of tasks necessary for their execution by providing appropriate inputs to the coordinators.

Implementation of the organizer is obtained by a syntax-directed translation schema which generates speech recognition algorithms and then another translation schema to organize the required tasks [21].

A translation schema is a simplified decision schema where the mapping from an input language is one-to-one to the single-output language, thus providing a translation of a command at the higher level to an execution sentence in the lower level. Feedback is selectively provided by the coordination level to complete the input strings. For the manipulator under consideration, the process was assumed to take place in a hospital environment as an aid to disabled patients, and therefore no learning was required at the organization level.

An English-like input language was assigned for the speech recognition and motion organization translation schema. Their associated grammars were generated to interpret these commands into the various tasks to be executed by the vision systems, mechanical arm, and sensors. Detailed information is given in Ref. [15].

5. CONCLUSIONS

This chapter has presented the early development of the hierarchically intelligent control and software organization for real-time end-point control of a general-purpose robotic manipulator. It was based on the *principle of increasing precision with decreasing intelligence* and was composed of three levels of intelligent controls. The design approach was developed to guarantee robustness of the control process through a frequency-domain-designed loop preserving the advantages of the approximately optimal feedback control approach. The introduction of high-level decision schemata and codes produced the coordinating and learning capabilities at the coordination and organization level of the hierarchy. Translation codes were used to implement speech recognition and task organizations at the organization level. Experimental work reported in Ref. [22] establishes the feasibility of implementation with microcomputers. Therefore, hierarchically intelligent control is a promising method of designing robots and qualifies as a good subject for the introductory chapter in a book on modern robotics and advanced automation.

The next chapters, written by various experts in the field, present the details for analysis and design of complete modern robotic systems by covering respectively the areas of mechanical design, control programming languages, vision and sensors, and legged locomotion. The information they provide complement each other and definitely provide more substance to the present chapter.

REFERENCES

[1] Agin, G. J., and Binford, T. O., Computer description of curved objects. *Proc. 3rd Int. Joint Conference on Artificial Intelligence*, Stanford, CA, August 1973.

[2] Bejczy, A. K., Remote manipulator system technology review. Tech. Rep. No. 760-77, Jet Propulsion Lab., Pasadena, CA, July 1972.

[3] Bejczy, A. K., New techniques for terminal phase control of manipulators. Tech. Rep. 760-98, Jet Propulsion Lab., Pasadena, CA, February 1974.

[4] Bejczy, A. K., Robot arm dynamics and control. Tech. Rep. 33-669, Jet Propulsion Lab., Pasadena, CA, February 1970.

[5] Bejczy, A. K., External state sensors and control tutorial workshop. Session 4 at *18th IEEE Conference on Decision and Control*, Ft. Lauderdale, FL, December 11, 1979.

[6] Birk, J., et al., Determining work piece orientation in a robot hand using vision. *Proc. 17th Conference on Decision and Control*, San Diego, CA, January 1979.

[7] Conant, R. C., Laws of information which govern systems. *IEEE Transactions on SMC, SMC-6*(4), 240–255, 1976.

[8] Feigenbaum, E. A., and Feldman, J. (eds.), *Computers and Thought*. New York: McGraw-Hill, 1963.

[9] Firschein, O., Fischier, M., Coles, S. L., and Tenenbaum, J. M., Forecasting and assessing the impact of artificial intelligence on society. Tech. Rep., Stanford Research Institute, Menlo Park, CA, February 1973.

[10] Freedy, A., Hull, F. C., Lucaccini, L. F., and Lyman, J., A computer-based learning system for remote manipulator control. *IEEE Trans. Systems Man and Cyber., SMC-1*(4), 356–364, 1971.

[11] Freedy, A., and Weltman, G., A prototype learning system as a potential controller for industrial robot arms. *Proc. Second International Symposium on Industrial Robots*, Chicago, IL, May 16–18, 1972.

[12] Fu, K. S., Learning control systems and intelligent control systems: An intersection of artificial intelligence and automatic control. *IEEE Trans. Automatic Control, AC-16*(1), 70–72, 1971.

[13] Fu, K. S. (ed.), *Pattern Recognition and Machine Learning*. New York: Plenum Press, 1971.

[14] Fu, K. S., *Sequential Methods in Pattern Recognition and Machine Learning*. New York; Academic Press, 1968.

[15] Fu, K. S., *Syntactic Methods in Pattern Recognition*. New York: Academic Press, 1974.

[16] Fu, K. S., Stochastic automata as models of learning systems. In J. T. Tou (ed.), *Computer and Information Sciences-II*. New York: Academic Press, 1967.

[17] Fu, K. S., On syntactic pattern recognition and stochastic languages. In S. Wantanabe (ed.), *Frontiers of Pattern Recognition*. New York: Academic Press, 1972.

[18] Fu, K. S., Learning techniques in system design—A brief review. *Fifth World Congress of IFAC*, Paris, June 12–16, 1972.

[19] Fu, K. S., and Fung, L. W., Decision making in fuzzy environment. Tech. Report TR-EE 73-22, Purdue University, West Lafayette, IN, May 1973.

[20] Fu, K. S., and McLaren, R. W., An application of stochastic automata to the synthesis of learning systems. Tech. Report TR-EE 65-17, Purdue University, West Lafayette, IN, September 1965.

[21] Graham, J. H., and Saridis, G. N. Linguistic design structures for hierarchical systems. *IEEE Trans. on SMC, SMC-12*(3), 325–333, 1982.

[22] Graupe, D., et al., A microprocessor system for multifunctional control of upper limb prostheses via EMG identification. *IEEE Trans. on Automatic Control, AC-23*, 538–544, 1978.

[23] Klinger, A., Natural language, linguistic processing and speech understanding: Recent

research and future goals. UCLA Tech. Report R-1377ARPA Report, Los Angeles, CA, December, 1973.

[24] Lee, S., and Saridis, G. N., The control of a prosthetic arm by EMG patterns classification. *IEEE Transaction on Automatic Control, AC-28*, (August), 1983.

[25] Luo, G-L, and Saridis, G. N., Robust compensation of optimal control for manipulators. *Proceedings of Conference on Decision and Control*, Orlando, FL, December 1982.

[26] Lyman, J., and Freedy, A., Summary of research activities on upper prosthesis control. *Bull Prosthetic Res., Winter*, 1973.

[27] McGhee, R. B., Control of legged locomotion systems. *Proceedings 1977, JACC*, San Francisco, June 1977.

[28] Mendel, J. M., and Fu, K. S. (eds.), *Adaptive, Learning, and Pattern Recognition Systems Theory and Applications*. New York: Academic Press, 1970.

[29] Minsky, M. L., *Artificial Intelligence*. New York: McGraw-Hill, 1972.

[30] Minsky, M. L., and Papert, S. A., Research on intelligent automata. Project MAC, Status Report II, MIT Cambridge, MA, September, 1967.

[31] Nikolic, Z. J., and Fu, K. S., An algorithm for learning without external supervision and its application to learning control systems. *IEEE Trans. on Automatic Control, AC-11*, (July), 1966.

[32] Nilsson, N. J., A mobile automaton: An application of artificial intelligence techniques. *Proc. International Joint Conference on Artificial Intelligence*, Washington, DC, May, 1969.

[33] Nilsson, N. J., Current artificial intelligence research at SRI. Oral Presentation, University of California, Los Angeles, May 1971.

[34] Nilsson, N. J., *Learning Machines*. New York: McGraw-Hill, 1965.

[35] Normal, M. F., Mathematical learning theory. In G. B. Dantzig and A. F. Veinott (eds.), *Mathematics of the Decision Sciences*. New York: Am. Math. Soc. Publications, 1968.

[36] Paul, R. P., *Robotic Manipulators*. Cambridge, MA: MIT Press, 1981.

[37] Rosen, C. A., and Nilsson, N. J., An intelligent automaton. 1967 IEEE International Convention Record, Part 9, New York, March 1967.

[38] Saridis, G. N., *Self-Organizing Control of Stochastic Systems*. New York: Marcel Dekker, 1977.

[39] Saridis, G. N., Self-organizing control and application to trainable manipulators and learning prostheses. *Proc. Sixth IFAC Congress*, Cambridge, MA, August, 1975.

[40] Saridis, G. N., Toward the realization of intelligent controls. *IEEE Proc. 67*, (8), 1979.

[41] Saridis, G. N., Graham, J. and Lee, G., An integrated syntactic approach and suboptimal control for manipulators and prostheses. *Proc. of 18th Conference on Decision and Control*, Ft. Lauderdale, FL, December, 1979.

[42] Saridis, G. N., and Lee, C. S. G., Computer controlled manipulators with visual inputs. *Optical Engineering, 18*, (5), 1979.

[43] Saridis, G. N., and Lee, C. S. G., On hierarchically intelligent control and management of traffic systems. *Computer Control of Urban Systems*, New York: ASCE Publication, 1979.

[44] Saridis, G. N., and Lee, C. S. G., Heuristic control of trainable manipulators. *Proceedings 1976 JACC*, pp. 712–716, West Lafayette, IN, July, 1976.

[45] Saridis, G. N., and Lee, C. S. G., Approximation theory of optimal control for trainable manipulators. *IEEE Transactions on SMC, SMC-8*, (3), 1979.

[46] Saridis, G. N., Intelligent robotic control. *IEEE Transactions on Automatic Control, AC-28*, (4), 1983.

[47] Saridis, G. N., and Stephanou, H. E., A hierarchical approach to the control of a prosthetic arm. *IEEE Transactions on SMC, SMC-7*, (6), 407–420, 1977.

[48] Shapiro, I. J., and Narendra, K. S., Use of stochastic automata for parameter self-optimization with multimodal performance criteria. *IEEE Transactions on Systems Science and Cybernetics, SSC-5*, (October), 1969; and Witten, I. H., Comments on 'Use of stochastic

automata for parameter self-optimization with multimodal performance criteria,' *IEEE Transactions on Systems, Man, and Cybernetics, SMC-2*, (April), 1972.

[49] Sklansky, J., Learning systems for automatic control. *IEEE Transactions Automatic Control, AC-11*, (1), 6–19, 1966.

[50] Slagle, J. R., *Artificial Intelligence: The Heuristic Programming Approach.* New York: McGraw-Hill, 1971.

[51] Tsypkin, Ya.Z., Generalized learning algorithms. *Automation and Remote Control, 1*, 86–92, 1970.

[52] Whitney, D. E., Resolved motion control of manipulators and human prosthesis. *IEEE Transactions Man-Machine Systems, MMS-10*, (2), 1969.

[53] Whitney, D. E., State space models of remote manipulation tasks. *Proc. 1969 International Joint Conference on Artificial Intelligence*, Washington, DC, 1969.

[54] Widrow, B., Generalization and information storage in networks of adaline 'neurons'. In M. C. Yovits, G. T. Jacobi, and C. D. Goldstein, (eds.), *Self-Organizing Systems.* Washington, DC: Spartan Books, 1962.

[55] Winston, P. E., Learning structural descriptions from examples. Tech. Report Project MAC TR-76, MIT, Cambridge, MA, 1970.

[56] Zadeh, L. A., Fuzzy Sets. *Inform. Control, 8*, (3), 338–355, 1965.

[57] Wu, G. H., and Paul, R. P., Manipulator compliance based on joint torque control. *Proceedings of 1980 CDC*, Albuquerque, NM, December, 1980.

Chapter 2

ROBOT ARM KINEMATICS AND DYNAMICS

C. S. G. Lee

ABSTRACT

This chapter presents the basic fundamentals in robot arm kinematics and dynamics. Kinematic equation and joint solution of a n-joint manipulator are derived. Three robot arm dynamics formulations suitable for robot arm control are briefly derived.

1. INTRODUCTION

An industrial robot is a general-purpose manipulator which consists of several rigid bodies (called links) connected in series by revolute or prismatic joints (see Figure 1). One end of the chain is attached to a supporting base while the other end is free and attached with a tool to manipulate objects or perform assembly tasks. The motion of the joints results in relative motion

Advances in Automation and Robotics, Volume 1, pages 21–63.

ISBN: 0-89232-399-X

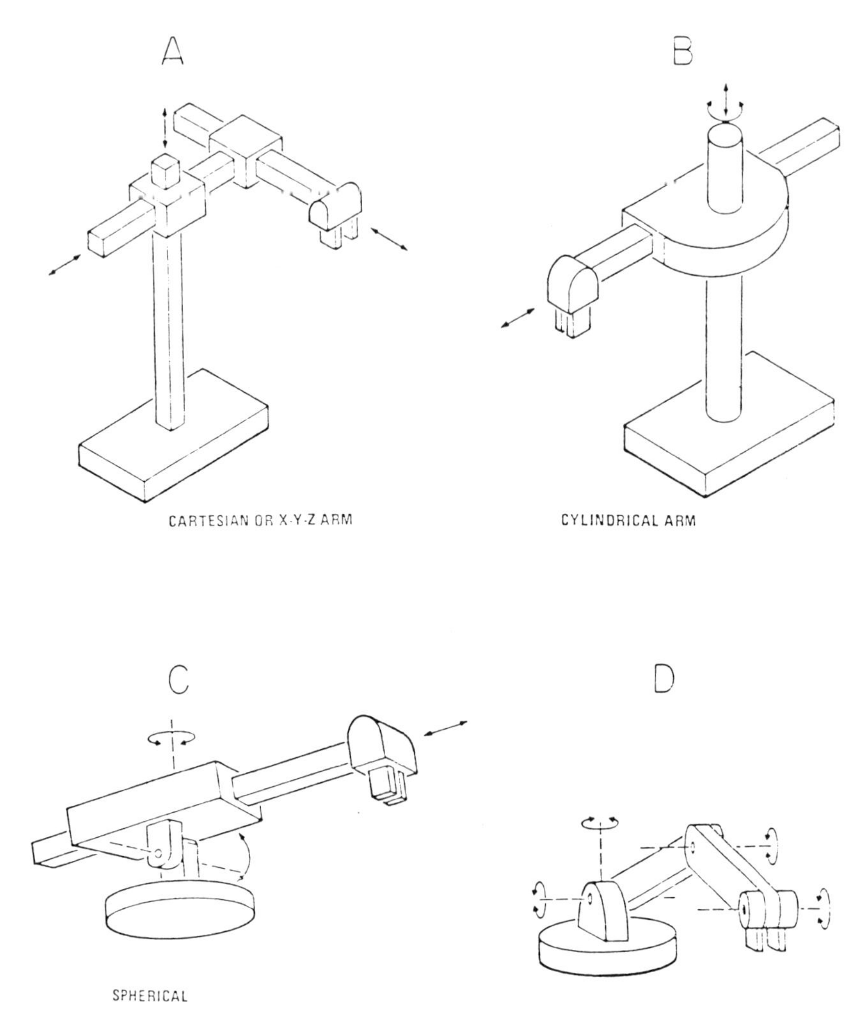

Figure 1. Various robot arm categories

of the links. Mechanically a robot is composed of an arm (or mainframe) and a wrist subassembly plus a tool. It is designed to reach a workpiece located within its work volume. The work volume is a sphere of influence of a robot whose arm can deliver the wrist subassembly unit to any point within the sphere. The arm subassembly typically consists of three-degree-of-freedom movement. The combination of the movements will place or position the wrist unit at the workpiece. The wrist subassembly unit usually consists of three rotary motions. The combination of these motions will orient the tool according to the configuration of the object to ease pickup.

Hence for a six-joint robot, the arm subassembly is the positioning mechanism, while the wrist subassembly is the orientation mechanism.

The word *robot* is derived from the Czech word *robota* meaning work. Webster's dictionary defines it as "an automatic device that performs functions ordinarily ascribed to human beings." According to this definition, a washing machine can be treated as a robot. We shall adopt a more restrictive definition used by the Robot Institute of America: "A robot is a reprogrammable multi-functional manipulator designed to move material, parts, tools, or specialized devices, through variable programmed motions for the performance of a variety of tasks." In short, a robot can be identified as a reprogrammable general-purpose manipulator with external sensors that can perform various assembly tasks. With this definition, a robot must be an autonomous system that possesses "intelligence." Such intelligence is normally due to the computer unit associated with its control system.

Presently there are many commercially available industrial robots which are widely used in simple material-handling, spot/arc-welding, and parts assembly tasks. The most widely used industrial robots are the Unimate 2000B and PUMA 260/550/560 series[1] robots developed by Unimation, Inc., the T^3 by Cincinnati Milacron, the Versatran by Prab, the ASEA robot, and the SIGMA by Olivetti of Italy. These robots exhibit their characteristics in motion and geometry. Mechanically they fall into one of the four basic motion-defining categories according to the motion of the arm subassembly (see Figure 1):

1. Cartesian coordinate (three linear axes)
2. Cylindrical coordinate (two linear axes and one rotary axis)
3. Spherical or polar coordinate (one linear axis and two rotary axes)
4. Revolute or articulated coordinate (three rotary axes)

Most automated manufacturing tasks are done by special-purpose machines which are designed to perform their prespecified functions in a manufacturing process. The inflexibility of these machines makes the computer-controlled manipulators more attractive and cost-effective in various manufacturing and assembly tasks. Present-day industrial robots, though controlled by a mini/microcomputer, are basically simple positional machines. They execute a given task by playing back prerecorded or preprogrammed sequences of motions which have been previously guided or taught by a user with a hand-held control/teach box. Moreover, the robot is equipped with little or no external sensors (both contact and noncontact) to obtain the vital information about the working environment. As a result, they are used mainly in simple material-handling tasks. Future research efforts should be directed toward improving the overall performance of the manipulator systems. One way is to study the basic fundamentals in robot arm kinematics and dynamics that are an integral part of robotics research.

Robot arm kinematics deals with the study of the geometry of motion of a robot arm with respect to a fixed-reference coordinate system as a function of time without regard to the forces/moments that cause the motion. Thus it deals with the spatial configuration of the robot as a function of time, in particular the relations between the joint-variable space and the position and orientation of a robot arm. The kinematics problem usually consists of two subproblems—the direct kinematics and the inverse kinematics problems (or the arm solution).

The direct kinematics problem is to find the position and orientation of the manipulator hand with respect to a reference coordinate system, given the joint angle vector $\boldsymbol{\vartheta}(t) = (\vartheta_1(t), \vartheta_2(t), \vartheta_3(t), \vartheta_4(t), \vartheta_5(t), \vartheta_6(t))^T$ and the link/joint parameters of the robot arm.[2] The inverse kinematics problem is to calculate the joint angle vector $\theta(t)$ given the position and orientation of the end effector with respect to the reference coordinate system and the link/joint parameters of the robot arm. Since the independent variables in a robot arm are the joint angles and a task is generally stated in terms of the base or world coordinate system, the inverse kinematics solution is used more frequently in computer applications. The direct kinematics results in a 4×4 homogeneous transformation matrix which relates the spatial configuration between neighboring links. These homogeneous transformation matrices are useful in deriving the dynamic equations of motion of a robot arm.

Robot arm dynamics deals with the mathematical formulations of the equations of motion of a robot arm. The dynamic equations of motion of a manipulator are a set of equations describing the dynamic behavior of the manipulator. Such equations of motion are useful for (i) simulating the robot arm motion in a computer, (ii) designing suitable control equations for a robot arm, and (iii) evaluating the kinematic design and structure of a robot arm.

2. ROBOT ARM KINEMATICS

In this section, vector and matrix algebra will be utilized to develop a systematic and generalized approach to describe and represent the location of the links of a robot arm with respect to a fixed reference frame. Since the links of a robot arm may rotate and/or translate with respect to a reference coordinate frame, a body-attached coordinate frame will be established at the joint for each link. This reduces the direct kinematics problem to finding a transformation matrix that relates the body-attached coordinate frame to the reference coordinate frame. A 3×3 rotation matrix is then used to describe the rotational operations of the body-attached frame with respect to the reference frame. Later the homogeneous coordinates will be used to represent position vectors in a three-dimensional space and the rotation matrices will be expanded to 4×4 homogeneous transformation matrices to include the translational operations of the body-attached coordinate frames. This matrix representation of a rigid mechanical link to describe the spatial geometry of a robot arm was first used by Denavit and Hartenberg [6]. The advantage of using the Denavit–Hartenberg representation of linkages is its algorithmic universality in deriving the kinematic equation of a robot arm.

2.1 Rotation Matrices

A 3×3 rotation matrix can be defined as a transformation matrix which operates on a position vector in a three-dimensional Euclidean space and maps its coordinates expressed in a rotated coordinate system $OUVW$ (body-attached frame) to a reference coordinate system $OXYZ$. From Figure 2, we are given two right-handed rectangular coordinate systems, namely the $OXYZ$ coordinate system with OX, OY, and OZ as its coordinate axes and the $OUVW$ coordinate system with OU, OV, and OW as its coordinate axes. Both coordinate systems have their origins coincide at the point O. The

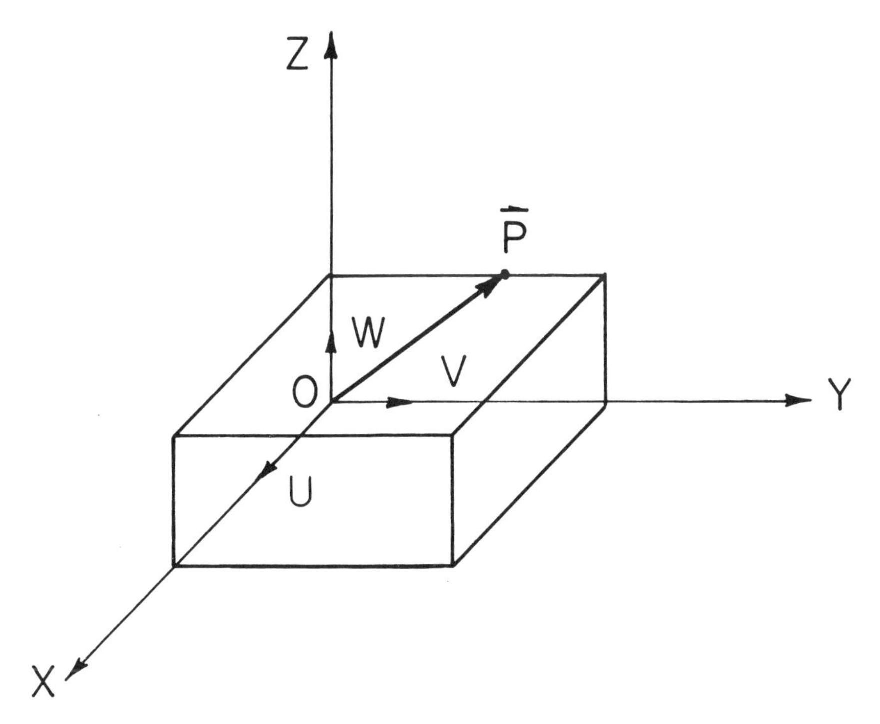

Figure 2. Coordinate systems for a rigid body.

$OXYZ$ coordinate system is fixed in the three-dimensional space and considered to be the reference frame. The $OUVW$ coordinate frame is rotating with respect to the reference frame $OXYZ$. Physically one can consider the $OUVW$ coordinate system to be a body-attached coordinate frame. That is, it is permanently and conveniently attached to the rigid body (e.g., and aircraft or a link of a robot arm) and moves together with it. Let $(\mathbf{i}_x, \mathbf{j}_y, \mathbf{k}_z)$ and $(\mathbf{i}_u, \mathbf{j}_v, \mathbf{k}_w)$ be the unit vectors along the coordinate axes of the $OXYZ$ and $OUVW$ systems, respectively. A point $\mathbf{p}$ in the space can be represented by its coordinates with respect to both coordinate systems. For ease of discussion, we shall assume that $\mathbf{p}$ is at rest and fixed with respect to the $OUVW$ coordinate frame. Then the point $\mathbf{p}$ can be represented by its coordinates with respect to the $OUVW$ and $OXYZ$ coordinate systems respectively as

$$\mathbf{p}_{uvw} = (p_u, p_v, p_w)^T \quad \text{and} \quad \mathbf{p}_{xyz} = (p_x, p_y, p_z)^T \tag{1}$$

where superscript T denotes transpose operation on matrices and vectors; $\mathbf{p}_{xyz}$ and $\mathbf{p}_{uvw}$ represent the same point $\mathbf{p}$ in the space with reference to different coordinate systems.

We would like to find a 3×3 transformation matrix $\mathbf{A}$ that will transform the coordinates of $\mathbf{p}_{uvw}$ to the coordinates expressed with respect to the $OXYZ$ coordinate system after the $OUVW$ coordinate system has been rotated. That is,

$$\mathbf{p}_{xyz} = \mathbf{A}\mathbf{p}_{uvw} \tag{2}$$

Note that physically the point $\mathbf{p}_{uvw}$ has been rotated together with the $OUVW$ coordinate system.

Recalling the definition of the components of a vector, we have

$$\mathbf{p}_{uvw} = p_u\mathbf{i}_u + p_v\mathbf{j}_v + p_w\mathbf{k}_w \tag{3}$$

and p_x, p_y, and p_z represent the components of $\mathbf{p}$ along the OX, OY, and OZ axes, respectively, or the projections of $\mathbf{p}$ onto the respective axes. Thus using the definition of a scalar product and Eq. (3):

$$\begin{aligned} p_x &= \mathbf{i}_x \cdot \mathbf{p} = \mathbf{i}_x \cdot \mathbf{i}_u \mathbf{p}_u + \mathbf{i}_x \cdot \mathbf{j}_v p_v + \mathbf{i}_x \cdot \mathbf{k}_w p_w \\ p_y &= \mathbf{j}_y \cdot \mathbf{p} = \mathbf{j}_y \cdot \mathbf{i}_u p_u + \mathbf{j}_y \cdot \mathbf{j}_v p_v + \mathbf{j}_y \cdot \mathbf{k}_w p_w \\ p_z &= \mathbf{k}_z \cdot \mathbf{p} = \mathbf{k}_z \cdot \mathbf{i}_u p_u + \mathbf{k}_z \cdot \mathbf{j}_v p_v + \mathbf{k}_z \cdot \mathbf{k}_w \mathbf{p}_w \end{aligned} \tag{4}$$

or, expressed in matrix form,

$$\begin{bmatrix} p_x \\ p_y \\ p_z \end{bmatrix} = \begin{bmatrix} \mathbf{i}_x \cdot \mathbf{i}_u & \mathbf{i}_x \cdot \mathbf{j}_v & \mathbf{i}_x \cdot \mathbf{k}_w \\ \mathbf{j}_y \cdot \mathbf{i}_u & \mathbf{j}_y \cdot \mathbf{j}_v & \mathbf{j}_y \cdot \mathbf{k}_w \\ \mathbf{k}_z \cdot \mathbf{i}_u & \mathbf{k}_z \cdot \mathbf{j}_v & \mathbf{k}_z \cdot \mathbf{k}_w \end{bmatrix} \begin{bmatrix} p_u \\ p_v \\ p_w \end{bmatrix} \tag{5}$$

and

$$A = \begin{bmatrix} \mathbf{i}_x \cdot \mathbf{i}_u & \mathbf{i}_x \cdot \mathbf{j}_v & \mathbf{i}_x \cdot \mathbf{k}_w \\ \mathbf{j}_y \cdot \mathbf{i}_u & \mathbf{j}_y \cdot \mathbf{j}_v & \mathbf{j}_y \cdot \mathbf{k}_w \\ \mathbf{k}_z \cdot \mathbf{i}_u & \mathbf{k}_z \cdot \mathbf{j}_v & \mathbf{k}_z \cdot \mathbf{k}_w \end{bmatrix} \tag{6}$$

Similarly, one can obtain the coordinates of $\mathbf{p}_{uvw}$ from the coordinates of $\mathbf{p}_{xyz}$:

$$\mathbf{p}_{uvw} = \mathbf{B}\mathbf{p}_{xyz}$$

$$\begin{bmatrix} p_u \\ p_v \\ p_w \end{bmatrix} = \begin{bmatrix} \mathbf{i}_u \cdot \mathbf{i}_x & \mathbf{i}_u \cdot \mathbf{j}_y & \mathbf{i}_u \cdot \mathbf{k}_z \\ \mathbf{j}_v \cdot \mathbf{i}_x & \mathbf{j}_v \cdot \mathbf{j}_y & \mathbf{j}_v \cdot \mathbf{k}_z \\ \mathbf{k}_w \cdot \mathbf{i}_x & \mathbf{k}_w \cdot \mathbf{j}_y & \mathbf{k}_w \cdot \mathbf{k}_z \end{bmatrix} \begin{bmatrix} p_x \\ p_y \\ p_z \end{bmatrix} \tag{7}$$

Since dot products are commutative, one can see that from Eqs. (6) and (7):

$$\mathbf{B} = \mathbf{A}^{-1} = \mathbf{A}^T \tag{8}$$

and

$$\mathbf{BA} = \mathbf{A}^T\mathbf{A} = \mathbf{A}^{-1}\mathbf{A} = \mathbf{I}_3 \tag{9}$$

The transformation is called an *orthogonal* transformation, and since the vectors in the dot products are all unit vectors, it is also called an *orthonormal* transformation.

The prime interest of developing the above transformation matrix is to find the rotation matrices that represent rotations of the $OUVW$ coordinate system about each of the three principal axes of the reference coordinate system $OXYZ$.

If the $OUVW$ coordinate system is rotated an α angle about the OX axis to arrive at a new location in the space, then the point $\mathbf{p}_{uvw}$ having coordinates $(p_u, p_v, p_w)^T$ with respect to the $OUVW$ system will have different coordinates $(p_x, p_y, p_z)^T$ with respect to the reference system $OXYZ$. The necessary transformation matrix $\mathbf{R}_{x,\alpha}$ is called the rotation matrix about the OX axis with α angle. $R_{x,\alpha}$ can be derived from the above transformation matrix concept, that is,

$$\mathbf{p}_{xyz} = \mathbf{R}_{x,\alpha} \cdot \mathbf{p}_{uvw} \tag{10}$$

and

$$\mathbf{R}_{x,\alpha} = \begin{bmatrix} \mathbf{i}_x \cdot \mathbf{i}_u & \mathbf{i}_x \cdot \mathbf{j}_v & \mathbf{i}_x \cdot \mathbf{k}_w \\ \mathbf{j}_y \cdot \mathbf{i}_u & \mathbf{j}_y \cdot \mathbf{j}_v & \mathbf{j}_y \cdot \mathbf{k}_w \\ \mathbf{k}_z \cdot \mathbf{i}_u & \mathbf{k}_z \cdot \mathbf{j}_v & \mathbf{k}_z \cdot \mathbf{k}_w \end{bmatrix} = \begin{bmatrix} 1 & 0 & 0 \\ 0 & \cos\alpha & -\sin\alpha \\ 0 & \sin\alpha & \cos\alpha \end{bmatrix} \tag{11}$$

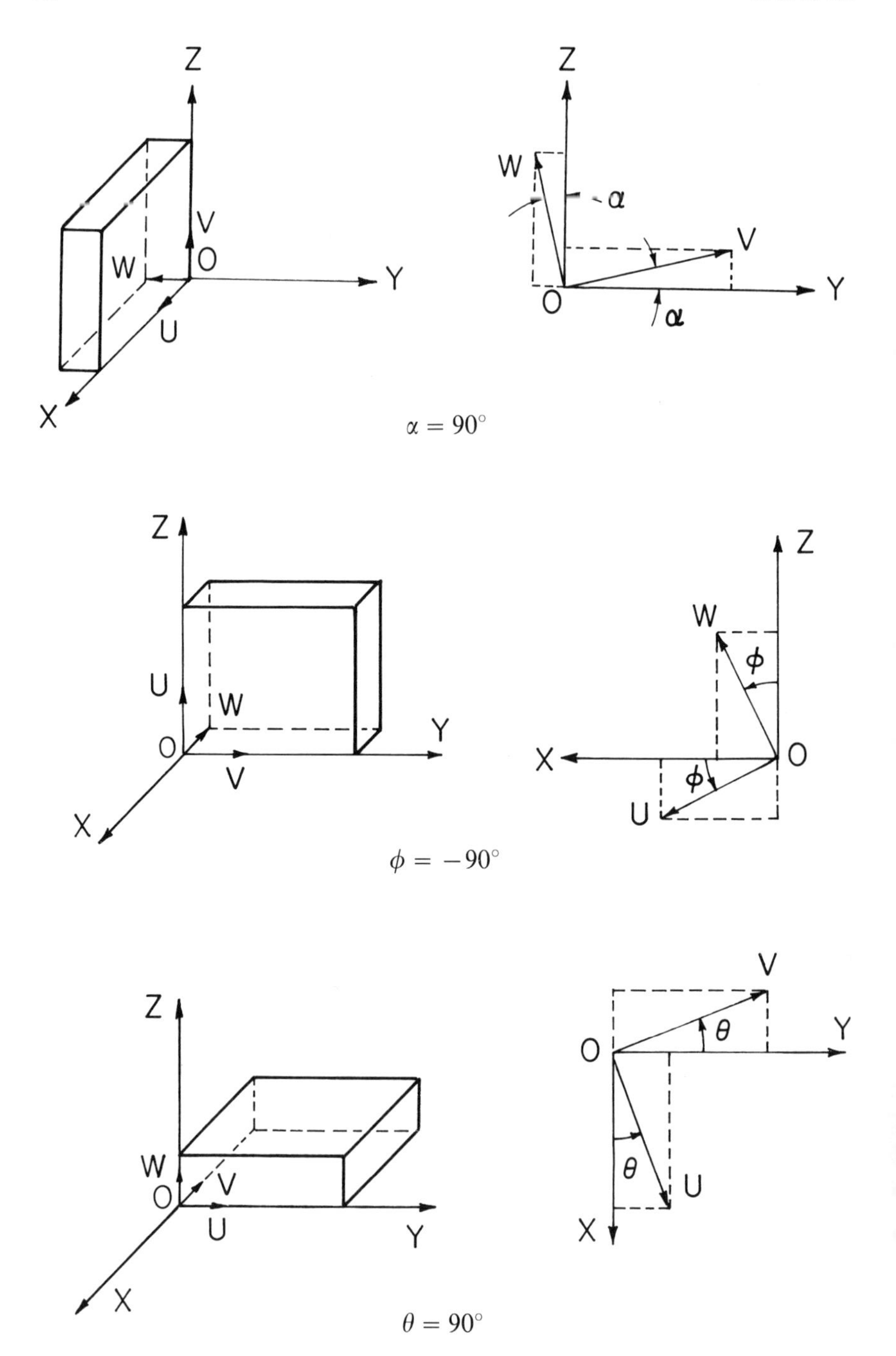

Figure 3. Rotating coordinate systems.

with $\mathbf{i}_x \equiv \mathbf{i}_u$. Similarly the 3×3 rotation matrices for rotation about the OY axis with φ angle and about the OZ axis with ϑ angle are respectively (see Figure 3)

$$\mathbf{R}_{y,\varphi} = \begin{bmatrix} \cos\varphi & 0 & \sin\varphi \\ 0 & 1 & 0 \\ -\sin\varphi & 0 & \cos\varphi \end{bmatrix}; \qquad \mathbf{R}_{z,\vartheta} = \begin{bmatrix} \cos\vartheta & -\sin\vartheta & 0 \\ \sin\vartheta & \cos\vartheta & 0 \\ 0 & 0 & 1 \end{bmatrix} \tag{12}$$

$\mathbf{R}_{x,\alpha}$, $\mathbf{R}_{y,\varphi}$, and $\mathbf{R}_{z,\vartheta}$ are called the *basic rotation matrices.* Other finite rotation matrices can be found from these matrices.

Example 1. Find the resultant rotation matrix that represents a rotation of φ angle about the OY axis followed by a rotation of ϑ angle about the OW axis followed by a rotation of α angle about the OU axis.

Solution:

$$\mathbf{R} = \mathbf{R}_{y,\varphi} \cdot \mathbf{R}_{w,\vartheta} \cdot \mathbf{R}_{u,\alpha} = \begin{bmatrix} C\varphi & 0 & S\varphi \\ 0 & 1 & 0 \\ -S\varphi & 0 & C\varphi \end{bmatrix} \begin{bmatrix} C\vartheta & -S\vartheta & 0 \\ S\vartheta & C\vartheta & 0 \\ 0 & 0 & 1 \end{bmatrix} \begin{bmatrix} 1 & 0 & 0 \\ 0 & C\alpha & -S\alpha \\ 0 & S\alpha & C\alpha \end{bmatrix}$$

$$= \begin{bmatrix} C\varphi C\vartheta & S\varphi S\alpha - C\varphi S\vartheta C\alpha & C\varphi S\vartheta S\alpha + S\varphi C\alpha \\ S\vartheta & C\vartheta C\alpha & -C\vartheta S\varphi \\ -S\varphi C\vartheta & S\varphi S\vartheta C\alpha + C\varphi S\alpha & C\varphi C\alpha - S\varphi S\vartheta S\alpha \end{bmatrix}$$

where $C\varphi \equiv \cos\varphi$; $S\varphi \equiv \sin\varphi$; $C\vartheta \equiv \cos\vartheta$; $S\vartheta \equiv \sin\vartheta$; $C\alpha \equiv \cos\alpha$; $S\alpha \equiv \sin\alpha$. ■

2.2 Homogeneous Coordinates and Transformation Matrix

Since a 3×3 rotation matrix does not give us any provision for translation and scaling, a fourth coordinate or component is introduced to a position vector $\mathbf{p} = (p_x, p_y, p_z)^T$ in a three dimensional space making it $\hat{\mathbf{p}} = (wp_x, wp_y, wp_z, w)^T$. We say that the position vector $\hat{\mathbf{p}}$ is expressed in homogeneous coordinates. In this section, we shall use a "hat" (that is, $\hat{\mathbf{p}}$) to indicate the representation of a Cartesian vector in homogeneous coordinates. Later, if no confusion exists, these "hats" will be lifted. The concept of homogeneous coordinate representation of points in a three-dimensional Euclidean space is useful in developing matrix transformations that include rotation, translation, scaling, and perspective transformation. In general, the representation of a n-component position vector by an $(n + 1)$-component vector is called homogeneous coordinate representation. In a homogeneous coordinate representation, the transformation of a n-dimensional vector is performed in the

$(n + 1)$-dimensional space and the physical n-dimensional vector is obtained by dividing the homogeneous coordinates by the fourth coordinate, w. Thus in a three-dimensional space, a position vector $\mathbf{p} = (p_x, p_y, p_z)^T$ is represented by an augmented vector $(wp_x, wp_y, wp_z, w)^T$ in the homogeneous coordinate representation. The physical coordinates are related to the homogeneous coordinates as follows:

$$p_x = \frac{wp_x}{w}, \qquad p_y = \frac{wp_y}{w}, \qquad p_z = \frac{wp_z}{w}$$

There is no unique homogeneous coordinates representation for a position vector in the three-dimensional space. For example,

$$\hat{\mathbf{p}}_1 = (w_1 p_x, w_1 p_y, w_1 p_z, w_1)^T$$

and

$$\hat{\mathbf{p}}_2 = (w_2 p_x, w_2 p_y, w_2 p_z, w_2)^T$$

are all homogeneous coordinates representing the same position vector $\mathbf{p} = (p_x, p_y, p_z)^T$. Thus one can view the the fourth component of the homogeneous coordinates, w, as a scale factor. If this coordinate is unity ($w = 1$), then the transformed homogeneous coordinates of a position vector are the same as the physical coordinates of the vector. In robotics applications, this scale factor will always be equal to 1, although it is commonly used in computer graphics as a universal scale factor taking on any positive values.

The homogeneous transformation matrix is a 4×4 matrix which maps a position vector expressed in homogeneous coordinates from one coordinate system to another coordinate system. A homogeneous transformation matrix can be considered to consist of four submatrices:

$$\mathbf{T} = \left[\begin{array}{c|c} \text{Rotation matrix} & \text{Position vector} \\ \hline \text{Perspective transf.} & \text{Scaling} \end{array}\right] = \left[\begin{array}{c|c} \mathbf{R}_{3\times 3} & \mathbf{p}_{3\times 1} \\ \hline \mathbf{t}_{1\times 3} & 1 \times 1 \end{array}\right]$$

The upper left 3×3 submatrix represents the rotation matrix; the upper right 3×1 submatrix represents the position vector of the origin of the rotated coordinate system with respect to the reference system; the lower left 1×3 submatrix represents perspective transformation; and the fourth diagonal element is the global scaling factor. The homogeneous transformation matrix can be used to explain the geometric relationship between the body-attached frame $OUVW$ and the reference coordinate system $OXYZ$.

If a position vector $\mathbf{p}$ in a three-dimensional space is expressed in homogeneous coordinates [i.e., $\hat{\mathbf{p}} = (p_x, p_y, p_z, 1)^T$], then using the transformation matrix concept, a 3×3 rotation matrix can be extended to a 4×4 homogeneous transformation matrix T_{rot} for pure rotation operations. Thus Eqs.

(11)–(12) expressed in homogeneous rotation matrix become

$$\mathbf{T}_{x,\alpha} = \begin{bmatrix} 1 & 0 & 0 & 0 \\ 0 & \cos\alpha & -\sin\alpha & 0 \\ 0 & \sin\alpha & \cos\alpha & 0 \\ 0 & 0 & 0 & 1 \end{bmatrix};$$

$$\mathbf{T}_{y,\varphi} = \begin{bmatrix} \cos\varphi & 0 & \sin\varphi & 0 \\ 0 & 1 & 0 & 0 \\ -\sin\varphi & 0 & \cos\varphi & 0 \\ 0 & 0 & 0 & 1 \end{bmatrix};$$

$$\mathbf{T}_{z,\vartheta} = \begin{bmatrix} \cos\vartheta & -\sin\vartheta & 0 & 0 \\ \sin\vartheta & \cos\vartheta & 0 & 0 \\ 0 & 0 & 1 & 0 \\ 0 & 0 & 0 & 1 \end{bmatrix}$$

These 4×4 rotation matrices $\mathbf{T}_{x,\alpha}$, $\mathbf{T}_{y,\varphi}$, and $\mathbf{T}_{z,\vartheta}$ are called the *basic homogeneous rotation matrices.*

The upper right 3×1 submatrix of the homogeneous transformation matrix has the effect of translating the $OUVW$ coordinate system which has parallel axes as the reference coordinate system $OXYZ$ but whose origin is at (dx, dy, dz) of the reference coordinate system

$$\mathbf{T}_{\text{tran}} = \begin{bmatrix} 1 & 0 & 0 & dx \\ 0 & 1 & 0 & dy \\ 0 & 0 & 1 & dz \\ 0 & 0 & 0 & 1 \end{bmatrix}$$

The above 4×4 transformation matrix T_{tran} is called the *basic homogeneous translation matrix.*

The lower left 1×3 submatrix of the homogeneous transformation matrix represents perspective transformation. They are useful for computer vision and the calibration of the camera model. At this moment, the elements of this submatrix are set to zero to indicate null perspective transformation.

The fourth diagonal element produces global scaling as in

$$\begin{bmatrix} 1 & 0 & 0 & 0 \\ 0 & 1 & 0 & 0 \\ 0 & 0 & 1 & 0 \\ 0 & 0 & 0 & s \end{bmatrix} \begin{bmatrix} x \\ y \\ z \\ 1 \end{bmatrix} = \begin{bmatrix} x \\ y \\ z \\ s \end{bmatrix}; \qquad s > 0$$

The physical Cartesian coordinates of the vector are

$$p_x = \frac{x}{s}, \qquad p_y = \frac{y}{s}, \qquad p_z = \frac{z}{s}, \qquad w = \frac{s}{s} = 1$$

Therefore, the fourth diagonal element in the homogeneous transformation matrix has the effect of globally reducing the coordinates if $s > 1$ and enlarging the coordinates if $0 < s < 1$.

In summary, a 4×4 homogeneous transformation matrix maps a vector expressed in homogeneous coordinates with respect to the $OUVW$ coordinate system to the reference coordinate system $OXYZ$. That is,

$$\mathbf{p}_{xyz} = \mathbf{T}\mathbf{p}_{uvw}, \qquad \text{with } w = 1$$

$$\mathbf{T} = \begin{bmatrix} n_x & s_x & a_x & p_x \\ n_y & s_y & a_y & p_y \\ n_z & s_z & a_z & p_z \\ 0 & 0 & 0 & 1 \end{bmatrix} = \begin{bmatrix} \mathbf{n} & \mathbf{s} & \mathbf{a} & \mathbf{p} \\ 0 & 0 & 0 & 1 \end{bmatrix} \tag{13}$$

2.3 Geometric Interpretation of Homogeneous Transformation Matrices

The geometric interpretation of a homogeneous transformation matrix may be of interest at this time. In general, a homogeneous transformation matrix for a three-dimensional space can be represented as in Eq. (13).

Let us choose a point $\mathbf{p}$ fixed in the $OUVW$ coordinate system and expressed in homogeneous coordinates as $(0, 0, 0, 1)^T$, that is $\mathbf{P}_{uvw}$ is the origin of the $OUVW$ coordinate system. Then the upper right 3×1 submatrix indicates the position of the origin of the $OUVW$ frame with respect to the $OXYZ$ reference coordinate frame. Next let us choose the point $\mathbf{p}$ to be $(1, 0, 0, 1)^T$. That is $\mathbf{P}_{uvw} \equiv \mathbf{i}_u$. Furthermore, we assume that the origins of both coordinate systems coincide at a point O. This has the effect of making the elements in the upper right 3×1 submatrix a null vector. Then the first column (or $\mathbf{n}$ vector) of the homogeneous transformation matrix represents the coordinates of the OU axis of $OUVW$ with respect to the $OXYZ$ coordinate system. Similarly choosing $\mathbf{p}$ to be $(0, 1, 0, 1)^T$ and $(0, 0, 1, 1)^T$ one can identify that the second (or $\mathbf{s}$ vector) and third column (or $\mathbf{a}$ vector) elements of the homogeneous transformation matrix represent the OV and OW axes, respectively, of the $OUVW$ coordinate system with respect to the reference coordinate system. Thus given a reference frame $OXYZ$ and a homogeneous transformation matrix $\mathbf{T}$, the column vectors of the rotation submatrix represent the principal axes of the $OUVW$ coordinate system with respect to the reference coordinate frame and one can draw the orientation of all the principal axes of the $OUVW$ coordinate frame with respect to the reference coordinate frame. The fourth column vector of the homogeneous transformation matrix represents the position of the origin of the $OUVW$ coordinate system with respect to the reference system. In other words, a homogeneous transformation matrix geometrically represent the *location* of a rotated and/or translated coordinate system (position and orientation) with respect to a reference coordinate system.

Since the inverse of a rotation submatrix is equivalent to its transpose, the row vectors of a rotation submatrix represents the principal axes of the references coordinate system with respect to the rotated coordinate system $OUVW$. However, the inverse of a homogeneous transformation matrix is *not* equivalent to its transpose. The position of the origin of the reference coordinate system with respect to the $OUVW$ coordinate system can only be found after the inverse of the homogeneous transformation matrix is determined. In general, the inverse of a homogeneous transformation matrix can be found to be

$$\mathbf{T}^{-1} = \begin{bmatrix} n_x & n_y & n_z & -\mathbf{n}^T\mathbf{p} \\ s_x & s_y & s_z & -\mathbf{s}^T\mathbf{p} \\ a_x & a_y & a_z & -\mathbf{a}^T\mathbf{p} \\ 0 & 0 & 0 & 1 \end{bmatrix} = \begin{bmatrix} & \mathbf{R}^T_{3\times 3} & & \begin{matrix} -\mathbf{n}^T\mathbf{p} \\ -\mathbf{s}^T\mathbf{p} \\ -\mathbf{a}^T\mathbf{p} \end{matrix} \\ 0 & 0 & 0 & 1 \end{bmatrix} \tag{14}$$

Thus from Eq. (14), the column vectors of the inverse of a homogeneous transformation matrix represent the principal axes of the reference system with respect to the rotated coordinate system $OUVW$ and the upper right 3×1 submatrix represents the position of the origin of the reference frame with respect to the $OUVW$ system. This geometric interpretation of the homogeneous transformation matrices is a very important concept and will be used frequently to provide insights to many robot arm kinematics problems.

Example 2. Find a homogeneous transformation $\mathbf{T}$ that represents a rotation of α angle about the OX axis, followed by a translation of a units along the OX axis, followed by a translation of d units along the OZ axis, followed by a rotation of ϑ angle about the OZ axis.

Solution:

$$\mathbf{T} = \mathbf{T}_{z,\vartheta}\mathbf{T}_{z,d}\mathbf{T}_{x,a}\mathbf{T}_{x,\alpha} = \begin{bmatrix} \cos\vartheta & -\sin\vartheta & 0 & 0 \\ \sin\vartheta & \cos\vartheta & 0 & 0 \\ 0 & 0 & 1 & 0 \\ 0 & 0 & 0 & 1 \end{bmatrix} \begin{bmatrix} 1 & 0 & 0 & 0 \\ 0 & 1 & 0 & 0 \\ 0 & 0 & 1 & d \\ 0 & 0 & 0 & 1 \end{bmatrix}$$

$$\times \begin{bmatrix} 1 & 0 & 0 & a \\ 0 & 1 & 0 & 0 \\ 0 & 0 & 1 & 0 \\ 0 & 0 & 0 & 1 \end{bmatrix} \begin{bmatrix} 1 & 0 & 0 & 0 \\ 0 & \cos\alpha & -\sin\alpha & 0 \\ 0 & \sin\alpha & \cos\alpha & 0 \\ 0 & 0 & 0 & 1 \end{bmatrix}$$

$$= \begin{bmatrix} \cos\vartheta & -\cos\alpha\sin\vartheta & \sin\alpha\sin\vartheta & a\cos\vartheta \\ \sin\vartheta & \cos\alpha\cos\vartheta & -\sin\alpha\cos\vartheta & a\sin\vartheta \\ 0 & \sin\alpha & \cos\alpha & d \\ 0 & 0 & 0 & 1 \end{bmatrix} \quad \blacksquare$$

2.4 Homogeneous Transformation Matrix Summary

From the previous discussion, there are two coordinate systems, the fixed reference coordinate frame $OXYZ$ and the moving (translating and rotating) coordinate frame $OUVW$. In order to describe the geometric relationship between these two coordinate systems, a 4×4 homogeneous transformation matrix is used. Homogeneous transformation matrices have the combined effects of rotation, translation, perspective, and global scaling when operating on position vectors expressed in homogeneous coordinates.

If these two coordinate systems are assigned to each link of a robot arm, say link $i - 1$ and link i, respectively, then the ink $i - 1$ coordinate system is the reference coordinate system and the link i coordinate system is the moving coordinate system, when joint i is activated. Using the $\mathbf{T}$ matrix, we can specify a point $\mathbf{p}_i$ at rest in link i and expressed in the link i (or $OUVW$) coordinate system in terms of the link $i - 1$ (or $OXYZ$) coordinate system as

$$\mathbf{p}_{i-1} = \mathbf{T} \cdot \mathbf{p}_i$$

where $\mathbf{T}$ = the 4×4 homogeneous transformation matrix relating the two coordinate systems;

$\mathbf{p}_i$ = the 4×1 augmented position vector $(x_i, y_i, z_i, 1)^T$ representing a point in the link i coordinate system expressed in homogeneous coordinates; and

$\mathbf{p}_{i-1}$ = the 4×1 augmented position vector $(x_{i-1}, y_{i-1}, z_{i-1})^T$ representing the same point $\mathbf{p}_i$ in terms of the link $i - 1$ coordinate system.

We shall continue to investigate the use of homogeneous transformation matrices in representing the links of a robot arm and relating adjacent link coordinate systems after we introduce the notation and parameters for links and joints of a robot arm.

2.5 Links, Joints, and Their Parameters

A mechanical manipulator consists of a sequence of rigid bodies, called links, connected by either revolute or prismatic joints. Each joint–link pair constitutes one degree of freedom. Hence for an n-degree-of-freedom manipulator, there are n joint–link pairs with link 0 (which is not considered to be part of the robot) attached to a supporting base where an inertial coordinate frame is usually established for this dynamic system and the last link is attached with a tool. The joints and links are numbered outward from the base; thus joint 1 is the point of connection between link 1 and the supporting base. Each link is connected to at most two others so that no closed loops are formed.

A joint axis (for joint i) is established at the connection of two links (see Figure 4). This joint axis will have two normals connected to it, one for each

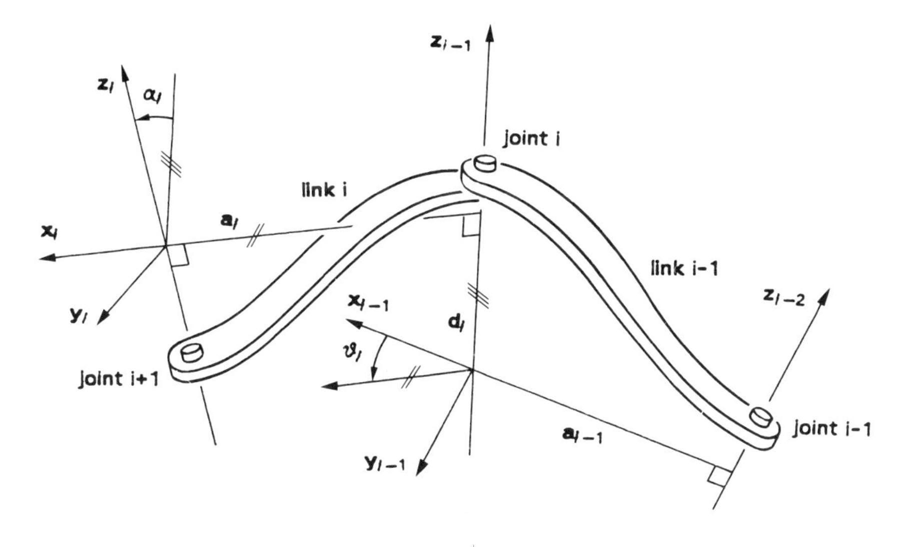

Figure 4. Link coordinate system and its parameters.

of the links. The relative position of two such connected links (links $i - 1$ and link i) is given by d_i, which is the distance measured along the joint axis between the normals. The joint angle ϑ_i between the normals is measured in a plane normal to the joint axis. Hence d_i and ϑ_i may be called the *distance* and the *angle* between the adjacent links, respectively. They determine the relative position of neighboring links.

A link i ($i = 1, \ldots, 6$) is connected to at most two other links (link $i - 1$ and link $i + 1$); thus two joint axes are established at both ends of the connection. The significance of links, from a kinematic perspective, is that they maintain a fixed configuration between their joints. It can be characterized by two parameters: the a_i and α_i. The a_i is the shortest distance measured along the common normal between the joint axes (i.e., the z_{i-1} and z_i axes for joint i and joint $i + 1$, respectively) and α_i is the angle between the joint axes measured in a plane perpendicular to a_i. Thus a_i and α_i may be called the *length* and the *twist angle* of the link i, respectively. They determine the structure of link i.

In summary there are four parameters a_i, α_i, d_i, and ϑ_i associated with each link of a manipulator. If a sign convention for each of these parameters has been established, then these parameters constitute a minimal sufficient set to completely determine the kinematic configuration of each link of a robot arm. It is worth noting that these four parameters come in pairs: the link parameters (a_i, α_i) determine the structure of the link; the joint parameters (d_i, ϑ_i) determine the relative position of neighboring links.

2.6 Denavit–Hartenberg Representation

In order to describe the translational and rotational relationship between adjacent links, Denavit and Hartenberg (D–H) [6] proposed a matrix method of systematically establishing an orthonormal coordinate system (body-attached frame) to each link of an articulated chain. The D–H representation results in a 4×4 homogeneous transformation matrix representing each link's coordinate system at the joint with respect to the previous link's coordinate system. Thus, through sequential transformations, the end effector expressed in the "hand coordinates" can be transformed and expressed in the "based coordinates," which is the inertial frame of this dynamic system.

An orthonormal Cartesian coordinate system (x_i, y_i, z_i) can be established for each link at its joint axis, where $i = 1, 2, 3, \ldots, n$ (n = number of degrees of freedom) plus the base coordinate frame.[3] Since a rotary joint has only one degree of freedom, each (x_i, y_i, z_i) coordinate frame of a robot arm corresponds to joint $(i + 1)$ and is fixed in link i. When the joint actuator activates joint i, link i will move with respect to link $i - 1$. Since the ith coordinate system is fixed in link i, then it moves together with the link i. Thus the nth coordinate frame moves with the hand (link n). The base coordinates are defined as the 0th coordinate frame (x_0, y_0, z_0), which is also the inertial coordinate frame of the robot arm. Thus for a PUMA 560 series robot arm, we have seen coordinate frames, namely, $(\mathbf{x}_0, \mathbf{y}_0, \mathbf{z}_0)$, $(\mathbf{x}_1, \mathbf{y}_1, \mathbf{z}_1), \ldots, (\mathbf{x}_6, \mathbf{y}_6, \mathbf{z}_6)$.

Every coordinate frame is determined and established on the basis of three rules: (a) The $\mathbf{z}_{i-1}$ axis lies along the axis of motion of the ith joint. (b) The $\mathbf{x}_i$ axis is normal to the $\mathbf{z}_{i-1}$ axis, pointing away from it. (c) The $\mathbf{y}_i$ axis completes the right-handed coordinate system as required.

By these rules, one is free to choose the location of coordinate frame 0 anywhere in the supporting base, as long as the $\mathbf{z}_0$ axis lies along the axis of motion of the first joint. The last coordinate frame (nth frame) can be placed anywhere in the hand, as long as the $\mathbf{x}_n$ axis is normal to the $\mathbf{z}_{n-1}$ axis.

The D–H representation of a rigid link depends on four geometric quantities associated with each link. These four quantities completely describe any revolute/prismatic joint. Referring to Figure 4, these four parameters are defined as follows:

- ϑ_i is the joint angle about the $\mathbf{z}_{i-1}$ axis from the $\mathbf{x}_{i-1}$ axis to the $\mathbf{x}_i$ axis (using the right-hand rule).
- d_i is the distance along the $\mathbf{z}_{i-1}$ axis from the origin of the $(i - 1)$th coordinate frame to the intersection of the $\mathbf{z}_{i-1}$ axis with the $\mathbf{x}_i$ axis.
- a_i is the offset distance along the $\mathbf{x}_i$ axis from the intersection of the $\mathbf{z}_{i-1}$ axis with the $\mathbf{x}_i$ axis to the origin of the ith frame (or shortest distance between the $\mathbf{z}_{i-1}$ and $\mathbf{z}_i$ axes).
- α_i is the offset angle about the $\mathbf{x}_i$ axis from the $\mathbf{z}_{i-1}$ axis to the $\mathbf{z}_i$ axis (using the right-hand rule.)

For a rotary joint, d_i, a_i, and α_i are the joint parameters and remain constants for a robot, while ϑ_i is the joint variable that changes when link i moves (or rotates) with link $i-1$. For a prismatic joint, ϑ_i, a_i, and α_i are the joint parameters and remain constants for a robot, while d_i is the joint variable. For the remaining of this chapter, *joint variable* refers to ϑ_i (or d_i), the varying quantity, and *joint parameters* refer to the remaining three geometric constant values (d_i, a_i, α_i), for a rotary joint, or (ϑ_i, a_i, α_i) for a prismatic joint.

With the above three basic rules for establishing an orthonormal coordinate system for each link and the geometric interpretation of the joint and link parameters, a procedure for establishing *consistent* orthonormal coordinate systems for a robot is outlined in Algorithm 1. An example of applying this algorithm to a PUMA robot arm is given in Figure 5.

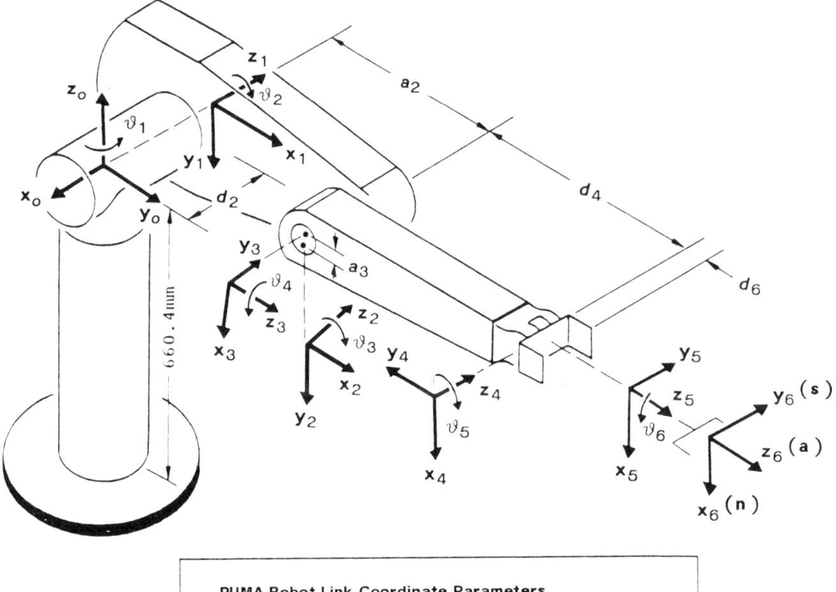

PUMA Robot Link Coordinate Parameters					
Joint i	ϑ_i	α_i	a_i	d_i	Range
1	90	-90	0	0	-160 to +160
2	0	0	431.8 mm	149.09 mm	-225 to +45
3	90	90	- 20.32 mm	0	-45 to +225
4	0	-90	0	433.07 mm	-110 to +170
5	0	90	0	0	-100 to +100
6	0	0	0	56.25 mm	-266 to +266

Figure 5. Establishing link coordinate systems for a PUMA robot.

Once the D–H coordinate system has been established for each link, a homogeneous transformation matrix can easily be developed relating the ith coordinate frame to the $(i-1)$th coordinate frame. Looking at Figure 4, it is obvious that a point $\mathbf{r}_i$ expressed in the ith coordinate system may be expressed in the $(i-1)$th coordinate system as $\mathbf{r}_{i-1}$ by performing the following successive transformations: (a) Rotate about the $\mathbf{z}_{i-1}$ axis an angle of ϑ_i to align the $\mathbf{x}_{i-1}$ axis with the $\mathbf{x}_i$ axis ($\mathbf{x}_{i-1}$ axis is parallel to $\mathbf{x}_i$). (b) Translate along the $\mathbf{z}_{i-1}$ axis a distance of d_i to bring the $\mathbf{x}_{i-1}$ and $\mathbf{x}_i$ axes into coincidence. (c) Translate along the $\mathbf{x}_i$ axis a distance of a_i to bring the two origins into coincidence. And, finally, (d) rotate about the $\mathbf{x}_i$ axis an angle of α_i to bring the two coordinate systems in coincidence.

Each of these four operations can be expressed by a basic homogeneous rotation/translation matrix, and the product of these four basic homogeneous transformation matrices yields a composite homogeneous transformation matrix, $\mathbf{A}_{i-1}$, known as the D–H transformation matrix for adjacent coordinate frames. Thus

---

ALGORITHM 1 (*Link Coordinate System Assignment*). Given an n-degree-of-freedom robot arm, this algorithm assigns an orthonormal coordinate system to each link of the robot arm at the joint axis. The labeling of the coordinate systems begins from the supporting base to the end effector of the robot arm. The relations between adjacent links can be represented by a 4×4 homogeneous transformation matrix. (Note that the assignment of coordinate systems is not unique.)

D1. [Establish the base coordinate system] Establish a right-handed orthonormal coordinate system $(\mathbf{x}_0, \mathbf{y}_0, \mathbf{z}_0)$ at the supporting base with the $\mathbf{z}_0$ axis lying along the axis of motion of joint one. The $\mathbf{x}_0$ and $\mathbf{y}_0$ axes can be conveniently established and are normal to the $\mathbf{z}_0$ axis.

D2. [Initialize and Loop] For each i, $i = 1, \ldots, n$, perform steps D3–D6.

D3. [Establish joint axis] Align the $\mathbf{z}_i$ with the axis of motion (rotary or sliding) of joint $i+1$.

D4. [Establish the origin of the ith coordinate system] Locate the origin of the ith coordinate system at the intersection of the $\mathbf{z}_i$ and $\mathbf{z}_{i-1}$ axes or at the intersection of common normal between the $\mathbf{z}_i$ and $\mathbf{z}_{i-1}$ axes and the $\mathbf{z}_i$ axis.

D5. [Establish $\mathbf{x}_i$ axis] Establish $\mathbf{x}_i = \pm(\mathbf{z}_{i-1} \times \mathbf{z}_i)/\|\mathbf{z}_{i-1} \times \mathbf{z}_i\|$ or along the common normal between the $\mathbf{z}_{i-1}$ and $\mathbf{z}_i$ axes when they are parallel.

D6. [Establish $\mathbf{y}_i$ axis] Assign $\mathbf{y}_i = +(\mathbf{z}_i \times \mathbf{x}_i)/\|\mathbf{z}_i \times \mathbf{x}_i\|$ to complete the right-handed coordinate system. (Extend the $\mathbf{z}_i$ and the $\mathbf{x}_i$ axes if necessary for steps D8–D11.)

D7. [Find joint and link parameters] For each i, $i = 1, \ldots, n$, perform steps D8–D11.

D8. [Find d_i] Now d_i is the distance from the origin of the $i-1$th coordinate system to the intersection of the $\mathbf{z}_{i-1}$ axis and the $\mathbf{x}_i$ axis along the $\mathbf{z}_{i-1}$ axis. It is the joint variable if joint i is prismatic.

D9. [Find a_i] Now a_i is the distance from the intersection of the $\mathbf{z}_{i-1}$ axis and the $\mathbf{x}_i$ axis to the origin of the ith coordinate system along the $\mathbf{x}_i$ axis.

D10. [Find ϑ_i] Now ϑ_i is the angle of rotation from the $\mathbf{x}_{i-1}$ axis to the $\mathbf{x}_i$ axis about the $\mathbf{z}_{i-1}$ axis. It is the joint variable if joint i is rotary.

D11. [Find α_i] Now α_i is the angle of rotation from the $\mathbf{z}_{i-1}$ axis to the $\mathbf{z}_i$ axis about the $\mathbf{x}_i$ axis.

$$\mathbf{A}_{i-1}^{i} = \mathbf{T}_{z,d}\mathbf{T}_{z,\vartheta}\mathbf{T}_{x,a}\mathbf{T}_{x,\alpha}$$

$$= \begin{bmatrix} 1 & 0 & 0 & 0 \\ 0 & 1 & 0 & 0 \\ 0 & 0 & 1 & d_i \\ 0 & 0 & 0 & 1 \end{bmatrix} \begin{bmatrix} \cos\vartheta_i & -\sin\vartheta_i & 0 & 0 \\ \sin\vartheta_i & \cos\vartheta_i & 0 & 0 \\ 0 & 0 & 1 & 0 \\ 0 & 0 & 0 & 1 \end{bmatrix} \begin{bmatrix} 1 & 0 & 0 & a_i \\ 0 & 1 & 0 & 0 \\ 0 & 0 & 1 & 0 \\ 0 & 0 & 0 & 1 \end{bmatrix}$$

$$\times \begin{bmatrix} 1 & 0 & 0 & 0 \\ 0 & \cos\alpha_i & -\sin\alpha_i & 0 \\ 0 & \sin\alpha_i & \cos\alpha_i & 0 \\ 0 & 0 & 0 & 1 \end{bmatrix}$$

$$= \begin{bmatrix} \cos\vartheta_i & -\cos\alpha_i\sin\vartheta_i & \sin\alpha_i\sin\vartheta_i & a\cos\vartheta_i \\ \sin\vartheta_i & \cos\alpha_i\cos\vartheta_i & -\sin\alpha_i\cos\vartheta_i & a\sin\vartheta_i \\ 0 & \sin\alpha_i & \cos\alpha_i & d_i \\ 0 & 0 & 0 & 1 \end{bmatrix} \tag{15}$$

and using Eq. (14), its inverse can be found to be

$$[\mathbf{A}_{i-1}^{i}]^{-1} = \mathbf{A}_{i-1}^{i} = \begin{bmatrix} \cos\vartheta_i & \sin\vartheta_i & 0 & -a_i \\ -\sin\vartheta_i\cos\alpha_i & \cos\alpha_i\cos\vartheta_i & \sin\alpha_i & -d_i\sin\alpha_i \\ \sin\vartheta_i\sin\alpha_i & -\sin\alpha_i\cos\vartheta_i & \cos\alpha_i & -d_i\cos\alpha_i \\ 0 & 0 & 0 & 1 \end{bmatrix} \tag{16}$$

where α_i, a_i, d_i are constants while ϑ_i is the joint variable for a revolute joint.

Using the $\mathbf{A}_{i-1}^{i}$ matrix, one can relate a point $\mathbf{p}_i$ at rest in link i and expressed in homogeneous coordinates with respect to the coordinate system i to the coordinate system $(i-1)$ established at link $(i-1)$ by

$$\mathbf{p}_{i-1} = \mathbf{A}_{i-1}^{i}\mathbf{p}_i$$

where $\mathbf{p}_{i-1} = (x_{i-1}, y_{i-1}, z_{i-1}, 1)^T$ and $\mathbf{p}_i = (x_i, y_i, z_i, 1)^T$.

Table 1. PUMA Coordinates Transformation Matrices

$$\mathbf{A}_{i-1}^{i} = \begin{bmatrix} C\vartheta_i & -C\alpha_i S\vartheta_i & S\alpha_i S\vartheta_i & a_i C\vartheta_i \\ S\vartheta_i & C\alpha_i C\vartheta_i & -S\alpha_i C\vartheta_i & a_i S\vartheta_i \\ 0 & S\alpha_i & C\alpha_i & d_i \\ 0 & 0 & 0 & 1 \end{bmatrix}$$

$$\mathbf{A}_0^1 = \begin{bmatrix} C_1 & 0 & -S_1 & 0 \\ S_1 & 0 & C_1 & 0 \\ 0 & -1 & 0 & 0 \\ 0 & 0 & 0 & 1 \end{bmatrix} \quad \mathbf{A}_1^2 = \begin{bmatrix} C_2 & -S_2 & 0 & a_2C_2 \\ S_2 & C_2 & 0 & a_2S_2 \\ 0 & 0 & 1 & d_2 \\ 0 & 0 & 0 & 1 \end{bmatrix} \quad \mathbf{A}_2^3 = \begin{bmatrix} C_3 & 0 & S_3 & a_3C_3 \\ S_3 & 0 & -C_3 & a_3S_3 \\ 0 & 1 & 0 & 0 \\ 0 & 0 & 0 & 1 \end{bmatrix}$$

$$\mathbf{A}_3^4 = \begin{bmatrix} C_4 & 0 & -S_4 & 0 \\ S_4 & 0 & C_4 & 0 \\ 0 & -1 & 0 & d_4 \\ 0 & 0 & 0 & 1 \end{bmatrix} \quad \mathbf{A}_4^5 = \begin{bmatrix} C_5 & 0 & S_5 & 0 \\ S_5 & 0 & -C_5 & 0 \\ 0 & 1 & 0 & 0 \\ 0 & 0 & 0 & 1 \end{bmatrix} \quad \mathbf{A}_5^6 = \begin{bmatrix} C_6 & -S_6 & 0 & 0 \\ S_6 & C_6 & 0 & 0 \\ 0 & 0 & 1 & d_6 \\ 0 & 0 & 0 & 1 \end{bmatrix}$$

where $C_i \equiv \cos\vartheta_i$; $S_i \equiv \sin\vartheta_i$; $C_{ij} \equiv \cos(\vartheta_i + \vartheta_j)$; $S_{ij} \equiv \sin(\vartheta_i + \vartheta_j)$

The six $\mathbf{A}_{i-1}^{i}$ transformation matrices for a PUMA robot have been found based on the coordinate systems established in Figure 5. These $\mathbf{A}_{i-1}^{i}$ matrices are listed in Table 1.

2.7 Kinematic Equations for Manipulators

The homogeneous matrix $\mathbf{T}_0^i$ which specifies the position and orientation of the end point of link i with respect to the base coordinate system is the chain product of successive coordinate transformation matrices of $\mathbf{A}_{i-1}^{i}$, expressed as

$$\mathbf{T}_0^i = \mathbf{A}_0^1\mathbf{A}_1^2 \cdots \mathbf{A}_{i-1}^{i} = \prod_{j=1}^{i} \mathbf{A}_{j-1}^{j}; \qquad \text{for } i = 1, 2, \ldots, n$$

$$= \begin{bmatrix} \mathbf{x}_i & \mathbf{y}_i & \mathbf{z}_i & \mathbf{p}_i \\ 0 & 0 & 0 & 1 \end{bmatrix} = \begin{bmatrix} \mathbf{R}_0^i & \mathbf{p}_0^i \\ 0 & 1 \end{bmatrix} \tag{17}$$

Specifically for $i = 6$, we obtain the $\mathbf{T}$ matrix, $\mathbf{T} = \mathbf{A}_0^6$, which specifies the position and orientation of the manipulator hand with respect to the base coordinate system. This $\mathbf{T}$ matrix is used so frequently in robot arm kinematics that it is called the "arm matrix." Consider the $\mathbf{T}$ matrix to be of the form

$$\mathbf{T} = \begin{bmatrix} \mathbf{x}_6 & \mathbf{y}_6 & \mathbf{z}_6 & \mathbf{p}_6 \\ 0 & 0 & 0 & 1 \end{bmatrix} = \begin{bmatrix} \mathbf{n} & \mathbf{s} & \mathbf{a} & \mathbf{p} \\ 0 & 0 & 0 & 1 \end{bmatrix} = \begin{bmatrix} n_x & s_x & a_x & p_x \\ n_y & s_y & a_y & p_y \\ n_z & s_z & a_z & p_z \\ 0 & 0 & 0 & 1 \end{bmatrix} \tag{18}$$

Here (see Figure 6)

- **n** is the normal vector of the hand; assuming parallel-jaw hand, it is orthogonal to the fingers of the robot arm.
- **s** is the sliding vector of the hand; it is pointing in the direction of the finger motion as the gripper opens and closes.
- **a** is the approach vector of the hand; it is pointing in the direction normal to the palm of the hand (i.e., normal to the tool mounting plate of the arm.)
- **p** is the position vector of the hand; it points from the origin of the base coordinate system to the origin of the hand coordinate system, which is usually located at the center point of the fully closed fingers.

The direct kinematics solution is therefore simply a matter of calculating $\mathbf{T} = \mathbf{A}_0^6$ by chain-multiplying the six $\mathbf{A}_{i-1}^i$ matrices or by evaluating each element in the **T** matrix. The arm matrix **T** for the PUMA robot arm shown in Figure 5 is found to be

$$\mathbf{T} = \mathbf{A}_0^1 \cdot \mathbf{A}_1^2 \cdot \mathbf{A}_2^3 \cdot \mathbf{A}_3^4 \cdot \mathbf{A}_4^5 \cdot \mathbf{A}_5^6 = \begin{bmatrix} n_x & s_x & a_x & p_x \\ n_y & s_y & a_y & p_y \\ n_z & s_z & a_z & p_z \\ 0 & 0 & 0 & 1 \end{bmatrix} \tag{19}$$

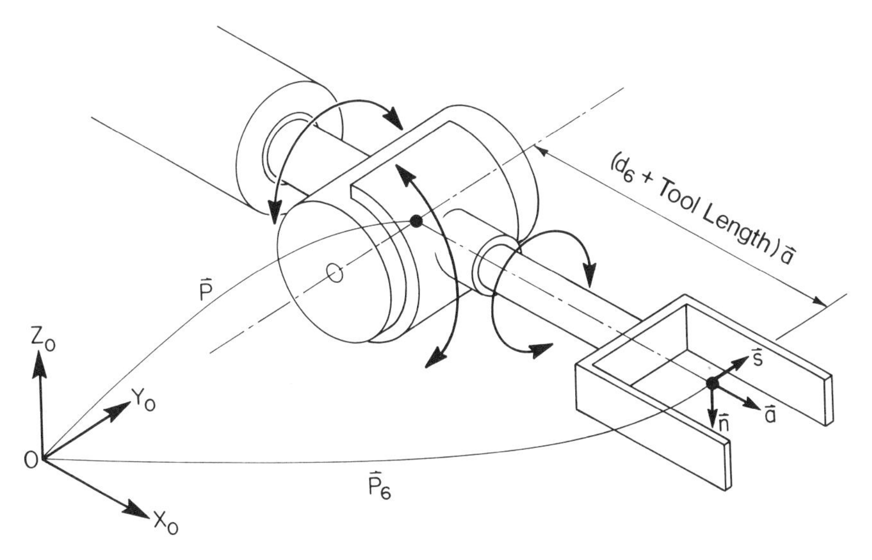

Figure 6. Hand coordinate system and [**n**, **s**, **a**].

where

$$\begin{aligned} n_x &= C_1[C_{23}(C_4C_5C_6 - S_4S_6) - S_{23}S_5C_6] - S_1[S_4C_5C_6 + C_4S_6] \\ n_y &= S_1[C_{23}(C_4C_5C_6 - S_4S_6) - S_{23}S_5C_6] + C_1[S_4C_5C_6 + C_4S_6] \\ n_z &= -S_{23}[C_4C_5C_6 - S_4S_6] - C_{23}S_5C_6 \end{aligned} \tag{20}$$

$$\begin{aligned} s_x &= C_1[-C_{23}(C_4C_5S_6 + S_4C_6) + S_{23}S_5S_6] - S_1[-S_4C_5S_6 + C_4C_6] \\ s_y &= S_1[-C_{23}(C_4C_5S_6 + S_4C_6) + S_{23}S_5S_6] + C_1[-S_4C_5S_6 + C_4C_6] \\ s_z &= S_{23}(C_4C_5S_6 + S_4C_6) + C_{23}S_5S_6 \end{aligned} \tag{21}$$

$$\begin{aligned} a_x &= C_1(C_{23}C_4S_5 + S_{23}C_5) - S_1S_4S_5 \\ a_y &= S_1(C_{23}C_4S_5 + S_{23}C_5) + C_1S_4S_5 \\ a_z &= -S_{23}C_4S_5 + C_{23}C_5 \end{aligned} \tag{22}$$

$$\begin{aligned} p_x &= C_1[d_6(C_{23}C_4S_5 + S_{23}C_5) + S_{23}d_4 + a_3C_{23} + a_2C_2] - S_1(d_6S_4S_5 + d_2) \\ p_y &= S_1[d_6(C_{23}C_4S_5 + S_{23}C_5) + S_{23}d_4 + a_3C_{23} + a_2C_2] + C_1(d_6S_4S_5 + d_2) \\ p_z &= d_6(C_{23}C_5 - S_{23}C_4S_5) + C_{23}d_4 - a_3S_{23} - a_2S_2 \end{aligned} \tag{23}$$

Example 3. A robot work station has been set up with a TV camera as shown in the accompanying diagram. The camera can see the origin of the base coordinate system where a six-joint robot is attached to. It can also see the center of an object (assumed to be a cube) to be manipulated by the robot. If a local coordinate system has been established at the center of the cube, this object as seen by the camera can be represented by a homogeneous transformation matrix $\mathbf{T}_1$. If the origin of the base coordinate system as seen by the camera can also be expressed by a homogeneous transformation matrix $\mathbf{T}_2$ and

$$\mathbf{T}_1 = \begin{bmatrix} 0 & 1 & 0 & 1 \\ 1 & 0 & 0 & 10 \\ 0 & 0 & -1 & 9 \\ 0 & 0 & 0 & 1 \end{bmatrix}; \qquad \mathbf{T}_2 = \begin{bmatrix} 1 & 0 & 0 & -10 \\ 0 & -1 & 0 & 20 \\ 0 & 0 & -1 & 10 \\ 0 & 0 & 0 & 1 \end{bmatrix}$$

then (1) what is the position of the center of the cube with respect to the base coordinate system; and (2) assuming that the cube is within the arm's reach, what is the orientation matrix $[\mathbf{n}, \mathbf{s}, \mathbf{a}]$ if you want the gripper (or finger) of the hand to be aligned with the $\mathbf{y}$ axis of the object and at the same time pick up the object from the top?

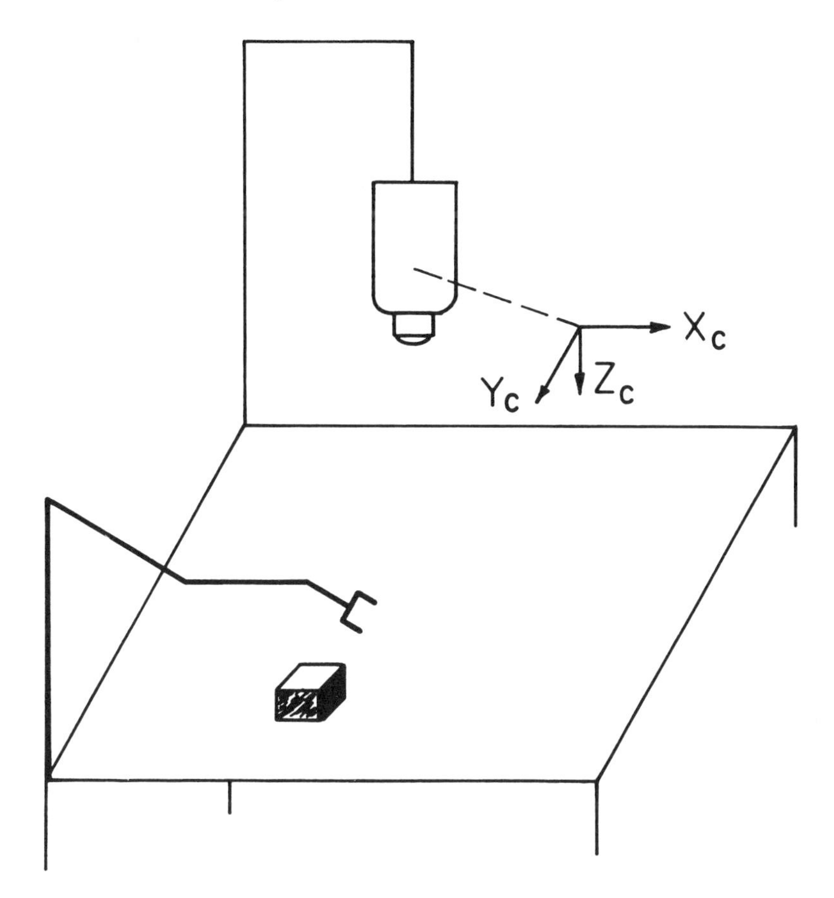

Example 3—Diagram

Solution:

$$\mathbf{T}_{\text{camera}}^{\text{cube}} \equiv \mathbf{T}_1 = \begin{bmatrix} 0 & 1 & 0 & 1 \\ 1 & 0 & 0 & 10 \\ 0 & 0 & -1 & 9 \\ 0 & 0 & 0 & 1 \end{bmatrix}$$

and

$$\mathbf{T}_{\text{camera}}^{\text{base}} \equiv \mathbf{T}_2 = \begin{bmatrix} 1 & 0 & 0 & -10 \\ 0 & -1 & 0 & 20 \\ 0 & 0 & -1 & 10 \\ 0 & 0 & 0 & 1 \end{bmatrix}$$

To find $\mathbf{T}_{\text{base}}^{\text{cube}}$, we use the "chain product" rule:

$$\mathbf{T}_{\text{base}}^{\text{cube}} = \mathbf{T}_{\text{base}}^{\text{camera}} \cdot \mathbf{T}_{\text{camera}}^{\text{cube}} = (\mathbf{T}_2)^{-1} \cdot \mathbf{T}_1$$

Using Eq. (14) to invert the $\mathbf{T}_2$ matrix, we obtain the resultant transformation matrix:

$$\mathbf{T}_{\text{base}}^{\text{cube}} = \begin{bmatrix} 1 & 0 & 0 & 10 \\ 0 & -1 & 0 & 20 \\ 0 & 0 & -1 & 10 \\ 0 & 0 & 0 & 1 \end{bmatrix} \begin{bmatrix} 0 & 1 & 0 & 1 \\ 1 & 0 & 0 & 10 \\ 0 & 0 & -1 & 9 \\ 0 & 0 & 0 & 1 \end{bmatrix} = \begin{bmatrix} 0 & 1 & 0 & 11 \\ -1 & 0 & 0 & 10 \\ 0 & 0 & 1 & 1 \\ 0 & 0 & 0 & 1 \end{bmatrix}$$

Therefore the cube is at a location $(11, 10, 1)^T$ from the base coordinate system. Its **x**, **y**, and **z** axes are parallel to the $-\mathbf{y}$, **x**, and **z** axes of the base coordinate system, respectively.

To find $[\mathbf{n}, \mathbf{s}, \mathbf{a}]$, we make use of

$$\mathbf{T}_0^6 = \begin{bmatrix} \mathbf{n} & \mathbf{s} & \mathbf{a} & \mathbf{p} \\ 0 & 0 & 0 & 1 \end{bmatrix}$$

where $\mathbf{p} = (11, 10, 1)^T$ from the above solution. From the above figure, we want the approach vector (**a**) to align with the negative direction of the OZ axis of the base coordinate system, $\mathbf{a} = (0, 0, -1)^T$; the **s** vector can be aligned in either direction of the **y** axis of $\mathbf{T}_{\text{base}}^{\text{cube}}$ (that is, $\mathbf{s} = (\pm 1, 0, 0)^T$); and the **n** vector can be obtained from the cross-product of **s** and **a**:

$$\mathbf{n} = \begin{vmatrix} \mathbf{i} & \mathbf{j} & \mathbf{k} \\ s_x & s_y & s_z \\ a_x & a_y & a_z \end{vmatrix} = \begin{vmatrix} \mathbf{i} & \mathbf{j} & \mathbf{k} \\ \pm 1 & 0 & 0 \\ 0 & 0 & -1 \end{vmatrix} = \begin{bmatrix} 0 \\ \pm 1 \\ 0 \end{bmatrix}$$

Therefore the orientation matrix $[\mathbf{n}, \mathbf{s}, \mathbf{a}]$ is found to be

$$[\mathbf{n}, \mathbf{s}, \mathbf{a}] = \begin{bmatrix} 0 & 1 & 0 \\ +1 & 0 & 0 \\ 0 & 0 & -1 \end{bmatrix} \quad \text{or} \quad \begin{bmatrix} 0 & -1 & 0 \\ -1 & 0 & 0 \\ 0 & 0 & -1 \end{bmatrix} \quad \blacksquare$$

2.8 The Inverse Kinematics Solution of a PUMA Robot Arm

Computer-based robots are usually servoed in the joint-variable space while the objects to be manipulated are usually expressed in the world coordinate system. In order to control the position and orientation of the end effector of a robot to reach its object, the inverse kinematics solution is more important. In other words, given the position and orientation of the end effector of a robot as $\mathbf{T}_0^6$, we would like to find the corresponding joint angles

$\vartheta = (\vartheta_1, \vartheta_2, \vartheta_3, \vartheta_4, \vartheta_5, \vartheta_6)^T$ of the robot so that the end effector can be positioned as desired.

In general, the inverse kinematics problem can be solved by matrix algebraic, iterative, or geometric approach. Pieper [24] presented the kinematics solution for any six-degree-of-freedom manipulator which has revolute or prismatic pairs for the first three joints and the joint axes of the last three joints intersect at a point. The solution can be expressed as a fourth-degree polynomial in one unknown, and closed-form solution for the remaining unknowns. Paul [21,23] presented a matrix algebraic approach using the 4×4 homogeneous transformation matrices in solving the kinematics solution for the same class of simple manipulators as discussed by Pieper. Uicker [30] and Milenkovic and Huang [19] presented iterative solutions for most industrial robots. The iterative solution often requires more computation, and it does not guarantee convergence to the correct solution, especially in the singular and degenerate cases.

A geometric approach will be briefly presented here to find the joint angle solution given the arm matrix. Detailed derivation can be found in Ref. [15]. This approach can be generalized to solve the inverse kinematics problem of most present-day industrial robots. The solution is calculated in two stages. First a position vector pointing from the shoulder to the wrist is derived. This is used to derive the solution of the first three joints. The last three joints are solved using the calculated values from the first three joints and the "orientation" submatrices of $\mathbf{T}_0^i$, $i = 4, 5, 6$.

2.8.1 Arm Solution for $(\vartheta_1, \vartheta_2, \vartheta_3)$ of a PUMA Robot Arm

Using the notation defined in the previous sections, a position vector $\mathbf{p}$ which points from the origin of the shoulder coordinate system $(\mathbf{x}_0, \mathbf{y}_0, \mathbf{z}_0)$ to the end point of the link 3 is derived as (see Figure 7):

$$\mathbf{p} = \mathbf{p}_6 - d_6\mathbf{a} = (p_x, p_y, p_z)^T \tag{24}$$

Recalling that

$$\mathbf{T}_0^i = \prod_{j=1}^{i} \mathbf{A}_{j-1}^j$$

and finding the position vector of $\mathbf{T}_0^4$, we have

$$\begin{bmatrix} p_x \\ p_y \\ p_z \end{bmatrix} = \begin{bmatrix} C_1(a_2C_2 + a_3C_{23} + d_4S_{23}) - d_2S_1 \\ S_1(a_2C_2 + a_3C_{23} + d_4S_{23}) + d_2C_1 \\ d_4C_{23} - a_3S_{23} - a_2S_2 \end{bmatrix} \tag{25}$$

In order to evaluate ϑ_1 for $-\pi \leq \vartheta_1 \leq \pi$, an arctangent function, arctan 2 (y/x), which returns $\tan^{-1}(y/x)$ adjusted to the proper quadrant will be used.

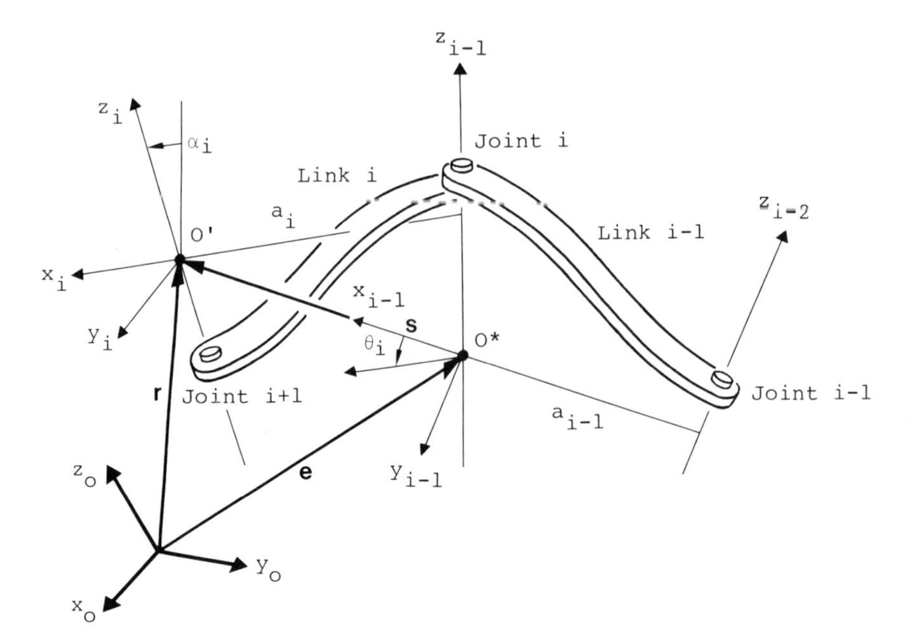

Figure 7. Relationship between O, O^*, O' frames.

It is defined as

$$\vartheta = \tan^{-1}\left(\frac{y}{x}\right) = \begin{cases} 0^\circ \leq \vartheta \leq 90^\circ, & \text{for } +x \text{ and } +y \\ 90^\circ \leq \vartheta \leq 180^\circ, & \text{for } -x \text{ and } +y \\ -180^\circ \leq \vartheta \leq -90^\circ, & \text{for } -x \text{ and } -y \\ -90^\circ \leq \vartheta \leq 0^\circ, & \text{for } +x \text{ and } -y \end{cases}$$

From Eq. (25), using the p_x and p_y component equations to solve for ϑ_1, we have

$$\vartheta_1 = \tan^{-1}\left[\frac{\pm p_y\sqrt{p_x^2 + p_y^2 - d_2^2} - d_2 p_x}{\pm p_x\sqrt{p_x^2 + p_y^2 - d_2^2} + d_2 p_y}\right]; \qquad -\pi \leq \vartheta_1 \leq \pi \qquad (26)$$

Because of the $\pm$ sign from the square root function, we have two solutions corresponding to two arm configurations—positive square root sign for left shoulder arm and negative square root sign for right shoulder arm.

From Eq. (25), squaring the components and sum them up, and using the equation $r\sin(\alpha + \beta) = r\sin\alpha\cos\beta + r\cos\alpha\sin\beta$, we have

$$\vartheta_3 = \tan^{-1}\left[\frac{wd_4 - a_3\sqrt{4a_2^2(a_3^2 + d_4^2) - w^2}}{d_4\sqrt{4a_2^2(a_3^2 + d_4^2) - w^2} + wa_3}\right]; \qquad -\pi \leq \vartheta_3 \leq \pi \qquad (27)$$

where $w = (p_x^2 + p_y^2 + p_z^2 - d_2^2 - a_2^2 - a_3^2 - d_4^2)/2a_2$. The positive square root sign of the above equation gives the below-arm configuration (elbow below the hand) and the negative square root sign is for the above-arm configuration (elbow above the hand).

From Eq. (25), using the p_z-component equation and again using the equation $r \sin(\alpha + \beta) = r \sin \alpha \cos \beta + r \cos \alpha \sin \beta$, we can obtain ϑ_2:

$$\vartheta_2 = \tan^{-1}\left[\frac{p_z\sqrt{p_x^2 + p_y^2 - d_2^2} + (a_2 + a_3C_3 + d_4S_3)(d_4C_3 - a_3S_3)}{p_z(d_4C_3 - a_3S_3) - (a_2 + a_3C_3 + d_4S_3)\sqrt{p_x^2 + p_y^2 - d_2^2}}\right]; \quad -\pi \leq \vartheta_2 \leq \pi \qquad (28)$$

The negative square root sign of the above equation gives the left-arm configuration and the positive square root sign for the right-arm configuration.

Knowing the first three joint angles $\vartheta_1, \vartheta_2, \vartheta_3$, we can evaluate the $\mathbf{T}_0^3$ matrix.

2.8.2 Arm Solution for $(\vartheta_4, \vartheta_5, \vartheta_6)$ of a PUMA Robot Arm

In order to find the solutions of the last three joint angles of a PUMA robot arm, we set these joints to meet the following criteria (see Figure 6):

1. Set joint 4 such that a rotation about joint 5 will align the axis of motion of joint 6 with the given approach vector (**a** or **T**)
2. Set joint 5 to align the axis of motion of joint 6 with the approach vector.
3. Set joint 6 to align the given orientation vector (or sliding vector or $\mathbf{y}_6$) and normal vector.

Mathematically, the above criteria respectively mean

$$\mathbf{z}_4 = \frac{\pm(\mathbf{z}_3 \times \mathbf{a})}{\|\mathbf{z}_3 \times \mathbf{a}\|}; \qquad \text{given } \mathbf{a} = (a_x, a_y, a_z)^T$$

$$\mathbf{a} = \mathbf{z}_5; \qquad \text{given } \mathbf{a} = (a_x, a_y, a_z)^T$$

$$\mathbf{s} = \mathbf{y}_6; \qquad \text{given } \mathbf{s} = (s_x, s_y, s_z)^T \text{ and } \mathbf{n} = (n_x, n_y, n_z)^T$$

From the above criteria, it can be shown that the following is true:

$$S_4 = -(\mathbf{x}_3 \cdot \mathbf{z}_4); \qquad C_4 = (\mathbf{y}_3 \cdot \mathbf{z}_4)$$

where $\mathbf{x}_3$ and $\mathbf{y}_3$ are the x and y column vector of $\mathbf{T}_0^3$, respectively. Thus

$$\vartheta_4 = \tan^{-1}\left[\frac{C_1a_y - S_1a_x}{C_1C_{23}a_x + S_1C_{23}a_y - S_{23}a_z}\right]; \qquad -\pi \leq \vartheta_4 \leq \pi \qquad (29)$$

To find ϑ_5, we use the criterion that aligns the axis of rotation of joint 6 with the approach vector (or $\mathbf{a} = \mathbf{z}_5$). Using the fact that

$$S_5 = \mathbf{x}_4 \cdot \mathbf{a}; \qquad C_5 = -(\mathbf{y}_4 \cdot \mathbf{a})$$

where $\mathbf{x}_4$ and $\mathbf{y}_4$ are the x and y column vector of $\mathbf{T}_0^4$, respectively, and $\mathbf{a}$ is the approach vector, then

$$\vartheta_5 = \tan^{-1}\left[\frac{(C_1C_{23}C_4 - S_1S_4)a_x + (S_1C_{23}C_4 + C_1S_4)a_y - C_4S_{23}a_z}{C_1S_{23}a_x + S_1S_{23}a_y + C_{23}a_z}\right]; \qquad -\pi \le \vartheta_5 \le \pi \quad (30)$$

If $\vartheta_5 \approx 0$, then the degenerate case occurs. This indicates that at this particular position/orientation configuration, a five-axis robot arm rather than a six-axis one would suffice.

Table 2. A PUMA Robot Arm Joint Angle Solutions

$$\vartheta_1 = \tan^{-1}\left[\frac{\pm p_y\sqrt{p_x^2 + p_y^2 - d_2^2} - d_2p_x}{\pm p_x\sqrt{p_x^2 + p_y^2 - d_2^2} + d_2p_y}\right]$$

$$\vartheta_2 = \tan^{-1}\left[\frac{p_z\sqrt{p_x^2 + p_y^2 - d_2^2} + (a_2 + a_3C_3 + d_4S_3)(d_4C_3 - a_3S_3)}{p_z(d_4C_3 - a_3S_3) - (a_2 + a_3C_3 + d_4S_3)\sqrt{p_x^2 + p_y^2 - d_2^2}}\right]$$

$$\vartheta_3 = \tan^{-1}\left[\frac{wd_4 - a_3\sqrt{4a_2^2(a_3^2 + d_4^2) - w^2}}{d_4\sqrt{4a_2^2(a_3^2 + d_4^2) - w^2} + wa_3}\right]$$

$$\vartheta_4 = \tan^{-1}\left[\frac{C_1a_y - S_1a_x}{C_1C_{23}a_x + S_1C_{23}a_y - S_{23}a_z}\right]$$

$$\vartheta_5 = \tan^{-1}\left[\frac{(C_1C_{23}C_4 - S_1S_4)a_x + (S_1C_{23}C_4 + C_1S_4)a_y - C_4S_{23}a_z}{C_1S_{23}a_x + S_1S_{23}a_y + C_{23}a_z}\right]$$

$$\vartheta_6 = \tan^{-1}\left[\frac{(-S_1C_4 - C_1C_{23}S_4)n_x + (C_1C_4 - S_1C_{23}S_4)n_y + (S_4S_{23})n_z}{(-S_1C_4 - C_1C_{23}S_4)s_x + (C_1C_4 - S_1C_{23}S_4)s_y + (S_4S_{23})s_z}\right]$$

where

$$w = \frac{p_x^2 + p_y^2 + p_z^2 - d_2^2 - a_2^2 - a_3^2 - d_4^2}{2a_2} \qquad -180^\circ \le \vartheta_1, \vartheta_2, \vartheta_3, \vartheta_4, \vartheta_5, \vartheta_6 \le 180^\circ$$

- Degenerate Case ($\vartheta_5 \approx 0$):
  ϑ_4 = current value of ϑ_4 or 0; and
  $(\vartheta_4 + \vartheta_6)$ = total angle required to align the orientation
- For a given arm configuration:
  $(\vartheta_1, \vartheta_2, \vartheta_3, \vartheta_4, \vartheta_5, \vartheta_6)$ is a set of solutions; and
  $(\vartheta_1, \vartheta_2, \vartheta_3, \vartheta_4 + \pi, -\vartheta_5, \vartheta_6 + \pi)$ is another set of solutions

Up to now, we have aligned the axis of joint 6 with the approach vector. Next we need to align the gripper to ease picking up the object. The criterion for doing this is to set $\mathbf{s} = \mathbf{y}_6$. Using the fact that

$$S_6 = \mathbf{y}_5 \cdot \mathbf{n}; \qquad C_6 = \mathbf{y}_5 \cdot \mathbf{s}$$

where $\mathbf{y}_5$ is the y column vector of $\mathbf{T}_0^5$ and $\mathbf{n}$ and $\mathbf{s}$ are the normal and sliding vectors of $\mathbf{T}_0^6$, respectively, then

$$\vartheta_6 = \tan^{-1}\left[\frac{(-S_1C_4 - C_1C_{23}S_4)n_x + (C_1C_4 - S_1C_{23}S_4)n_y + (S_4S_{23})n_z}{(-S_1C_4 - C_1C_{23}S_4)s_x + (C_1C_4 - S_1C_{23}S_4)s_y + (S_4S_{23})s_z}\right]; \quad -\pi \le \vartheta_6 \le \pi \tag{31}$$

If the degenerate case occurs, then $(\vartheta_4 + \vartheta_6)$ = total angle required to align the sliding vector ($\mathbf{s}$).

In summary, there are four solutions to the inverse kinematics problem—two for the right-shoulder-arm configuration and two for the left-shoulder-arm configuration. For each arm configuration, Eqs. (26)–(31) give one set of solutions $(\vartheta_1, \vartheta_2, \vartheta_3, \vartheta_4, \vartheta_5, \vartheta_6)$, and $(\vartheta_1, \vartheta_2, \vartheta_3, \vartheta_4 + \pi, -\vartheta_5, \vartheta_6 + \pi)$ give another set of solutions. The solution of the joint angles of a PUMA robot arm (given the $\mathbf{T}_0^6$ matrix) is obtained and tabulated in Table 2 for convenience.

3. ROBOT ARM DYNAMICS

Various approaches are available to formulate the robot arm dynamics, such as the Lagrange–Euler (L–E), the Newton–Euler (N–E), the recursive Lagrange–Euler (R–L) and the generalized d'Alembert principle formulations (G–D). In this section, only the L–E, the N–E, and the G–D formulations will be discussed, and the equations of motion of a six-linked manipulator will be derived. The derivations follow closely the work done by Bejczy [2], Paul [21,22], Luh [18], and Lee [15].

The derivation of the dynamic model of a manipulator based on the L–E method is simple and systematic. The resulting dynamic equations of motion, excluding the dynamics of the electronic control device and the gear friction, are a set of second-order coupled nonlinear differential equations. Bejczy [2], Based on the 4×4 homogeneous transformation matrix representation of kinematic chain and the Lagrangian formulation, has shown that the dynamic equations of motion for a six-joint manipulator (Stanford arm) are highly nonlinear and consist of inertia loading, coupling reaction forces between joints (Coriolis and centrifugal), and gravity loading effects. It has been recognized that the dynamic equations of motion as formulated are computationally inefficient [28,29], and real-time control based on the "complete" dynamic model has been found difficult to achieve if not impossible [10,28,29]. It takes about 8 seconds (FORTRAN simulation) to compute the move between two adjacent set points on a preplanned trajectory for a six-joint robot arm. To improve the speed of computation, simplified sets of equations have been used by other investigators. In general, these "approximate" models simplify the underlying physics by neglecting second-order terms such as the Coriolis and centrifugal reaction terms [2,22]. These approximate models, when used in control purposes, result in suboptimal dynamic performance restricting arm movement to low speeds. At high arm speeds the neglected terms become significant, making accurate position control of the robot arm impossible.

An approach which has the advantage of both speed and accuracy was based on the N–E vector formulation [18]. The derivation is simple but messy and involves vector cross-product terms. The resulting dynamic equations, excluding the dynamics of the control device and the gear friction, are a set of forward and backward recursive equations. This set of recursive equations can be applied to the robot links sequentially. The forward recursion propagates kinematics information (such as angular velocities, angular accelerations, linear accelerations, total forces, and moments exerted at the center of mass of each link) from the base reference frame (inertial frame) to the end effector. The backward recursion propagates the forces and moments exerted on each link from the end effector of the manipulator to the base reference frame. Because of the nature of the formulation and the method of systematically computing the torques, computations are much simpler, which makes it possible to achieve a short computing time. This algorithm takes about 3 ms to compute the feedback joint torques per trajectory set point using a PDP II/45 computer [15]. This enables us to implement real-time control of a robot arm in the joint-variable space.

The inefficiency of the equations of motion as formulated by the L–E method comes mainly from the 4×4 homogeneous transformation matrices. To improve the computation time, Hollerbach [10] exploited the recursive nature of the Lagrangian formulations. However, the recursive equations destroy the "structure" of the dynamic model which is useful to provide insight for the design of controller. For control analysis, one would like to obtain a set of closed-form differential equations that describe the dynamic behavior of a manipulator and in which the interaction and coupling reaction forces can be easily identified so that an appropriate controller can be designed to compensate their effects. To obtain such efficient set of equations of motion, an approach based on the generalized d'Alembert principle has been proposed [3,16]. This method used rotation matrices and relative position vectors between joints to increase the computational efficiency.

The computation of the applied torques from the generalized d'Alembert equations of motion is of order $O(n^3)$, while the L–E formulation is of order $O(n^4)$ and the N–E method is of order $O(n)$.

3.1 Lagrange–Euler Formulation [2,15,22].

The following derivation of the equations of motion for an n-degree-of-freedom manipulator is based on the homogeneous transformation matrices developed in the previous sections.

3.1.1 *Kinetic Energy Calculation*

Consider a position vector expressed in homogeneous coordinates, $\mathbf{r}_i = (x_i, y_i, z_i, 1)^T$, which points from the ith coordinate system to a differential mass dm located in the ith link. The kinetic energy dK_i of this differential mass is $\frac{1}{2}\mathrm{Tr}(\mathbf{v}_0^i(\mathbf{v}_0^i)^T)dm$,[4] which equals

$$dK_i = \frac{1}{2} \sum_{j=1}^{i} \sum_{k=1}^{i} \mathrm{Tr}\left\{ \frac{\partial \mathbf{T}_0^i}{\partial \vartheta_j} \mathbf{r}_i(\mathbf{r}_i)^T \, \mathrm{dm} \left[\frac{\partial \mathbf{T}_0^i}{\partial \vartheta_k} \right]^T \dot{\vartheta}_j \dot{\vartheta}_k \right\} \tag{32}$$

If each link is integrated over its entire mass and the kinetic energies of all links are summed, then we obtain the kinetic energy of the robot arm with respect to the inertial frame:

$$\mathrm{K.E.} = \sum_{i=1}^{n} \int dK_i = \sum_{i=1}^{n} \left\{ \frac{1}{2} \mathrm{Tr} \left\{ \sum_{j=1}^{i} \sum_{k=1}^{i} \frac{\partial \mathbf{T}_0^i}{\partial \vartheta_j} \mathbf{J}_i \left[\frac{\partial \mathbf{T}_0^i}{\partial \vartheta_k} \right]^T \dot{\vartheta}_j \dot{\vartheta}_k \right\} \right\} \tag{33}$$

where $\mathbf{J}_i$ can be expressed as

$$\mathbf{J}_i = \int \mathbf{r}_i \mathbf{r}_i^T \, \mathrm{dm} = \begin{bmatrix} \int x_i^2 \, \mathrm{dm} & \int x_i y_i \, \mathrm{dm} & \int x_i z_i \, \mathrm{dm} & \int x_i \, \mathrm{dm} \\ \int x_i y_i \, \mathrm{dm} & \int y_i^2 \, \mathrm{dm} & \int y_i z_i \, \mathrm{dm} & \int y_i \, \mathrm{dm} \\ \int x_i z_i \, \mathrm{dm} & \int y_i z_i \, \mathrm{dm} & \int z_i^2 \, \mathrm{dm} & \int z_i \, \mathrm{dm} \\ \int x_i \, \mathrm{dm} & \int y_i \, \mathrm{dm} & \int z_i \, \mathrm{dm} & \int \mathrm{dm} \end{bmatrix} \tag{34}$$

If we use inertia tensor l_{ij} which is defined as

$$l_{ij} = \int \left[\delta_{ij} \left(\sum_k x_k^2 \right) - x_i x_j \right] \mathrm{dm} \tag{35}$$

then $\mathbf{J}_i$ can be expressed in inertia tensor as

$$\mathbf{J}_i = \begin{bmatrix} \dfrac{-l_{xx} + l_{yy} + l_{zz}}{2} & -l_{xy} & -l_{xz} & m_i \bar{x}_i \\ -l_{xy} & \dfrac{l_{xx} - l_{yy} + l_{zz}}{2} & -l_{yz} & m_i \bar{y}_i \\ -l_{xz} & -l_{yz} & \dfrac{l_{xx} + l_{yy} - l_{zz}}{2} & m_i \bar{z}_i \\ m_i \bar{x}_i & m_i \bar{y}_i & m_i \bar{z}_i & m_i \end{bmatrix} \tag{36}$$

or as

$$\mathbf{J}_i = m_i \cdot \begin{bmatrix} \frac{-k_{i11}^2 + k_{i22}^2 + k_{i33}^2}{2} & k_{i12}^2 & k_{i13}^2 & \bar{x}_i \\ k_{i12}^2 & \frac{k_{i11}^2 - k_{i22}^2 + k_{i33}^2}{2} & k_{i23}^2 & \bar{y}_i \\ k_{i13}^2 & k_{i23}^2 & \frac{k_{i11}^2 + k_{i22}^2 - k_{i33}^2}{2} & \bar{z}_i \\ \bar{x}_i & \bar{y}_i & \bar{z}_i & 1 \end{bmatrix} \tag{37}$$

where $\bar{\mathbf{r}}_i = (\bar{x}_i, \bar{y}_i, \bar{z}_i, 1)^T$ is the location of the center of mass of link i with respect to the ith coordinate frame, and k_{ijk} is the radius of gyration of link i about the $\boldsymbol{j} - \boldsymbol{k}$ axes.

3.1.2 Potential Energy Calculation

The total potential energy of the robot arm is the sum of the potential energy of each link expressed in the base coordinate frame:

$$\text{P.E.} = \sum_{i=1}^{n} P_i = \sum_{i=1}^{6} -m_i \mathbf{g} \mathbf{T}_0^i \bar{\mathbf{r}}_i \tag{38}$$

where $\mathbf{g}$ is the gravity row vector $= (g_x, g_y, g_z, 0)$ and $|\mathbf{g}| = 9.8062 \text{ m/s}^2$.

3.1.3 Lagrange–Euler Equations of Motion

From Eqs. (33) and (38), we can form the Lagrangian function $L =$ K.E. $-$ P.E. and, applying the Lagrange–Euler formulation to the Lagrangian function, we obtain the necessary generalized torque τ_i for joint i to drive the ith link of the robot arm:

$$\begin{aligned} \tau_i = \frac{d}{dt}\left(\frac{\partial L}{\partial \dot{\vartheta}_i}\right) - \frac{\partial L}{\partial \vartheta_i} &= \sum_{k=i}^{n} \sum_{j=1}^{k} \operatorname{Tr}\left\{\frac{\partial \mathbf{T}_0^k}{\partial \vartheta_j} \mathbf{J}_k \left(\frac{\partial \mathbf{T}_0^k}{\partial \vartheta_i}\right)^T\right\} \ddot{\vartheta}_j \\ &+ \sum_{r=i}^{n} \sum_{j=1}^{r} \sum_{k=1}^{r} \operatorname{Tr}\left\{\frac{\partial^2 \mathbf{T}_0^r}{\partial \vartheta_j \, \partial \vartheta_k} \mathbf{J}_r \left(\frac{\partial \mathbf{T}_0^r}{\partial \vartheta_i}\right)^T\right\} \dot{\vartheta}_j \dot{\vartheta}_k \\ &- \sum_{j=i}^{n} m_j \mathbf{g} \left\{\frac{\partial \mathbf{T}^{jo}}{\partial \vartheta_i}\right\} \bar{\mathbf{r}}_j, \qquad \text{for } i = 1, 2, \ldots, n \end{aligned} \tag{39}$$

or expressed in matrix form explicitly,

$$\sum_{k=1}^{n} D_{ik} \ddot{\vartheta}_k + \sum_{k=1}^{n} \sum_{m=1}^{n} H_{ikm} \dot{\vartheta}_k \dot{\vartheta}_m + G_i = \tau_i, \qquad \text{for } i = 1, 2, \ldots, n \tag{40}$$

Or

$$\mathbf{D}(\vartheta) + \mathbf{h}(\vartheta, \dot{\vartheta}) + \mathbf{c}(\vartheta) = \tau \tag{41}$$

where

$$D_{ik} = \sum_{j=\max(l,k)}^{n} \mathrm{Tr}\left\{\frac{\partial \mathbf{T}_0^j}{\partial \vartheta_k} \mathbf{J}_j \left(\frac{\partial \mathbf{T}_0^j}{\partial \vartheta_i}\right)^T\right\}, \qquad \text{for } i = 1, 2, \ldots, n \tag{42}$$

$$H_{ikm} = \sum_{j=\max(l,k,m)}^{n} \mathrm{Tr}\left\{\frac{\partial^2 \mathbf{T}_0^j}{\partial \vartheta_k\, \partial \vartheta_m} \mathbf{J}_j \left(\frac{\partial \mathbf{T}_0^j}{\partial \vartheta_i}\right)^T\right\}, \qquad \text{for } i, k, m = 1, 2, \ldots, n \tag{43}$$

$$G_i = \sum_{j=1}^{n} \left(-m_j \mathbf{g}\left\{\frac{\partial \mathbf{T}_0^j}{\partial \vartheta_i}\right\} \bar{\mathbf{r}}_j\right), \qquad \text{for } i = 1, 2, \ldots, n \tag{44}$$

The coefficients G_i, D_{ik} and H_{ikm} in Eqs. (42)–(44) are functions of both the joint variables and inertial parameters of the manipulator, and Bejczy called them *the dynamic coefficients of the manipulator*. The physical meaning of these dynamic coefficients can easily be seen from the Lagrange–Euler equations of motion given by Eq. (40):

1. The coefficients G_i are the gravity loading terms due to the links. They are defined by Eq. (44).
2. The coefficients D_{ik} are related to the acceleration of the joint variables; they are defined by Eq. (42). In particular, for $i = k$, D_{ii} is related to the acceleration of joint i where the driving torque τ_i acts, while for $i \neq k$, D_{ik} is related to the reaction torque (or force) induced by the acceleration of joint k and acting at joint i, or vice versa. Since the inertia matrix is symmetric and $\mathrm{Tr}(\mathbf{A}) = \mathrm{Tr}(\mathbf{A}^T)$, it can been shown that $D_{ik} = D_{kj}$.
3. The coefficients H_{ikm} are related to the velocity of the joint variables; they are defined by Eq. (43). The last two indices km are related to the velocities of joints k and m whose velocities induce a reaction torque (or force) at joint i. Thus, the first index i is always related to the joint where the velocity-induced reaction torques (or forces) are "felt." In particular, for $k = m$, H_{ikk} is related to the centrifugal force generated by the angular velocity of joint k and "felt" at joint i, while for $k \neq m$, H_{ikm} is related to the Coriolis force generated by the velocities of joints k and m and "felt" at joint i.

Because of its matrix structure, this formulation is appealing from a control viewpoint in that it gives a set of state equations as in Eq. (41). This form allows one to design a control law that compensates all the nonlinear effects easily. Computationally, however, these equations of motion are extremely inefficient as compared with other formulations.

3.2 Newton–Euler Formulation [1.18]

In the previous section, a set of highly coupled nonlinear second-order differential equations of motion have been derived using the Lagrange–Euler approach. Using these equations to compute the joint torques from the given joint positions, velocities and accelerations for each trajectory set point in real time has been a computational bottleneck in the open-loop control problem. This is mainly due to the inefficiency of the Lagrange–Euler equations of motion which utilize the 4×4 homogeneous transformation matrices. Recently an efficient recursive Newton–Euler formulation has been proposed to derive the equations of motion for an open kinematic chain. This formulation when applied to a robot arm results in a set of forward and backward recursive equations with "messy" vector cross-product terms. The most significant of this formulation is the computation time of the applied torques could be reduced tremendously so that real-time control is possible. The derivation is based on a set of mathematical equations that describe the kinematic relationship of the moving links of a robot arm with respect to the base coordinate system.

With reference to Figure 7, and recalling that an orthonormal coordinate system $(\mathbf{x}_{i-1}, \mathbf{y}_{i-1}, \mathbf{z}_{i-1})$ is established at joint i, then the coordinate system $(\mathbf{x}_0, \mathbf{y}_0, \mathbf{z}_0)$ is our base coordinate system while the coordinate systems $(\mathbf{x}_{i-1}, \mathbf{y}_{i-1}, \mathbf{z}_{i-1})$ and $(\mathbf{x}_i, \mathbf{y}_i, \mathbf{z}_i)$ are attached to link $i-1$ with origin O^* and link i with origin O', respectively. The origin O' is located by a position vector $\mathbf{r}$ with respect to the origin O and by a position vector $\mathbf{s}$ with respect to the origin O^*. The origin O^* is located by a position vector $\mathbf{e}$ with respect to the base coordinate system.

Let $\mathbf{v}_e$ and $\boldsymbol{\omega}_e$ be the linear and angular velocity of the coordinate system $(\mathbf{x}_{i-1}, \mathbf{y}_{i-1}, \mathbf{z}_{i-1})$ with respect to the base coordinate system $(\mathbf{x}_0, \mathbf{y}_0, \mathbf{z}_0)$, respectively. Let $\boldsymbol{\omega}_r$ and $\boldsymbol{\omega}_s$ be the angular velocity of the coordinate system $(\mathbf{x}_i, \mathbf{y}_i, \mathbf{z}_i)$ with respect to $(\mathbf{x}_0, \mathbf{y}_0, \mathbf{z}_0)$ and $(\mathbf{x}_{i-1}, \mathbf{y}_{i-1}, \mathbf{z}_{i-1})$, respectively. Then the linear velocity $\mathbf{v}_r$ and angular velocity ω_r of the coordinate system $(\mathbf{x}_i, \mathbf{y}_i, \mathbf{z}_i)$ with respect to the base coordinate system are respectively [5]

$$\mathbf{v}_r = \frac{d^*\mathbf{s}}{dt} + \boldsymbol{\omega}_e \times \mathbf{s} + \mathbf{v}_e \tag{45}$$

$$\boldsymbol{\omega}_r = \boldsymbol{\omega}_e + \boldsymbol{\omega}_s \tag{46}$$

where $d^*(\quad)/dt$ denotes the time derivative with respect to the moving coordinate system $(\mathbf{x}_{i-1}, \mathbf{y}_{i-1}, \mathbf{z}_{i-1})$. The linear acceleration $\dot{\mathbf{v}}_r$ and angular acceleration $\dot{\boldsymbol{\omega}}_r$ of the coordinate system $(\mathbf{x}_i, \mathbf{y}_i, \mathbf{z}_i)$ with respect to the base coordinate system are respectively

$$\dot{\mathbf{v}}_r = \frac{d^{*2}\mathbf{s}}{dt^2} + \quad \times \mathbf{s} + 2\boldsymbol{\omega}_e \times \frac{d^*\mathbf{s}}{dt} + \boldsymbol{\omega}_e \times (\boldsymbol{\omega}_e \times \mathbf{s}) + \dot{\mathbf{v}}_e \tag{47}$$

$$\dot{\omega}_r = \dot{\omega}_e + \dot{\omega}_s \tag{48}$$

and

$$\dot{\omega}_s = \frac{d^*\omega_s}{dt} + \omega_e \times \omega_s \tag{49}$$

Based on Eqs. (45)–(49), the Newton–Euler equations of motion for manipulators having all rotary joints can be derived and are listed below. Detailed derivation of the equations of motion can be found in Ref. [18]:

***Forward Equations*: for $i = 1, 2, \ldots, n$**

$$\omega_i = \mathbf{R}_i^{i-1}(\omega_{i-1} + \dot{\vartheta}_i \mathbf{z}_{i-1}) \tag{50}$$

$$\boldsymbol{\alpha}_i = \mathbf{R}_i^{i-1}(\boldsymbol{\alpha}_{i-1} + \omega_{i-1} \times \dot{\vartheta}_i \mathbf{z}_{i-1} + \ddot{\vartheta}_i \mathbf{z}_{i-1}) \tag{51}$$

$$\mathbf{a}_i = \omega_i \times (\omega_i \times \mathbf{r}_i) + \alpha_i \times \mathbf{r}_i + \mathbf{R}_i^{i-1}\mathbf{a}_{i-1} \tag{52}$$

$$\bar{\mathbf{a}}_i = \omega_i \times (\omega_i \times \bar{\mathbf{r}}_i) + \boldsymbol{\alpha}_i \times \bar{\mathbf{r}}_i + \mathbf{a}_i \tag{53}$$

***Backward Equations*: for $i = n, n-1, \ldots, 1$**

$$\mathbf{f}_i = m_i \bar{\mathbf{a}}_{\underset{1}{i}} + \mathbf{R}_i^{i+1}\mathbf{f}_{i+1} = \mathbf{F}_i + \mathbf{R}_i^{i+1}\mathbf{f}_{i+1} \tag{54}$$

$$\mathbf{n}_i = \mathbf{I}_i \boldsymbol{\alpha}_i + \omega_i \times (\mathbf{I}_i \omega_i) + m_i(\bar{\mathbf{r}}_i + \mathbf{r}_i) \times \bar{\mathbf{a}}_i + \mathbf{r}_i \times \mathbf{R}_i^{i+1}\mathbf{f}_{i+1} + \mathbf{R}_i^{i+1}\mathbf{n}_{i+1} \tag{55}$$

$$\tau_i = (\mathbf{R}_i^{i-1}\mathbf{z}_0)^T \mathbf{n}_i \tag{56}$$

where $\mathbf{z}_0 = (0, 0, 1)^T$ and the "usual" initial conditions are $\boldsymbol{\omega}_0 = 0$, $\dot{\mathbf{v}}_0 = g\mathbf{z}_0$, $\dot{\omega}_0 = 0$, and $g = 9.8062\ m/s^2$, and $\mathbf{f}_{n+1}$ and $\mathbf{n}_{n+1}$ are external force and moment exerted on the hand, respectively. The variables in Eqs. (50)–(56) are defined as

ω_i = the angular velocity of link i with respect to the ith coordinate system;

$\boldsymbol{\alpha}_i$ = the angular acceleration of link i with respect to the ith coordinate system;

$\mathbf{r}_i$ = the origin of the ith frame with respect to the $(i-1)$th frame;

$\bar{\mathbf{r}}_i$ = the center of mass of link i with respect to the ith frame;

$\mathbf{a}_i$ = the linear acceleration of link i with respect to the ith coordinate system;

$\bar{\mathbf{a}}_i$ = the linear acceleration of the center of mass of link i with respect to the ith coordinate system;

$\mathbf{R}_{i-1}^i$ = the rotation matrix that maps position vectors from the ith coordinate system to the $(i-1)$th coordinate system;

$\mathbf{I}_i$ = the inertia about center of mass of link i with respect to the ith coordinate system;

$\mathbf{F}_i$ = the total external force exerted on link i with respect to the ith coordinate system;

$\mathbf{N}_i$ = the total moment exerted on link i with respect to the ith coordinate system;

$\mathbf{f}_i$ = the force exerted on link i by link $i-1$ with respect to the ith coordinate system,

$\mathbf{n}_i$ = the moment exerted on link i by link $i-1$ with respect to the ith coordinate system;

τ_i = the torque exerted on link i.

3.3 Generalised d'Alembert Equations of Motion [16]

Computationally, the L–E equations of motion are inefficient due to the 4×4 homogeneous matrix manipulations, while the efficiency of the N–E formulation can be seen from the vector formulation and its recursive nature. Because of its recursive computation, it does not provide much insight for the design of controllers. In order to obtain an efficient set of closed form equations of motion, one can utilize the relative position vector and rotation matrix representation to describe each link's kinematics information, obtain the kinetic and potential energies of the robot arm to form the Lagrangian function and apply the Lagrange–Euler formulation to obtain the equations of motion. The following section outlines the derivative of the equations of motion for manipulators having all the rotary joints. Detailed derivation can be found in Ref. [16].

Assuming that the links of a robot arm are rigid bodies, the kinetic energy of the ith link can be expressed as the summation of the kinetic energies due to the translational and rotational effects at its center of mass and summing all the kinetic energy of each link, the total kinetic energy of the robot arm equals

$$\text{K.E.} = \sum_{i=1}^{n} \left[\tfrac{1}{2} m_i(\mathbf{v}_i \cdot \mathbf{v}_i) + \tfrac{1}{2}(\mathbf{R}_i^0 \omega_i)^T \mathbf{I}_i (\mathbf{R}_i^0 \omega_i)\right] \tag{57}$$

where $\mathbf{v}_i$ = the linear velocity of the center of mass of link i with respect to the base coordinate frame;

ω_i = the angular velocity of link i with respect to the base coordinate system;

$\mathbf{R}_i^0 \omega_i$ = the angular velocity of link i with respect to the ith coordinate system; and

$\mathbf{I}_i$ = the inertia tensor matrix of link i about its center of mass expressed in the ith coordinate system.

The potential energy of the robot arm equals to the sum of all the potential energies of each link:

$$\text{P.E.} = \sum_{i=1}^{n} P_i = \sum_{i=1}^{n} -\mathbf{g} \cdot m_i \bar{\mathbf{r}}_i \tag{58}$$

where m_i = mass of link i, $\bar{\mathbf{r}}_i$ is the position vector of the center of mass of link i from the base coordinate frame, $\mathbf{g} = (g_x, g_y, g_z)^T$, and $|\mathbf{g}| = 9.8062\ m/s^2$.

From Eqs. (57)–(58), we can form the Legrangian function, L = K.E. − P.E., and applying the Lagrange–Euler formulation to the Lagrangian function, we obtain the generalized torque for joint i to drive the link i:

$$\sum_{j=1}^{n} D_{ij}\ddot{\vartheta}_j + H_i^{\text{tran}}(\vartheta, \dot{\vartheta}) + H_i^{\text{rot}}(\vartheta, \dot{\vartheta}) + G_i = \tau_i; \qquad i = 1, 2, \ldots, n \tag{59}$$

where

$$D_{ij} = D_{ij}^{\text{rot}} + D_{ij}^{\text{tran}} = \sum_{s=j}^{n} \{(\mathbf{R}_s^0 \mathbf{z}_{i-1})^T \mathbf{I}_s (\mathbf{R}_s^0 \mathbf{z}_{j-1})\}$$
$$+ \sum_{s=j}^{n} [m_s\{\mathbf{z}_{j-1} \times (\bar{\mathbf{r}}_s - \mathbf{p}_{j-1})\} \cdot \{\mathbf{z}_{j-1} \times (\bar{\mathbf{r}}_s - \mathbf{p}_{i-1})\}]; \quad i \le j \tag{60}$$

$$H_i^{\text{tran}}(\vartheta, \dot{\vartheta}) = \sum_{s=1}^{n}\left[m_s \left[\sum_{k=1}^{s-1} \left\{ \left\{ \left(\sum_{p=1}^{k} \dot{\vartheta}_p \mathbf{z}_{p-1} \right) \times \left(\left(\sum_{q=1}^{k} \dot{\vartheta}_q \mathbf{z}_{q-1} \right) \times \mathbf{p}_k^* \right) \right\} \right.\right.\right.$$
$$\left.\left. + \left\{ \sum_{p=2}^{k} \left(\left(\sum_{q=1}^{p-1} \dot{\vartheta}_q \mathbf{z}_{p-1} \right) \times \dot{\vartheta}_p \mathbf{z}_{p-1} \right) \times \mathbf{p}_k^* \right\} \right\} \right]$$
$$\left. \cdot \{\mathbf{z}_{i-1} \times (\bar{\mathbf{r}}_s - \mathbf{p}_{i-1})\} \right]$$
$$+ \sum_{s=i}^{n} \left[m_s \left\{ \left\{ \left(\sum_{p=1}^{s} \dot{\vartheta}_p \mathbf{z}_{p-1} \right) \times \left(\left(\sum_{q=1}^{s} \dot{\vartheta}_q \mathbf{z}_{q-1} \right) \times \bar{\mathbf{c}}_s \right) \right\} \right.\right.$$
$$\left. + \left\{ \sum_{p=2}^{s} \left(\left(\sum_{q=1}^{p-1} \dot{\vartheta}_q \mathbf{z}_{q-1} \right) \times \dot{\vartheta}_p \mathbf{z}_{p-1} \right) \times \bar{\mathbf{c}}_s \right\} \right\}$$
$$\left. \cdot \{\mathbf{z}_{i-1} \times (\bar{\mathbf{r}}_s - \mathbf{p}_{i-1})\} \right] \tag{61}$$

$$H_i^{\text{rot}}(\vartheta, \dot{\vartheta}) = \sum_{s=i}^{n} \left[(\mathbf{R}_s^0 \mathbf{z}_{i=1})^T \mathbf{I}_s \left\{ \sum_{j=1}^{s} \left(\dot{\vartheta}_j \mathbf{R}_s^0 \mathbf{z}_{j-1} \times \left(\sum_{k=j+1}^{s} \dot{\vartheta}_k \mathbf{R}_s^0 \mathbf{z}_{k-1} \right) \right) \right\} \right.$$
$$\left. + \left\{ \mathbf{R}_s^0 \mathbf{z}_{i-1} \times \left(\sum_{p=1}^{s} \dot{\vartheta}_p \mathbf{R}_s^0 \mathbf{z}_{p-1} \right) \right\}^T \mathbf{I}_s \left(\sum_{q=1}^{s} \dot{\vartheta}_q \mathbf{R}_s^0 \mathbf{z}_{q-1} \right) \right] \tag{62}$$

$$G_i = -\mathbf{g} \cdot \left[\mathbf{z}_{i-1} \times \sum_{j=1}^{n} m_j (\bar{\mathbf{r}}_j - \mathbf{p}_{i-1}) \right] \tag{63}$$

where $\bar{\mathbf{c}}_s$ is the position vector of the center of mass of link s from the $(s-1)$th coordinate frame with reference to the base coordinate frame; $\mathbf{p}_s^*$ is the position vector of the sth coordinate from the $(s-1)$th coordinate frame with reference to the base coordinate frame; $\mathbf{p}_s$ is the position vector of the sth coordinate with reference to the base coordinate frame; and $\mathbf{z}_{i-1}$ is the axis of the motion of joint i.

The dynamic coefficients D_{ij} and G_i are functions of both the joint variables and inertial parameters of the manipulator, while the H_i^{tran} and H_i^{rot} are functions of the joint variables, the joint velocities, and inertial parameters of the manipulator. These coefficients have the following physical interpretation:

1. The elements of the D_{ij} matrix are related to the link inertias of the manipulator. Equation (60) reveals the acceleration effects of joint j acting on joint i where the driving torque τ_i acts. The first term of Eq. (60) indicates the inertial effects of moving link j on joint i due to the *rotational* motion of link j, and vice versa. If $i = j$, it is the effective inertias felt at joint i due to the rotational motion of link i; while if $i \neq j$, it is the pseudoproducts of inertia of link j felt at joint i due to the rotational motion of link j. The second term has the same physical meaning except it is due to the *translational* motion of link j acting on point i.
2. The $H_i^{\text{tran}}(\vartheta, \dot{\vartheta})$ is related to the velocities of the joint variables. Equation (61) represents the combined centrifugal and Coriolis reaction torques felt at joint i due to the velocities of joints p and q resulted from the *translational* motion of links p and q. The first and third terms of Eq. (61) constitute the centrifugal and Coriolis reaction forces from all the links below link s and link s, respectively, in the kinematic chain due to the translational motion of the links. If $p = q$, then it represents the centrifugal reaction forces felt at joint i. If $p \neq q$, then it indicate the Coriolis forces acting on joint i. The second and fourth terms of Eq. (61) indicates the Coriolis reaction forces contributed from the links below link s and link s, respectively, due to the translational motion of the links.
3. The $H_i^{\text{rot}}(\vartheta, \dot{\vartheta})$ is also related to the velocities of the joint variables. Similar to the $H_i^{\text{tran}}(\vartheta, \dot{\vartheta})$, Eq. (62) reveals the combined centrifugal and Coriolis reaction torques felt at joint i due to the velocities of joints p and q resulted from the *rotational* motion of links p and q. The first term of Eq. (62) indicates purely the Coriolis reaction forces of joints p and q acting on joint i due to the rotational motion of the links. The second term is the combined centrifugal and Coriolis reaction forces acting on joint i. If $p = q$, then it indicates the centrifugal reaction

forces felt at joint i, while if $p \neq q$, then it represents the Coriolis forces acting on joint i due to the rotational motion of the links.

4. The coefficient G_i in Eq. (63) represents the gravity effects acting on joint i from the links above joint i.

Since these coefficients are used quite often in designing a feedback controller for the manipulator, it would be useful to evaluate the computational complexities of these coefficients in Eqs. (60)–(63). An example of using these coefficients in designing a feedback controller is the computed torque technique which is widely used in various institutions [2]. At first sight, Eqs. (60)–(63) seem to require a large amount of cimputations. However, most of the cross-product terms can be computed very fast. As an indication of their computational complexities, a block diagram explicitly showing the procedure calculating these coefficients for every set point in the trajectory in terms of multiplication and addition operations is shown in Figure 8. Table 3 summarizes the computational complexities of the L–E, N–E, and G–D equations of motion in terms of required mathematical operations per trajectory set point.

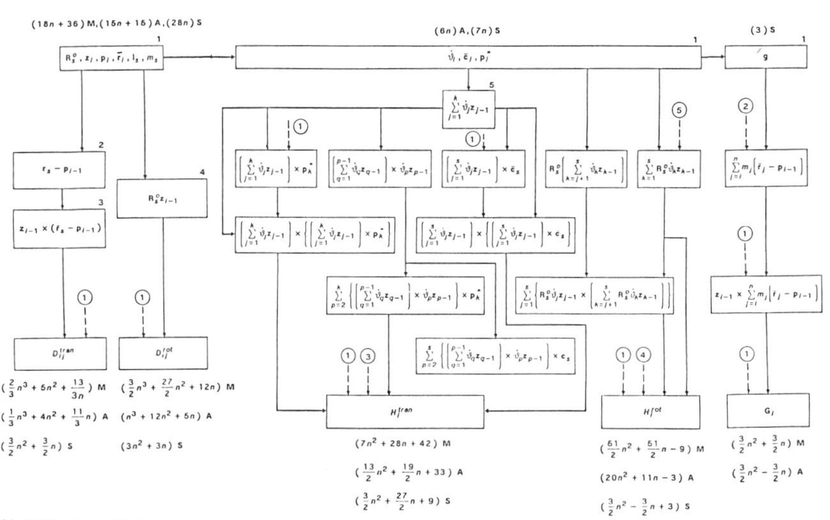

Figure 8. Computational procedure for D_{ij}, H_i^{tran}, H_i^{rot}, and G_i.

Table 3 Comparison of Various Robot Arm Dynamics Formulations*

Approach	*Lagrange–Euler*	*Newton–Euler*	*Generalized d'Alembert*
Multiplications	$\frac{128}{3}n^4 + \frac{512}{3}n^3 + \frac{739}{3}n^2 + \frac{160}{3}n$	$132n$	$\frac{13}{6}n^3 + \frac{105}{2}n^2 + \frac{268}{3}n + 69$
$n = 6$	101,348	792	2963
Additions	$\frac{98}{3}n^4 + \frac{781}{6}n^3 + \frac{559}{3}n^2 + \frac{245}{6}n$	$111n - 4$	$\frac{4}{3}n^3 + 44n^2 + \frac{146}{3}n + 45$
$n = 6$	77,405	662	2209
Kinematics Representation	4×4 Homogeneous Matrices	Rotation Matrices and Position Vectors	Rotation Matrices and Position Vectors
Equations of Motion	Closed-form Differential Equations	Recursive Equations	Closed-form Differential Equations

Note: *Here n number of degrees-of-freedom of the robot arm.

4. CONCLUSION

Basic fundamentals in robot arm kinematics and dynamics have been discussed in this chapter. The link and joint parameters of a manipulator are defined and a 4×4 homogeneous transformation matrix is introduced to describe the location of a link with respect to a fixed coordinate frame. The forward kinematic equation for a n-joint robot arm is derived. The inverse kinematics problem was introduced and solved using a geometric approach. This approach can be used to derive the inverse solution for other robot arms with rotary joints.

Three different formulations for robot arm dynamics have been presented and discussed. Comparisons among these formulations are tabulated in Table 3. The L–E equations of motion can be put in a well-structured form, but computationally it is very difficult to utilize for real-time control purposes unless the equations of motion are simplified. The N–E formulation results in a very efficient set of recursive equations, but they are very difficult to use for deriving advanced control laws in state space. The G–D equations of motion give fairly well-"structured" equations at the expense of higher computations. In addition to having faster computation time than the L–E equations of motion, the G–D equations of motion explicitly indicate the contributions of the translational and rotational effects of the links. Such information is useful for control analysis in obtaining an appropriate approximate model of the manipulator. To briefly summarize the results, a user is able to choose between a formulation which is highly structured but computationally inefficient (L–E), a formulation which has efficient computations at the expense of the "structure" of the equations of motion (N–E), and a formulation which retains the "structure" of the problem with only a moderate computing penalty (G–D).

ACKNOWLEDGMENT

This work was supported in part by the National Science Foundation Grant ECS-8106954. Any opinions, findings, and conclusions or recommendations expressed in this publication are those of the authors and do not necessarily reflect the views of the funding agency.

NOTES

1. PUMA is a trademark of Unimation, Inc.
2. Throughout this chapter, matrices are represented in upper-case boldface alphabets while vectors are in lower-case boldface alphabets.
3. Here, $(\mathbf{x}_i, \mathbf{y}_i, \mathbf{z}_i)$ actually represent are the unit vectors along the principal axes of the coordinate frame i, respectively, but are used here to denote the coordinate frame i.
4. Here, $\mathrm{Tr}\mathbf{A} = \sum_{i=1}^{n} a_{ii}$.

REFERENCES

[1] Armstrong, W. M., Recursive solution to the equations of motion of an n-link manipulator. *Proceedings of the 5th World Congress, Theory of Machines, Mechanisms*, Vol. 2, July 1979, pp 1343–1346.

[2] Bejczy, A. K., Robot arm dynamics and control. Technical Memo 33-669. Jet Propulsion Laboratory, February 1974.

[3] Chace, M. A., and Bayazitoglu, Y. O., Development and application of a generalized d'Alembert force for multifreedom mechanical systems. *Transactions of ASME, Journal of Engineering for Industry*, Series B, *93*, (Feb.), 317–327, 1971.

[4] Chase, M. A., Vector analysis of linkages, *Journal of Engineering for Industry, Trans. of ASME, 85*, Series B, 289–297.

[5] Crandall, S. H., Karnopp, D. C., Kurtz, E. F., Jr., and Pridmore-Brown, D. C., *Dynamics of Mechanical and Electromechanical Systems*. New York: McGraw-Hill, 1968, p. 177.

[6] Denavit, J., and Hartenberg, R. S., A kinematic notation for lower-pair mechanisms based on matrices. *Journal of Applied Mechanics*, June, 215–221, 1955.

[7] Duffy, J., and Rooney, J., A foundation for a unified theory of analysis of spatial mechanisms. *Trans. of ASME, Journal of Engineering for Industry, 97*, (4), Series B, 1159–1164, 1975.

[8] Duffy, J., *Analysis of Mechanisms and Robot Manipulator*. New York: John Wiley & Sons, 1980.

[9] Hartenberg, R. S., and Denavit, J., *Kinematic Synthesis of Linkages*. New York: McGraw-Hill Book Company, 1964.

[10] Hollerbach, J. M., A recursive Langrangian formulation of manipulator dynamics and a comparative study of dynamics formulation complexity. *IEEE Trans. on System, Man, and Cybernetics, SMC-10*, (11), 730–736, 1980.

[11] Huston, R. L., and Kelly, F. A., The development of equations of motion of single-arm robots. *IEEE Transactions on Systems, Man, and Cybernetics, SMC-12*, (3), 259–266, 1982.

[12] Huston, R. L. Passerello, C. E., and Harlow, M. W., Dynamics of multirigid body system. *J. Appl. Mech., 45*, 889–894, 1978.

[13] ICAM-Robotics Application Guide, Technical Report AFWAL-TR-80-4042 Volume II, Air Force Wright Aeronautical Labs., Air Force Systems Command, Wright-Patterson Air Force Base, Ohio.

[14] Kohli, D., and Soni, A. H., Kinematic analysis of spatial mechanisms via successive screw displacements. *Journal of Engineering for Industry, Trans, of ASME, 2*, Series B, 739–747, 1975.

[15] Lee, C. S. G., Robot arm kinematics, dynamics, and control. *Computer*, 15, (12), 62–80, 1982.

[16] Lee, C. S. G., Lee, B. H., and Nigam, R., Development of generalized d'Alembert equations of motion for mechanical manipulators. *Proceedings of the 22nd Conference on Decision and Control*, San Antonio, Texas, Dec. 14–16, 1983, pp 1205–1210.

[17] Lewis, R. A., Autonomous manipulation on a robot: Summary of manipulator software functions. Technical Memo 33-679, Jet Propulsion Lab, March 15, 1974.

[18] Luh, J. Y. S., Walker, M. W., and Paul, R. P., On-line computational scheme for mechanical manipulators. *Transactions of ASME, Journal of Dynamic Systems, Measurements, and Control, 120*, (June), 69–76, 1980.

[19] Milenkovic, V., and Huang, B., Kinematics of major robot linkages, *Proceedings of the 13th International Symposium on Industrial Robots*, April 17–19, 1983, Chicago, Illinois, pp. 16–31 to 16–47.

[20] Orin, D. E., McGhee, R. B., Vukobratovic, M., and Hartoch, G., Kinematic and kinetic analysis of open-chain linkages utilizing Newton–Euler methods. *Math, Biosc, 43*, 107–130, 1979.

[21] Paul, R. P., *Robot Manipulators: Mathematics, Programming and Control.* Cambridge, MA: MIT Press, 1981.

[22] Paul, R. P., Modeling, trajectory calculation and servoing of a computer controlled arm. Stanford Artificial Intelligence Laboratory, A.I. Memo 177, November 1972.

[23] Paul, R. P., Shimano, B. E., and Mayer, G., Kinematic control equations for simple manipulators. *IEEE Transactions of Systems, Man, and Cybernetics, SMC-11*, (6), 449–455, 1981.

[24] Pieper, D. L., The kinematics of manipulators under computer control. Computer Science Department, Stanford University, Artificial Intelligence Project Memo No. 72, October 1968.

[25] Scheinman, V. D., Design of a computer manipulator. Stanford Artificial Intelligence lab. Memo AIM-92, June 1969.

[26] Silver, W. M., On the equivalence of the lagrangian and Newton–Euler dynamics for manipulators. *The International Journal of Robotics Research, 1*, (2), 60–70, 1982.

[27] Stepanenko, Y., and Vukobratovic, M., Dynamics of articulated open-chain active mechanisms. *Math. Biosc., 28*, 137–170, 1976.

[28] Turney, J. L., Mudge, T. N., and Lee, C. S. G., Equivalence of two formulations for robot arm dynamics, SEL Report 142, ECE Department, University of Michigan, December 1980.

[29] Turney, J. L., Mudge, T. N., and Lee, C. S. G., Connection between formulations of robot arm dynamics with applications to simulation and control. CRIM Technical Report No. RSD-TR-4-82, the University of Michigan, April 1982.

[30] Uicker, J. J., Jr., Denavit, J., and Hartenberg, R. S., An iterative method for the displacement analysis of spatial mechanisms. *Trans. of ASME, Journal of Applied Mechanics, 31*, Series E, 309–314, 1964.

[31] Uicker, J. J., On the dynamic analysis of spatial linkages using 4 × 4 matrices. Ph.D. dissertation, Northwestern University, August 1965.

[32] Walker, M. W., and Orin, D. E., Efficient dynamic computer simulation of robotic mechanisms. *Transactions of ASME, Journal of Dynamic Systems, Measurement, and Control, 104*, (Sept.), 205–211, 1982.

[33] Yang A. T., and Freudenstein, F., Application of dual number quarternian algebra to the analysis of spatial mechanisms. *Trans. of ASME, Journal of Applied Mechanics, 31*, Series E, 1964.

[34] Yang, A. T., Displacement analysis of spatial five-link mechanisms using (3 × 3) matrices with dual-number elements. *Trans. of ASME, J. of Engineering for Industry, 91*, (1), Series B, 152–157, 1969.

[35] Yuan, M. S. C., and Freudenstein, F., Kinematic analysis of spatial mechanisms by means of screw coordinates (two parts). *Trans. of ASME, J. of Engineering for Industry, 93*, (1) 61–73, 1971.

Chapter 3

ON THE ROBOTIC MANIPULATOR CONTROL

C. S. G. Lee

ABSTRACT

The dynamic performance of robot manipulators is directly linked to the dynamic models and the associated controller. This chapter presents various control methods for industrial robots. It presents several existing control methods from simple servomechanism to advanced controls such as adaptive control with identification algorithm both in joint-variable and Cartesian coordinates.

1. INTRODUCTION

A mechanical manipulator can be modeled as an open-loop kinematic chain with several rigid links connected in series by either revolute or prismatic joints driven by actuators. One end of the chain is attached to a supporting

Advances in Automation and Robotics, Volume 1, pages 65–115.

ISBN: 0-89232-399-X

base while the other end is equipped with a tool (the end effector) to manipulate objects or perform assembly tasks. The motion of the joints results in relative motion of the links. The use of such manipulators in industrial manufacturing offers significant advantages, including improved product quality, cost savings, reliability, tolerance of working environments unacceptable to humans, and an adaptability to various tasks through simple reprogramming. These industrial robots are basically position-controlled machines performing such tasks as material handling (pick and place), spraying paint, spot/arc welding, loading and unloading numerical control machines.

Recently there has been a surge of interest in improving the overall performance of industrial robots, and in particular investigating efficient control strategies. The purpose of robot arm control is to maintain a prescribed motion for the arm along a desired time-based arm trajectory by applying corrective compensation torques to the actuators to adjust for any deviations of the arm from the trajectory. In general, the control problem consists of (a) obtaining dynamic models of the manipulator and (b) using these models to determine control laws or strategies to achieve the desired system response and performance. The current industrial approach to robot arm control system design treats each joint of the robot arm as a simple servomechanism. Such modeling is inadequate because it neglects the changes of the motion and configuration of the whole arm mechanism. These changes in the parameters of the controlled system are significant enough to render conventional feedback control strategies ineffective. The result is reduced servo response speed and damping, which limits the precision and speed of the end effector. Any significant performance gain in this and other areas of robot arm control requires the consideration of elaborate dynamic models sophisticated control techniques, and the exploitation of computer architecture.

From the control analysis point of view, the movement of a robot arm is usually performed in two distinct control phases. The first is the gross motion control in which the arm moves from an initial position/orientation to the vicinity of the desired target position/orientation along a planned trajectory. The second is the fine motion control in which the end effector of the arm dynamically interacts with the object using sensory feedback information from the sensor to complete the task. This chapter focuses on the gross motion control and discusses various control methods that are available for servoing computer-based industrial robots.

2. ROBOT ARM DYNAMICS

A priori information needed for robot arm control analysis is a set of closed form differential equations describing the dynamic behavior of the manipulator. Various approaches are available to derive the robot arm dynamics, and three formulations of robot arm dynamics suitable for robot arm control have been discussed in the previous chapter. We shall utilize those equations of motion and briefly present a state space formulation of the motion equations for control purpose.

2.1 Equations of Motion in State Space

In general, excluding the dynamics of the electronic control device, gear friction and backlash, the necessary generalized applied torque τ_i for joint i to drive the ith link of the robot arm can be obtained from the Lagrange–Euler formulation as [2, 9, 13, 19,]

$$\tau^i = \sum_{k=i}^{n} \sum_{j=1}^{k} \mathrm{Tr}\left\{\frac{\partial \mathbf{T}_0^k}{\partial q_j} \mathbf{J}_k \left(\frac{\partial \mathbf{T}_0^k}{\partial q_i}\right)^T\right\} \ddot{q}_j + \sum_{r=i}^{n} \sum_{j=1}^{r} \sum_{k=1}^{r} \mathrm{Tr}\left\{\frac{\partial^2 \mathbf{T}_0^r}{\partial q_j \partial q_k} \mathbf{J}_r \left(\frac{\partial \mathbf{T}_0^r}{\partial q_i}\right)^T\right\} \dot{q}_j \dot{q}_k$$
$$- \sum_{j=i}^{n} m_j \mathbf{g} \left\{\frac{\partial \mathbf{T}_0^j}{\partial q_i}\right\} \bar{\mathbf{r}}_j, \qquad \text{for } i = 1, 2, \ldots, n \tag{1}$$

where $\dot{q}_i$ and $\ddot{q}_i$ are the angular velocity and angular acceleration of joint i, respectively, and q_i is the generalized coordinate of the manipulator and indicates its angular position. If joint i is rotary, then $q_i \equiv \vartheta_i$; and if joint i is prismatic, then $q_i \equiv d_i$. $\mathbf{T}_0^i$ is a 4×4 homogeneous link transformation matrix which relates the spatial relationship between two coordinate frames (the ith and the base coordinate frames), $\bar{\mathbf{r}}_i$ is the position of the center of mass of link i with respect to the ith coordinate system, $\mathbf{g} = (g_x \, g_y, g_z, 0)$ is the gravity row vector and $|\mathbf{g}| = 9.8062 \; m/s^2$, the T superscript on vectors and matrices indicates the transpose operation, n is the number of degrees of freedom of the robot arm, and $\mathbf{J}_i$ is the inertial matrix of link i about the ith coordinate frame and can be expressed as

$$\mathbf{J}_i = \begin{bmatrix} \dfrac{-I_{xx} + I_{yy} + I_{zz}}{2} & -I_{xy} & -I_{xz} & m_i\bar{x}_i \\ -I_{xy} & \dfrac{I_{xx} - I_{yy} + I_{zz}}{2} & -I_{yz} & m_i\bar{y}_i \\ -I_{xz} & -I_{yz} & \dfrac{I_{xx} + I_{yy} - I_{zz}}{2} & m_i\bar{z}_i \\ m_i\bar{x}_i & m_i\bar{y}_i & m_i\bar{z}_i & m_i \end{bmatrix} \tag{2}$$

where m_i is the mass of link i and the elements of the inertia tensor I_{ij} are defined as

$$I_{ij} = \int \left[\delta_{ij}\left(\sum_k x_k^2\right) - x_i x_j\right] \mathrm{dm} \tag{3}$$

where δ_{ij} is the Kronecker delta, I_{xx}, I_{yy}, I_{zz} are called the moments of inertia about the $\mathbf{x}_i$, $\mathbf{y}_i$, $\mathbf{z}_i$ axes, respectively, $I_{ij}(i \neq j)$ are called the products of inertia, and dm is the differential mass.

Equation (1) can be expressed in matrix form explicitly as

$$\sum_{k=1}^{n} C_{ik}\ddot{q}_k + \sum_{k=1}^{n}\sum_{m=1}^{n} C_{ikm}\dot{q}_k\dot{q}_m + C_i = \tau_i, \qquad \text{for } i = 1, 2, \ldots, n \tag{4}$$

where

$$C_{ik} = \sum_{j=\max(i,k)}^{n} \operatorname{Tr}\left\{\frac{\partial \mathbf{T}_0^j}{\partial q_k}\mathbf{J}_j\left(\frac{\partial \mathbf{T}_0^j}{\partial q_i}\right)^T\right\}, \qquad \text{for } i, k = 1, 2, \ldots, n \tag{5}$$

$$C_{ikm} = \sum_{j=\max(i,k,m,)}^{n} \operatorname{Tr}\left\{\frac{\partial^2 \mathbf{T}_0^j}{\partial q_k\,\partial q_m}\mathbf{J}_j\left(\frac{\partial \mathbf{T}_0^j}{\partial q_i}\right)^T\right\}, \qquad \text{for } i, k, m = 1, 2, \ldots, n \tag{6}$$

$$C_i = \sum_{j=i}^{n}\left(-m_j\mathbf{g}\left\{\frac{\partial \mathbf{T}_0^j}{\partial q_i}\right\}\bar{\mathbf{r}}_j\right), \qquad \text{for } i = 1, 2, \ldots, n \tag{7}$$

Equation (4) can be rewritten in a more compact vector matrix notation as

$$\mathbf{D}(\mathbf{q})\ddot{\mathbf{q}}(t) + \mathbf{h}(\mathbf{q}, \dot{\mathbf{q}}) + \mathbf{c}(\mathbf{q}) = \boldsymbol{\tau}(t) \tag{8}$$

where $\boldsymbol{\tau}(t)$ is an $n \times 1$ applied torque vector for joint actuators, $\mathbf{q}(t)$ is the angular positions, $\dot{\mathbf{q}}(t)$ is the angular velocities, $\ddot{\mathbf{q}}(t)$ is an $n \times 1$ acceleration vector, $\mathbf{c}(\mathbf{q}(t))$ is an $n \times 1$ gravitational force vector, $\mathbf{h}(\mathbf{q}(t), \dot{\mathbf{q}}(t))$ is an $n \times 1$ Coriolis and centrifugal force vector, and $\mathbf{D}(\mathbf{q})$ is an $n \times n$ acceleration-related matrix.

Defining a $2n$-dimensional state vector of a manipulator as

$$\begin{aligned}\mathbf{x}^T(t) &= (\mathbf{q}^T(t), \dot{\mathbf{q}}^T(t)) \triangleq (q_1(t), \ldots, q_n(t), \dot{q}_1(t), \ldots, \dot{q}_n(t))\\ &= (\mathbf{x}_1^T(t), \mathbf{x}_2^T(t)) \triangleq (x_1(t), x_2(t), \ldots, x_{2n}(t))\end{aligned} \tag{9}$$

and an n-dimensional input vector as

$$\mathbf{u}^T(t) = (\tau_1(t), \tau_2(t), \ldots, \tau_n(t)) \tag{10}$$

Equation (8) can be expressed in state space representation as

$$\dot{\mathbf{x}}(t) = \boldsymbol{f}(\mathbf{x}(t), \mathbf{u}(t)) \tag{11}$$

where $\mathbf{f}(\cdot)$ is a $2n \times 1$ vector-valued function and continuously differentiable. Since $\mathbf{D}(\mathbf{q})$ is always nonsingular, the above equation can be expressed explicitly as

$$\begin{aligned}\dot{\mathbf{x}}_1(t) &= \mathbf{x}_2(t)\\ \dot{\mathbf{x}}_2(t) &= \mathbf{f}_2(\mathbf{x}(t)) + \mathbf{b}(\mathbf{x}_1(t))\mathbf{u}(t)\end{aligned} \tag{12}$$

where $\mathbf{f}_2(\mathbf{x})$ is an $n \times 1$ vector-valued function,

$$\mathbf{f}_2(\mathbf{x}) \equiv -\mathbf{D}^{-1}(\mathbf{x}_1)[\mathbf{h}(\mathbf{x}_1, \mathbf{x}_2) + \mathbf{c}(\mathbf{x}_1)] \tag{13}$$

and $\mathbf{b}(\mathbf{x}_1)$ is equivalent to the matrix $\mathbf{D}^{-1}(\mathbf{x}_1)$.

3. ROBOT ARM CONTROL

As stated earlier, the control problem requires the determination of the joint torques to be generated at each joint actuator for each set point on a planned trajectory in real time. Though the control problem may be stated in this rather simple manner, the solution for manipulator control is complicated by the arm's gravitational loading and inertial forces and the coupling reaction forces between joints. This section focuses on the study of control algorithms which utilize the above robot arm dynamic model(s) to efficiently control the manipulator.

Considering the robot arm control as a path/trajectory tracking problem (see Figure 1), the motion control can be classified into three major classes for discussion:

1. *Joint motion controls* [2,4,6,10,12,15,18,20,21,26]
   - Joint servomechanism—PUMA[2] robot arm control scheme [15]
   - Computed torque technique [2,12,18,20]
   - Near minimum-time control [6]
   - Variable structure control [26]
   - Nonlinear feedback control [4,10,21]

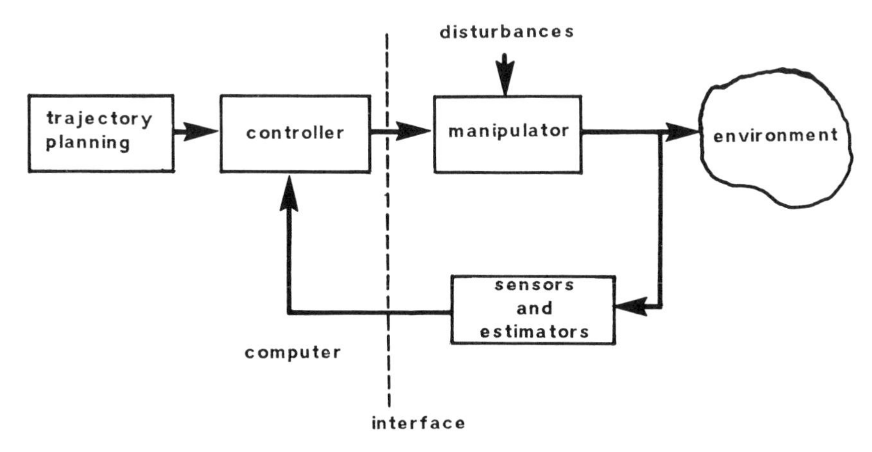

Figure 1. General robot arm control block diagram.

2. Resolved motion controls [7,23–25]

   - Resolved motion rate control [23,24]
   - Resolved motion acceleration control [17]
   - Resolved motion force control [25]

3. Adaptive controls [3,8,11,14]

   - Model referenced adaptive control [3]
   - Self-tuning adaptive control [8]
   - Adaptive control using perturbation theory [11]
   - Resolved motion adaptive control [14]

For these control methods, we assume that the desired motion is specified by a time-based path/trajectory of the manipulator either in joint coordinates or Cartesian coordinates. Each of the above control methods will be briefly described in the next several sections.

3.1 PUMA Robot Arm Control Strategy

This section discusses the computer control technique employed in the PUMA series robot arms. Like most industrial robot arms, the PUMA robot arm system employs the control technique used in the computer numerical control systems.

The current industrial practice treats each joint of the robot arm as a simple servomechanism. For the PUMA robot arm, the controller consists of an LSI-11/02 and six 6503 microprocessors each with a joint encoder, a digital-to-analog converter (DAC), and a current amplifier. The control structure is hierarchically arranged. At the top of the system hierarchy is the LSI-11/02 microcomputer, which serves as a supervisory computer. At the lower level are the six 6503 microprocessors—one for each degree of freedom (see Figure 2). The LSI-11/02 computer performs two major functions: (i) on-line user interaction and subtask scheduling from the user's VAL[3] commands and (ii) subtask coordination with the six 6503 microprocessors to carry out the command. The on-line interaction with the user includes parsing, interpreting, and decoding the VAL commands in addition to reporting appropriate error messages to the user. Once a VAL command has been decoded, various internal routines are called to perform scheduling and coordination functions. These functions, which reside in the EPROM of the LSI-11/02 computer, include the following: (i) Coordinate systems transformations (e.g., from the world coordinates *XYZOAT* to the joint coordinates $v_1, v_2, \ldots, v_6$, or vice versal). (ii) Joint-interpolated trajectory planning; this involves sending incremental location updates corresponding to each set

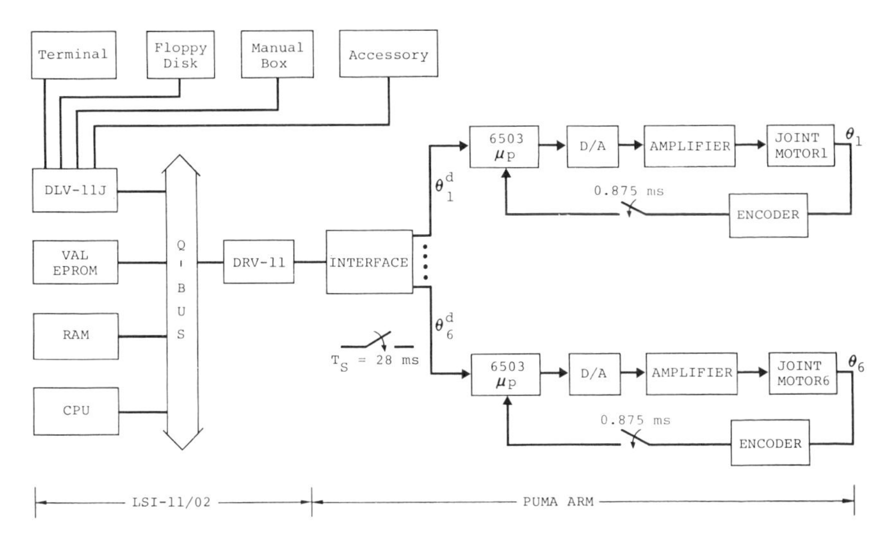

Figure 2. PUMA robot arm control scheme.

point to each joint every 28 ms. (iii) Acknowledging from the 6503 microprocessors that each axis of motion has completed its required incremental motion. (iv) Two set-point lookaheads to perform the continuous-path interpolation if the robot is in the continuous-path mode.

At the lower level in the system hierarchy is the joint controller, which consists of a digital servo board, an analog servo board, and power amplifiers. The 6503 microprocessor is an integral part of the joint controller which directly controls each axis of motion. Each microprocessor resides on a digital servo board with its EPROM and DAC. It communicates with the LSI-11/02 computer through a Unimation-designed interface board which functions as a demultiplexer that routes set-points information to each joint controller. The interface board is in turn connected to a 16-bit DEC parallel interface board (DRV-11) which transmits the data to and from the Q-Bus of the LSI-11/02 (see Figure 2). The microprocessor computes the error signal and sends it to the analog servo board which has an analog compensator designed for each joint motor. The feedback gain of the compensator is tuned to run at a specific "VAL speed." There are two servo loops for each joint control (see Figure 2). The outer loop provides position error information and is updated by the 6503 microprocessor about every 0.875 ms. The inner loop consists of analog devices and a compensator with derivative feedback to put damping on the velocity variable. Both servo loop gains are constant and tuned to perform as a "critically damped joint system" at the specific VAL speed. The main functions of the microprocessor include the following:

1. Every 28 ms, receive and acknowledge set points from the LSI-11/02 computer and perform interpolation between the current joint value and the desired joint value.
2. Every 0.875 ms, read the register value which stores the incremental values from the encoder mounted at each axis of rotation.
3. Update the error-actuating signals derived from the joint-interpolated set points and the values from the axis encoders.
4. Convert the error-actuating signal to voltage using the DACs, and send the voltage to the analog servo board which moves the joint.

It can be seen that the PUMA robot control scheme is basically a proportional plus integral plus derivative control method (PID). One of the main disadvantages of this control scheme is that the feedback gains are constants and perspecified. It does not have the capability of updating the feedback gains under varying payloads. Since an industrial robot is a highly nonlinear system, the inertial loading, the coupling between joints, and the gravity effects are all position-dependent terms. Furthermore, at high speeds the inertial loading term can change drastically. Thus, the above control scheme using constant feedback gains to control a nonlinear system does not perform well under varying speeds and payloads. In fact, the PUMA moves with noticeable vibrations at reduced speeds. One solution to the problem is the use of digital control in which the applied torques to the robot arm are computed by digital computer or microprocessor based on an appropriate dynamic model of the arm. A version of this method is discussed in the next section.

3.2 Computed Torque Technique

One of the basic control schemes is the computed torque technique based on the Lagrauge–Euler (L–E) [2,18–20] or the Newton–Euler (N–E) equations of motion [12,16]. Basically, the computed torque technique is a feedforward control and has a feedforward and a feedback component. The feedforward component compensates the interaction forces among all the various joints, and the feedback component computes the necessary correction torques to compensate any deviations from the desired trajectory. It assumes that one can accurately compute the counterparts of $\mathbf{D}(\mathbf{q})$, $\mathbf{h}(\mathbf{q}, \dot{\mathbf{q}})$, and $\mathbf{c}(\mathbf{q})$ in the L–E equations of motion [Eq. (8)] to minimize their nonlinear effects and use a proportional plus derivative control to servo the joint motors. Thus, the structure of the control law has the form

$$\tau(t) = \mathbf{D}_a(\mathbf{q})[\ddot{\mathbf{q}}^d(t) + \mathbf{K}_v(\dot{\mathbf{q}}^d(t) - \dot{\mathbf{q}}(t)) + \mathbf{K}_p(\mathbf{q}^d(t) - \mathbf{q}(t))] + \mathbf{h}_a(\mathbf{q}, \dot{\mathbf{q}}) + \mathbf{c}_a(\mathbf{q}) \quad (14)$$

where $\mathbf{K}_v$ and $\mathbf{K}_p$ are $n \times n$ derivative and position feedback gain matrices, respectively, and n is the number of degree of freedom of a manipulator.

Substituting τ from Eq. (14) into Eq. (8), we have

$$\mathbf{D}(\mathbf{q})\ddot{\mathbf{q}}(t) + \mathbf{h}(\mathbf{q}, \dot{\mathbf{q}}) + \mathbf{c}(\mathbf{q}) = \mathbf{D}_a(\mathbf{q})[\ddot{\mathbf{q}}^d(t) + \mathbf{K}_v(\dot{\mathbf{q}}^d(t) - \dot{\mathbf{q}}(t)) + \mathbf{K}_p(\mathbf{q}^d(t) - \mathbf{q}(t))] + \mathbf{h}_a(\mathbf{q}, \dot{\mathbf{q}}) + \mathbf{c}_a(\mathbf{q}) \quad (15)$$

If $\mathbf{D}_a(\mathbf{q})$, $\mathbf{h}_a(\mathbf{q}, \dot{\mathbf{q}})$, $\mathbf{c}_a(\mathbf{q})$ are equal to $\mathbf{D}(\mathbf{q})$, $\mathbf{h}(\mathbf{q}, \dot{\mathbf{q}})$, $\mathbf{c}(\mathbf{q})$, respectively, then Eq. (15) reduces to

$$\mathbf{D}(\mathbf{q})[\ddot{\mathbf{e}}(t) + \mathbf{K}_v\dot{\mathbf{e}}(t) + \mathbf{K}_p\mathbf{e}(t)] = 0 \quad (16)$$

where $\mathbf{e}(t) \triangleq \mathbf{q}^d(t) - \mathbf{q}(t)$ and $\dot{\mathbf{e}}(t) \triangleq \dot{\mathbf{q}}^d(t) - \dot{\mathbf{q}}(t)$.

Since $\mathbf{D}(\mathbf{q})$ is always nonsingular, $\mathbf{K}_p$ and $\mathbf{K}_v$ can be chosen such that the characteristic roots of Eq. (16) have negative real part, then the position error vector $\mathbf{e}(t)$ approaches zero asymptotically.

The computation of the joint torques based on the complete L–E equations of motion [Eq. (14)] is very inefficient. As a result, Paul [20] concluded that real-time closed-loop digital control is impossible or very difficult. It requires 2000 and 1500 floating-point additions and multiplications, respectively, to compute all the joint torques per set point for a Stanford arm. Because of this reason, it is common to simplify Eq. (14) by neglecting the velocity-related coupling term $\mathbf{h}_a(\mathbf{q}, \dot{\mathbf{q}})$ and the off-diagonal elements of the acceleration-related matrix $\mathbf{D}_a(\mathbf{q})$. In this case, the structure of the control law has the form

$$\tau(t) = \mathrm{diag}[\mathbf{D}_a(\mathbf{q})][\ddot{\mathbf{q}}^d(t) + \mathbf{K}_v(\dot{\mathbf{q}}^d(t) - \dot{\mathbf{q}}(t)) + \mathbf{K}_p(\mathbf{q}^d(t) - \mathbf{q}(t))] + \mathbf{c}_a(\mathbf{q}) \quad (17)$$

A computer simulation study has been conducted to show the effects of neglecting these terms when the controller [as in Eq. (17)] is based on the simplified L–E equations of motion and the robot arm is moving at high speeds.

An analogous control law in the joint-variable space can be derived from the N–E equations of motion to servo a robot arm. The control law is computed recursively using the N–E equations of motion. Using a PDP-11/45 computer, the feedback control equations can be computed within 3 ms for a trajectory set point if all the complex trigonometric functions are implemented as table lookup [15]. This certainly is quite acceptable for the time delay in the servo loop and thus allows one to perform real-time control on a mechanical manipulator with all its dynamics taken into consideration. The recursive control law can be obtained by substituting $\ddot{q}_i$ into the N–E equations of motion to obtain the necessary joint torque for each actuator:

$$\ddot{q}_i(t) = \ddot{q}_i^d(t) + \sum_{j=1}^{n} K_v^{ij}(\dot{q}_j^d(t) - \dot{q}_j(t)) + \sum_{j=1}^{n} K_p^{ij}(q_j^d(t) - q_j(t)) \quad (18)$$

where K_v^{ij} and K_p^{ij} are the derivative and position feedback gains for joint i, respectively, and $e_j(t) = q_j^d(t) - q_j(t)$ is the position error for joint j. The physical interpretation of putting Eq. (18) into the N–E recursive equations can be viewed as follows:

1. The first term will generate the desired torque for each joint if there is no modeling error and the physical system parameters are known. However, there are errors due to backlash, gear friction, uncertainty about the inertia parameters, and time delay in the servo loop so that deviation from the desired joint trajectory will be inevitable.
2. The remaining terms, in the N–E equations of motion, will generate the correction torque to compensate for small deviations from the desired joint trajectory.

The above recursive control law is a proportional plus derivative control and has the effect of compensating the inertial loading, coupling effects, and the gravity loading of the links. In order to achieve a "critically damped" system for each joint subsystem (which in turn loosely implies that the whole system behaves as a "critically damped" system), the feedback gain matrices $\mathbf{K}_p$ and $\mathbf{K}_v$ can be chosen as in Paul [19].

In summary, the computed torque technique is a feedforward compensation control. Based on a complete L–E equations of motion, the joint torques can be computed in $O(n^4)$ time. The analogous control law derived from the N–E equations of motion can be computed in $O(n)$ time. One of the main drawbacks of this control technique is that the convergence of the position error vector depends on the dynamic coefficients of $\mathbf{D}(\mathbf{q})$, $\mathbf{h}(\mathbf{q}, \dot{\mathbf{q}})$, and $\mathbf{c}(\mathbf{q})$ in the equations of motion.

3.3 Near-Minimum-Time Control

For most manufacturing tasks, it is desirable to move the manipulators at their highest speed to minimize the task cycle time. This prompted Kahn [6] to investigate the time-optimal control problem for mechanical manipulators. The objective of minimum-time control is to transfer the end effector of a manipulator from an initial position to a specified desired position in minimum time.

Let us briefly discuss the time-optimal control for a six-link manipulator. The state space representation of the equations of motion of a six-link robot are given as in Eq. (11). At the initial time $t = t_0$ the system is assumed to be in the initial state $\mathbf{x}(t_0) = \mathbf{x}_0$, and at the final minimum time $t = t_f$ the system is required to be in the desired final state $\mathbf{x}(t_f) = \mathbf{x}_f$. Furthermore, the admissible controls of the system are assumed to be bounded and satisfy the constraints

$$|u_i| \leq (u_i)_{\max}, \qquad \text{for all } t \tag{19}$$

Then the time-optimal control problem is to find an admissible control which transfers the system from the initial state $\mathbf{x}_0$ to the final state $\mathbf{x}_f$ while minimizing the performance index in Eq. (20) and subject to the constraints of Eq. (11),

$$J = \int_{t_0}^{t_f} dt = t_f - t_0 \tag{20}$$

Using the Pontryagin minimum principle [7], an optimal control which minimizes the above functional, J, must minimize the Hamiltonian. In terms of the optimal state vector $\mathbf{x}^*(t)$, the optimal control vector $\mathbf{u}^*(t)$, and the optimal adjoint variables $\mathbf{p}^*(t)$, and the Hamiltonian function

$$H(\mathbf{x}, \mathbf{p}, \mathbf{u}) = \mathbf{p}^T\mathbf{f}(\mathbf{x}, \mathbf{u}) + 1 \tag{21}$$

the necessary conditions for $\mathbf{u}^*(t)$ to be an optimal control are

$$\dot{\mathbf{x}}^*(t) = \frac{\partial H(\mathbf{x}^*, \mathbf{p}^*, \mathbf{u}^*)}{\partial \mathbf{p}}, \qquad \text{for all } t \in [t_0, t_f] \tag{22}$$

$$\dot{\mathbf{p}}^*(t) = -\frac{\partial H(\mathbf{x}^*, \mathbf{p}^*, \mathbf{u}^*)}{\partial \mathbf{x}}, \qquad \text{for all } t \in [t_0, t_f] \tag{23}$$

$$H(\mathbf{x}^*, \mathbf{p}^*, \mathbf{u}^*) \leq H(\mathbf{x}^*, \mathbf{p}^*, \mathbf{u}), \qquad \text{for all } t \in [t_0, t_f] \tag{24}$$

and for all admissible controls.

Obtaining $\mathbf{u}^*(t)$ from Eqs. (21)–(24), the optimization problem reduces to a two-point boundary value problem (TPBVP) with boundary conditions on the state $\mathbf{x}(t)$ at the initial and final times. Due to the nonlinearity of the equations of motion, a numerical solution is usually the only approach to this problem. However, the numerical solution only computes the control function and does not accommodate any system disturbances. In addition, the solution is optimal for the special initial and final conditions. Hence, the computations of the optimal control have to be performed for each manipulator motion. Furthermore, in practice, the numerical procedures do not provide an acceptable solution for the control of mechanical manipulators. Therefore, as an alternative to the numerical solution, Kahn [6] proposed an approximation to the optimal control, which results in a near-minimum-time control.

The suboptimal feedback control was obtained by approximating the nonlinear system [Eq. (12)] by a linear system and analytically found an optimal control for the linear system. The linear system is obtained by a change of variables followed by linearization of the equations of motion. A transformation is used to decouple the controls in the linearized system. Defining a new set of dependent variables $\xi_i(t)$, $i = 1, 2, \ldots, 2n$, the equations of motion can be transformed, using the new state variables, to

$$\xi_i(t) = x_i(t) - x_i(t_f), \qquad i = 1, 2, \ldots, n$$
$$\xi_i(t) = x_i(t), \qquad i = n+1, \ldots, 2n \tag{25}$$

Because of this change of variables, the control problem becomes one of moving the system from an initial state $\xi(t_0)$ to the origin of the ξ space.

In order to obtain the linearized system, Eq. (25) is substituted into Eq. (12) and Taylor series expansion is used to linearize the system about the origin of the ξ space. In addition, all sine and cosine functions of ξ_i are replaced by their series representations. As a result, the linearized equations of motion are

$$\dot{\xi}(t) = \mathbf{A}\xi(t) + \mathbf{B}\mathbf{v}(t) \tag{26}$$

where $\xi^T(t) = (\xi_1, \xi_2, \ldots, \xi_n)$ and $\mathbf{v}(t)$ is related to $\mathbf{u}(t)$. Although Eq. (26) is linear, the control functions $\mathbf{v}(t)$ are coupled. By properly selecting a set of basis vectors from the linearly independent columns of the controllability matrices of $\mathbf{A}$ and $\mathbf{B}$ to decouple the control function, a new set of equations with no coupling in control variables can be obtained:

$$\dot{\zeta}(t) = \mathbf{A}\zeta(t) + \mathbf{B}\mathbf{v}(t) \tag{27}$$

Using a three-link manipulator as an example and applying the above equations to it, we can obtain a three-double-integrator system with unsymmetric bounds on controls:

$$\dot{\zeta}_{2i-1}(t) = v_i$$
$$\dot{\zeta}_{2i}(t) = \zeta_{2i-1}, \qquad i = 1, 2, 3 \tag{28}$$

where $v_i^- \leq v_i \leq v_i^+$ and

$$v_i^+ = (u_i)_{\max} + c_i$$
$$v_i^- = -(u_i)_{\max} + c_i \tag{29}$$

From this point on, the solution to the time-optimal control and switching surfaces problem can be solved by the usual procedures. The linearized and decoupled suboptimal control [Eqs. (28) and (29)] generally results in response timesand trajectories which are reasonably close to the time-optimal solutions. However, this control method is usually too complex to be used for manipulatorswith four or more degrees of freedom, and it neglects the effect of unknown external loads.

3.4 Variable Structure Control

In 1978, Young [26] proposed to use the theory of variable structure systems for the control of manipulators. The variable structure systems (VSS) are a class of systems with discontinuous feedback control. For the last 20 years, the theory of variable structure systems has found numerous applications in control of various processes in steel, chemical, and aerospace industries. The main feature of VSS is that it has the sliding mode on the switching surface. While in the sliding mode, the system remains insensitive to parameter variations and disturbances and its trajectories lie in the switching surface. It is this insensitivity property of VSS that enables us to eliminate the interactions among the joints of the manipulator.The sliding phenomena do not depend on the system parameter and have a stable property. Hence the theory of VSS can be used to design a variable structure controller (VSC) which induces sliding mode and in which lie the robot arm's trajectories. Such design of variable structure controller does not require accurate dynamic modeling of the manipulator; only the bounds of the model parameters are sufficient to construct the controller.

The variable structure control differs from the time-optimal control in that the variable structure controller induces sliding mode in which the trajectories of the system lie. Furthermore, the system is insensitive to system parameter variations in the sliding mode.

Let us consider the variable structure control for a six-link manipulator. From Eq. (9), defining the state vector $\mathbf{x}^T(t)$ as

$$\mathbf{x}^T = (q_1, \ldots, q_6, \dot{q}_1, \ldots, \dot{q}_6) = (p_1, \ldots, p_6, v_1, \ldots, v_6) = (\mathbf{p}^T, \mathbf{v}^T) \quad (30)$$

and introducing the position error vector $\mathbf{e}_1(t) = \mathbf{p}(t) - \mathbf{p}^d$ and the velocity error vector $\mathbf{e}_2(t) = v(t)$ (with $\mathbf{v}^d = 0$), we have changed the tracking problem to a regulator problem. The error equations of the system become

$$\begin{aligned} \dot{\mathbf{e}}_1(t) &= \mathbf{v}(t) \\ \dot{\mathbf{v}}(t) &= \mathbf{f}_2(\mathbf{e}_1 + \mathbf{p}^d, \mathbf{v}) + \mathbf{b}(\mathbf{e}_1 + \mathbf{p}^d)\mathbf{u}(t) \end{aligned} \quad (31)$$

For the regulator system problem in Eq. (31), a variable structure control $\mathbf{u}(\mathbf{p}, \mathbf{v})$ can be constructed as

$$u_i(\mathbf{p}, \mathbf{v}) = \begin{cases} u_i^+(\mathbf{p}, \mathbf{v}); & \text{if } s_i(e_i, v_i) > 0; \quad i = 1, \ldots, 6 \\ u_i^-(\mathbf{p}, \mathbf{v}); & \text{if } s_i(e_i, v_i) < 0; \quad i = 1, \ldots, 6 \end{cases} \quad (32)$$

where $s_i(e_i, v_i)$ are the switching surfaces and found to be

$$s_i(e_i, v_i) = c_i e_i + v_i; \qquad c_i > 0; \quad i = 1, \ldots, 6 \quad (33)$$

and the synthesis of the control reduces to choosing the feedback controls as in Eq. (32) so that the sliding mode occurs on the intersection of the switching planes. By solving the algebraic equations of the switching planes,

$$s_i(e_i, v_i) = 0; \qquad i = 1, \ldots, 6 \quad (34)$$

a unique control exists and is found to be

$$\mathbf{u}_{\mathrm{eq}} = -\mathbf{D}(\mathbf{p})(\mathbf{f}(\mathbf{p}, \mathbf{v}) + \mathbf{C}\mathbf{v}) \tag{35}$$

where $\mathbf{C} \equiv \mathrm{diag}[c_1, c_2, \ldots, c_6]$. Then the sliding mode is obtained from Eq. (33) as

$$\dot{e}_j = -c_j e_j, \qquad i = 1, \ldots, 6 \tag{36}$$

The above equation represents six uncoupled first-order linear systems; each represents one degree of freedom of the manipulator when the system is in sliding mode. As we can see, the controller [Eq. (35)] forces the manipulator into sliding mode and the interactions among the joints are completely eliminated. The dynamics of the manipulator in the sliding mode depends only on the parameters c_i which are design parameters. With the choice of $c_i > 0$, we can obtain the asymptotic stability of the system in sliding mode and make a speed adjustment of the motion in sliding mode by varying the parameters c_i.

In summary, the variable structure control eliminates the nonlinear interactions among the joints by forcing the system into sliding mode. However, the controller produces a discontinuous feedback control signal that changes sign rapidly. The effects of such control signals to the physical control device of the manipulator should be taken into consideration for any applications to robot arm control. A more detailed discussion of designing multiinput controller for a VSS can be found in Ref. [26].

3.5 Nonlinear Feedback Control

Nonlinear feedback control methods [4,10,21] have been proposed by many researchers. The use of nonlinear feedback components to minimize the effects of the nonlinear coupling terms in a nonlinear control system and transform the nonlinear system to a linear system that can be controlled using state feedback is not new to control practitioners, but it is a good approach to control multijoint robot arms. There is a substantial body of nonlinear control theory which may allow one to design a near-optimal control for mechanical manipulators. Most of the existing robot control algorithms emphasize nonlinear compensations of the interactions among the links, as in the computed torque technique. Hemani and Camana [4] applied the nonlinear feedback control technique to a simple locomotion system which has a particular class of nonlinearity (sine, cosine, and polynomial) and obtained decoupled subsystems, postural stability, and desired periodic trajectories. Their approach is different from the method of linear system decoupling where the system to be decoupled must be linear. Saridis and Lee [21] proposed an iterative algorithm for sequential improvement of a nonlincar suboptimal control law. It provides an approximate optimal control for a manipulator. To achieve such a high quality of control this method also requires a considerable amount of computational time.

Another approach of using nonlinear feedback component was proposed to control a six-link robot arm system [10]. Using the nonlinear feedback component, the highly coupled nonlinear system was transformed to a quasi-linear system which was controlled via state feedback. Basically the control system consists of two servo loops. The inner loop consists of nonlinear feedback to minimize the nonlinear effects while the outer loop controls the quasi-linearized control system with switching functions that stablize the control system. The reformulation of the robot arm dynamic model to accommodate the nonlinear feedback control is briefly described here.

Since

$$\frac{dC_{ik}}{dt} = \sum_{m=1}^{n} C_{ikm}\dot{q}_m + \sum_{m=1}^{n} C_{kim}\dot{q}_m \tag{37}$$

where C_{ik}, C_i, and C_{ikm} are defined in Eqs. (5)–(7). The L–E equations of motion of a manipulator [Eq. (4)] can be rewritten as

$$\tau_i = \frac{d}{dt}\left[\sum_{k=1}^{n} C_{ik}\dot{q}_k\right] + C_i - \sum_{k=1}^{n}\sum_{m=1}^{n} C_{kim}\dot{q}_k\dot{q}_m, \qquad i = 1, \ldots, n \tag{38}$$

It should be noted that $C_{ikm} \neq C_{kim}$. If we let $q_i^d(t)$ be the desired angular position, and the angular position error $r_i(t) = q_i(t) - q_i^d(t)$, by substituting $r_i(t)$ and $\dot{r}_i(t)$ into Eq. (38), we have

$$\tau_i = \frac{d}{dt}\left[\sum_{k=1}^{n} D_{ik}\dot{r}_k\right] + D_i - \sum_{k=1}^{n}\sum_{m=1}^{n} D_{kjm}\dot{r}_k\dot{r}_m \tag{39}$$

where

$$D_{ik}(r_1, \ldots, r_n) \triangleq C_{ik}(q_1 - q_1^d, \ldots, q_n - q_n^d)$$

$$D_i(r_1, \ldots, r_n) \triangleq C_i(q_1 - q_1^d, \ldots, q_n - q_n^d)$$

$$D_{kim}(r_1, \ldots, r_n) \triangleq C_{kim}(q_1 - q_1^d, \ldots, q_n - q_n^d)$$

In other words, the control system is to be driven to the origin of the r_i space.

From Eq. (39), we can consider the input torque τ_i to be consisted of a feedback σ_i and a control u_i. That is, $\sigma_i + u_i = \tau_i$, $1 \leq i \leq n$. The feedback control component σ_i can be chosen as

$$\sigma_i = D_i - \sum_{k=1}^{n}\sum_{m=1}^{n} D_{kim}\dot{r}_k\dot{r}_m \tag{40}$$

With this feedback control, the control u_i can be found from Eqs. (39) and (40):

$$u_i + \sigma_i = \tau_i \quad \text{and} \quad u_i = \frac{d}{dt}\left[\sum_{k=1}^{n} D_{ik}\dot{r}_k\right] \tag{41}$$

Defining the generalized momenta p_i as

$$p_i = \sum_{k=1}^{n} D_{ik}\dot{r}_k, \qquad 1 \leq i \leq n; \quad \text{or} \quad \mathbf{p} = \mathbf{D}\dot{\mathbf{r}} \tag{42}$$

Eqs. (41) and (42) become

$$\begin{aligned} \dot{\mathbf{p}}(t) &= \mathbf{u}(t) \\ \dot{\mathbf{r}}(t) &= \mathbf{D}^{-1}(\mathbf{r})\mathbf{p}(t) \end{aligned} \tag{43}$$

where

$$\mathbf{p}(t) = (p_1, \ldots, p_n)^T, \qquad \mathbf{u}(t) = (u_1, \ldots, u_n)^T, \qquad \mathbf{r}(t) = (r_1, \ldots, r_n)^T$$

Equation (43) can be expressed in a matrix vector form as

$$\dot{\mathbf{x}}(t) = \mathbf{A}(\mathbf{x})\mathbf{x}(t) + \mathbf{B}\mathbf{u}(t) \tag{44}$$

where

$$\mathbf{x}(t) = (p_1, \ldots, p_n, r_1, \ldots, r_n)^T = \begin{bmatrix} \mathbf{p} \\ \mathbf{r} \end{bmatrix}, \qquad \mathbf{x} \in \mathbf{R}^{2n}$$

$$\mathbf{A}_{2n \times 2n}(\mathbf{x}) = \left[\begin{array}{c|c} 0 & 0 \\ \hline \mathbf{D}^{-1}(\mathbf{x}) & 0 \end{array}\right]_{2n \times 2n}; \qquad \mathbf{B}_{2n \times n} = \left[\begin{array}{ccccc} 1 & 0 & 0 & \cdot & 0 \\ 0 & 1 & 0 & \cdot & 0 \\ \cdot & \cdot & \cdot & \cdot & \cdot \\ \cdot & \cdot & \cdot & \cdot & \cdot \\ 0 & 0 & \cdot & \cdot & 1 \\ \hline 0 & 0 & \cdot & \cdot & 0 \\ \cdot & \cdot & \cdot & \cdot & \cdot \\ \cdot & \cdot & \cdot & \cdot & \cdot \\ 0 & 0 & \cdot & \cdot & 0 \end{array}\right] = \begin{bmatrix} \mathbf{I}_n \\ 0 \end{bmatrix}$$

Using the recursive computation technique in Ref. [5], the nonlinear feedback component in Eq. (40) can be computed in $O(n)$ time. Hence the implementation of the nonlinear feedback is straightforward.

As a result of the nonlinear feedback, the control system becomes quasi-linearized as in Eq. (43). Treating Eq. (44) as a linear system, a linear quadratic controller with switching functions for stability can be designed. The resulting control system is shown in Figure 3. A more detailed discussion of this controller design can be found in Ref. [10].

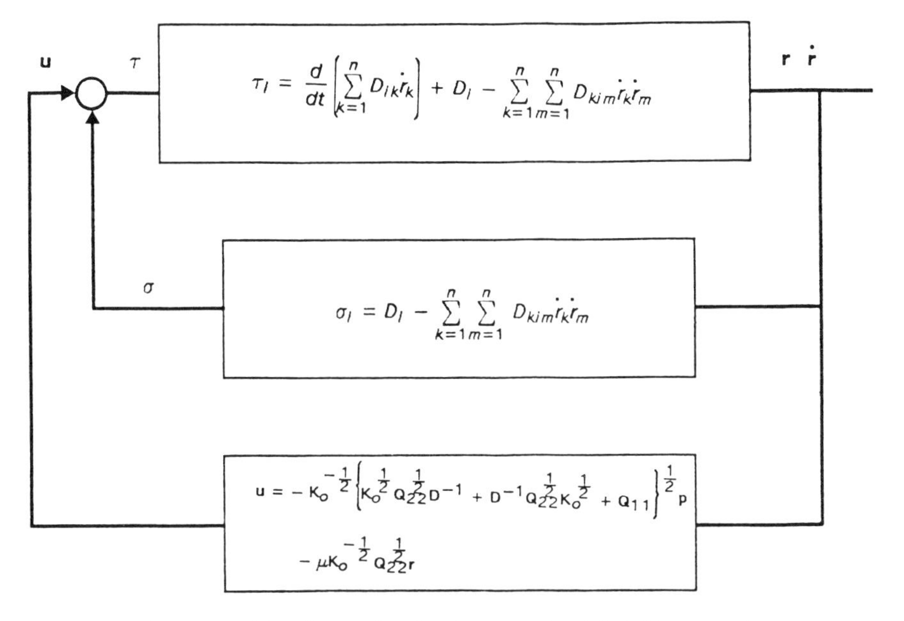

Figure 3. Nonlinear feedback control with LQ controller.

In summary, nonlinear feedback control minimizes the effects of the nonlinear coupling terms in a nonlinear control system and transforms the nonlinear system to a linear system that can be controlled using state feedback. However, it is worth noting that the nonlinear feedback control assumes perfect cancellation of the nonlinear terms in the system, which hardly can be achieved in practice, and the computation of nonlinearity by a computer must be carried out rapidly for sufficient control.

4. ROBOT ARM CONTROL—RESOLVED MOTION CONTROL

In the last section, several control methods are used to control a mechanical manipulator in the joint-variable space to follow a joint-interpolated trajectory. For most applications, resolved motion control, which commands the manipulator hand to move in a desired Cartesian direction in a coordinated position and rate control, may be more appropriate and desired. *Resolved motion* means that the motions of the various joint motors are combined and resolved into separately controllable hand motions along the world coordinate axes. This implies that several joint motors must run simultaneously at different time-varying rates in order to achieve desired coordinated hand motion along any one world coordinate axis. This enables the user to specify the direction and speed along any arbitrarily oriented path for the manipulator to follow. This motion control greatly simplifies the specification of the sequence of motion for completing a task because users are usually more adapted to the Cartesian coordinate system than the manipulator's joint angle coordinates.

In general, the desired motion of a manipulator is specified in terms of a time-based hand trajectory in Cartesian coordinates, while the servo control system requires the reference inputs specified in joint coordinates. The mathematical relationship between these two coordinate systems is important in designing efficient control in the Cartesian space. We shall briefly describe the basic kinematic theory relating these two coordinate systems for a six-link robot arm that will lead us to understand various resolved motion control methods that are described in the following sections.

The location of the manipulator hand with respect to a fixed reference coordinate system can be realized by establishing an orthonormal coordinate frame at the hand (the hand coordinate frame) [9,19] (see Figure 4). The problem of finding the location of the hand is reduced to finding the position and orientation of the hand coordinate frame with respect to the inertial frame of the manipulator. This can be conveniently achieved by a 4×4

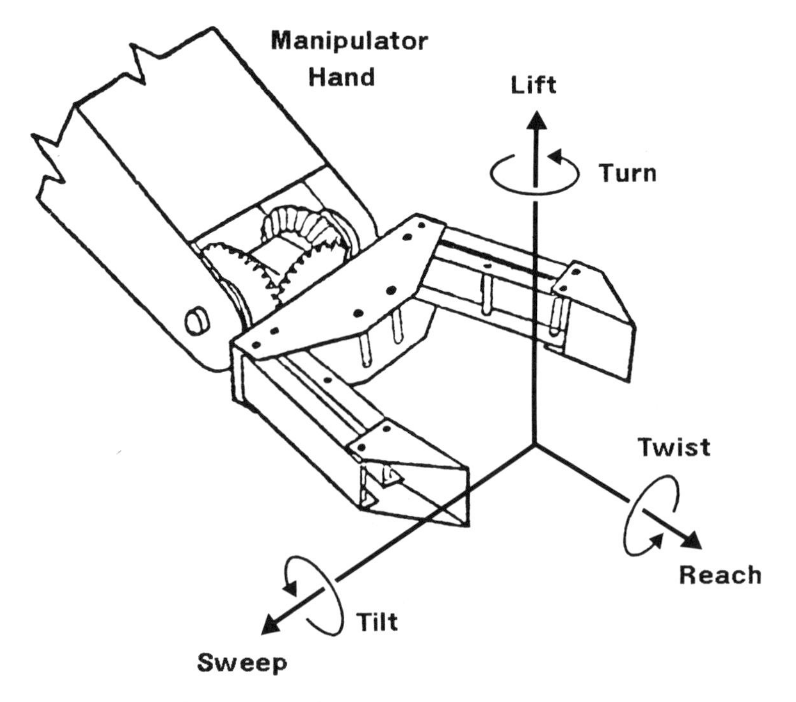

Figure 4. The hand coordinate system.

homogeneous transformation matrix [9,19]:

$$\mathbf{T}_{\text{base}}^{\text{hand}}(t) = \left[\begin{array}{c|c} \begin{matrix} \textit{Rotation} \\ \textit{matrix} \end{matrix} & \begin{matrix} \textit{Position} \\ \textit{vector} \end{matrix} \\ \hline 0 \quad 0 \quad 0 & 1 \end{array}\right] = \begin{bmatrix} n_x(t) & s_x(t) & a_x(t) & p_x(t) \\ n_y(t) & s_y(t) & a_y(t) & p_y(t) \\ n_z(t) & s_z(t) & a_z(t) & p_z(t) \\ 0 & 0 & 0 & 1 \end{bmatrix}$$

$$= \begin{bmatrix} \mathbf{n}(t) & \mathbf{s}(t) & \mathbf{a}(t) & \mathbf{p}(t) \\ 0 & 0 & 0 & 1 \end{bmatrix} \tag{45}$$

where $\mathbf{p}$ is the position vector of the hand and $\mathbf{n}$, $\mathbf{s}$, $\mathbf{a}$ are the unit vectors along the principal axes of the hand coordinate frame describing the orientation of the hand. Instead of using the rotation submatrix $[\mathbf{n}, \mathbf{s}, \mathbf{a}]$ to describe the orientation, we can use three Euler angles, yaw ($\alpha(t)$), pitch ($\beta(t)$), and roll ($\gamma(t)$), which are defined as rotations of the hand coordinate frame about the $\mathbf{x}_0$, $\mathbf{y}_0$ and $\mathbf{z}_0$ of the reference frame, respectively. One can obtain the elements of $[\mathbf{n}, \mathbf{s}, \mathbf{a}]$ from the Euler rotation matrix resulting from a rotation of α angle about the $\mathbf{x}_0$ axis, then a rotation of β angle about the $\mathbf{y}_0$ axis, and a rotation of γ angle about the $\mathbf{z}_0$ axis of the reference frame. Thus

$$\mathbf{R}_{\text{base}}^{\text{hand}}(t) = \begin{bmatrix} n_x(t) & s_x(t) & a_x(t) \\ n_y(t) & s_y(t) & a_y(t) \\ n_z(t) & s_z(t) & a_z(t) \end{bmatrix}$$

$$= \begin{bmatrix} C\gamma & -S\gamma & 0 \\ S\gamma & C\gamma & 0 \\ 0 & 0 & 1 \end{bmatrix} \cdot \begin{bmatrix} C\beta & 0 & S\beta \\ 0 & 1 & 0 \\ -S\beta & 0 & C\beta \end{bmatrix} \cdot \begin{bmatrix} 1 & 0 & 0 \\ 0 & C\alpha & -S\alpha \\ 0 & S\alpha & C\alpha \end{bmatrix}$$

$$= \begin{bmatrix} C\gamma C\beta & -S\gamma C\alpha + C\gamma S\beta S\alpha & S\gamma S\alpha + C\gamma S\beta C\alpha \\ S\gamma C\beta & C\gamma C\alpha + S\gamma S\beta S\alpha & -C\gamma S\alpha + S\gamma S\beta C\alpha \\ -S\beta & C\beta S\alpha & C\beta C\alpha \end{bmatrix} \tag{46}$$

where $\sin\alpha \equiv S\alpha$; $\cos\alpha \equiv C\alpha$; $\sin\beta \equiv S\beta$; $\cos\beta \equiv C\beta$; $\sin\gamma \equiv S\gamma$; and $\cos\gamma \equiv C\gamma$.

Let us define the position $[\mathbf{p}(t)]$, orientation $[\Phi(t)]$, linear velocity $[\mathbf{v}(t)]$, and angular velocity $[\Omega(t)]$ vectors of the manipulator hand with respect to the reference frame respectively:

$$\begin{aligned} \mathbf{p}(t) &\triangleq (p_x(t), p_y(t), p_z(t))^T, \qquad \Phi(t) \triangleq (\alpha(t), \beta(t), \gamma(t))^T \\ \mathbf{v}(t) &\triangleq (v_x(t), v_y(t), v_z(t))^T, \qquad \Omega(t) \triangleq (\omega_x(t), \omega_y(t), \omega_z(t))^T \end{aligned} \tag{47}$$

The linear velocity of the hand with respect to the reference frame is equal to the time derivative of the position of the hand:

$$\mathbf{v}(t) = \frac{d\mathbf{p}(t)}{dt} = \dot{\mathbf{p}}(t) \tag{48}$$

Since the inverse of a direction cosine matrix is equivalent to its transpose, the instantaneous angular velocities of the hand coordinate frame about the principal axes of the reference frame can be obtained from Eq. (46) as

$$\mathbf{R}\frac{d\mathbf{R}^T}{dt} = -\frac{d\mathbf{R}}{dt}\mathbf{R}^T = -\begin{bmatrix} 0 & -\omega_z & \omega_y \\ \omega_z & 0 & -\omega_x \\ -\omega_y & \omega_x & 0 \end{bmatrix}$$

$$= \begin{bmatrix} 0 & -S\beta\dot{\alpha} + \dot{\gamma} & -S\gamma C\beta\dot{\alpha} - C\gamma\dot{\beta} \\ S\beta\dot{\alpha} - \dot{\gamma} & 0 & C\gamma C\beta\dot{\alpha} - S\gamma\dot{\beta} \\ S\gamma C\beta\dot{\alpha} + C\gamma\dot{\beta} & -C\gamma C\beta\dot{\alpha} + S\gamma\dot{\beta} & 0 \end{bmatrix} \quad (49)$$

From the above equation, the relation between the $(\omega_x(t), \omega_y(t), \omega_z(t))^T$ and $(\dot{\alpha}(t), \dot{\beta}(t), \dot{\gamma}(t))^T$ can be found by equating the nonzero elements in the matrices:

$$\begin{bmatrix} \omega_x(t) \\ \omega_y(t) \\ \omega_z(t) \end{bmatrix} = -\begin{bmatrix} -C\gamma C\beta & S\gamma & 0 \\ -S\gamma C\beta & -C\gamma & 0 \\ S\beta & 0 & -1 \end{bmatrix}\begin{bmatrix} \dot{\alpha}(t) \\ \dot{\beta}(t) \\ \dot{\gamma}(t) \end{bmatrix} \quad (50)$$

Its inverse relation can be found easily:

$$\begin{bmatrix} \dot{\alpha}(t) \\ \dot{\beta}(t) \\ \dot{\gamma}(t) \end{bmatrix} = \sec\beta\begin{bmatrix} C\gamma & S\gamma & 0 \\ -S\gamma C\beta & C\gamma C\beta & 0 \\ C\gamma S\beta & S\gamma S\beta & C\beta \end{bmatrix}\begin{bmatrix} \omega_x(t) \\ \omega_y(t) \\ \omega_z(t) \end{bmatrix} \quad (51)$$

or expressed in vector matrix form:

$$\dot{\Phi}(t) \triangleq [\mathbf{S}(\Phi)]\Omega(t) \quad (52)$$

Based on the moving coordinate frame concept [24], the linear and angular velocities of the hand can be obtained from the velocities of the lower joints:

$$\begin{bmatrix} \mathbf{v}(t) \\ \Omega(t) \end{bmatrix} = [\mathbf{N}(\mathbf{q})]\dot{\mathbf{q}}(t) = [\mathbf{N}_1(\mathbf{q}), \mathbf{N}_2(\mathbf{q}), \ldots, \mathbf{N}_6(\mathbf{q})]\dot{\mathbf{q}}(t) \quad (53)$$

where $\dot{\mathbf{q}}(t) = (\dot{q}_1, \ldots, \dot{q}_6)^T$ is the joint velocity vector of the manipulator and $\mathbf{N}(\mathbf{q})$ is a 6×6 matrix whose ith column vector $\mathbf{N}_i(\mathbf{q})$ can be found from Ref. [24],

$$\mathbf{N}_i(\mathbf{q}) = \begin{cases} \begin{bmatrix} \mathbf{z}_{i-1} \times (\mathbf{p} - \mathbf{p}_{i-1}) \\ \mathbf{z}_{i-1} \end{bmatrix}, & \text{if joint } i \text{ is rotational} \\ \begin{bmatrix} \mathbf{z}_{i-1} \\ 0 \end{bmatrix}, & \text{if joint } i \text{ is translational} \end{cases} \quad (54)$$

where $\times$ indicates cross product, $\mathbf{p}_{i-1}$ is the position of the origin of the $(i-1)$th coordinate frame with respect to the reference frame, $\mathbf{z}_{i-1}$ is the unit vector along the axis of motion of joint i, and $\mathbf{p}$ is the position of the hand with respect to the reference coordinate frame.

If the inverse Jacobian matrix exists at $\mathbf{q}(t)$, then the joint velocities $\dot{\mathbf{q}}(t)$ of the manipulator can be computed from the hand velocities using Eq. (53):

$$\dot{\mathbf{q}}(t) = [\mathbf{N}^{-1}(\mathbf{q})]\begin{bmatrix}\mathbf{v}(t)\\ \Omega(t)\end{bmatrix} \tag{55}$$

Given the desired linear and angular velocities of the hand, this equation computes the joint velocities and indicates the rates at which the joint motors must be maintained in order to achieve a steady hand motion along the desired Cartesian direction.

The accelerations of the hand can be obtained by taking the time derivative of the velocity vector in Eq. (53):

$$\begin{bmatrix}\dot{\mathbf{v}}(t)\\ \dot{\Omega}(t)\end{bmatrix} = [\dot{\mathbf{N}}(\mathbf{q}, \dot{\mathbf{q}})]\dot{\mathbf{q}}(t) + [\mathbf{N}(\mathbf{q})]\ddot{\mathbf{q}}(t) \tag{56}$$

where $\ddot{\mathbf{q}}(t) = (\ddot{q}_1(t), \ldots, \ddot{q}_6(t))^T$ is the joint acceleration vector of the manipulator. Substituting $\dot{\mathbf{q}}(t)$ from Eq. (55) into Eq. (56) gives

$$\begin{bmatrix}\dot{\mathbf{v}}(t)\\ \dot{\Omega}(t)\end{bmatrix} = [\dot{\mathbf{N}}(\mathbf{q}, \dot{\mathbf{q}})][\mathbf{N}^{-1}(\mathbf{q})]\begin{bmatrix}\mathbf{v}(t)\\ \Omega(t)\end{bmatrix} + [\mathbf{N}(\mathbf{q})]\ddot{\mathbf{q}}(t) \tag{57}$$

and the joint accelerations $\ddot{\mathbf{q}}(t)$ can be computed from the hand velocities and accelerations as

$$\ddot{\mathbf{q}}(t) = [\mathbf{N}^{-1}(\mathbf{q})]\begin{bmatrix}\dot{\mathbf{v}}(t)\\ \dot{\Omega}(t)\end{bmatrix} - [\mathbf{N}^{-1}(\mathbf{q})][\dot{\mathbf{N}}(\mathbf{q}, \dot{\mathbf{q}})][\mathbf{N}^{-1}(\mathbf{q})]\begin{bmatrix}\mathbf{v}(t)\\ \Omega(t)\end{bmatrix} \tag{58}$$

The above kinematic relations between the joint coordinates and the Cartesian coordinates will be used in the next section for various resolved motion control methods and in deriving the resolved motion equations of motion of the manipulator hand in Cartesian coordinates.

4.1 Resolved Motion Rate Control

Resolved motion rate control (RMRC) [2324] means that the motions of the various joint motors are combined and run simultaneously at different time-varying rates in order to achieve steady hand motion along any one world coordinate axis. The mathematics that relates the world coordinates, such as lift (p_x), sweep (p_y), reach (p_z), yaw (α), pitch (β), and roll (γ) to the joint angle coordinate of a six-link manipulator is inherently nonlinear and can be expressed through a nonlinear vector valued function as

$$\mathbf{x}(t) = \mathbf{f}(\mathbf{q}(t)) \tag{59}$$

where $\mathbf{f}(\mathbf{q})$ is a 6×1 vector-valued function, and

$$\mathbf{x}(t) = \text{world coordinates} = (p_x, p_y, p_z, \alpha, \beta, \gamma)^T$$

and

$$\mathbf{q}(t) = \text{generalized coordinates} = (q_1, q_2, \ldots, q_n)^T$$

The relationship between the linear and angular velocities and the joint velocities of a six-link manipulator is given by Eq. (53).

For a more general discussion, if we assume that the manipulator has m degrees of freedom while the world coordinates of interest are of dimension n, then the joint angles and the world coordinates are related by a complex nonlinear function as in Eq. (59).

If we differentiate Eq. (59) with respect to time, we have

$$\frac{dx(t)}{dt} = \dot{\mathbf{x}}(t) = \mathbf{N}(\mathbf{q})\dot{\mathbf{q}}(t) \tag{60}$$

where $\mathbf{N}(\mathbf{q})$ is the Jacobian matrix with respect to $\mathbf{q}(t)$; that is,

$$[N_{ij}] = \frac{\partial f_i}{\partial q_j}, \qquad \textit{for } 1 \leq i \leq n, \quad 1 \leq j \leq m \tag{61}$$

We see that if we work with rate control, the relationship is linear as indicated by Eq. (60). When $\mathbf{x}(t)$ and $\mathbf{q}(t)$ are of the same dimension, that is, $m = n$, then the manipulator is nonredundant and the Jacobian matrix can be inverted at a particular nonsingular position $\mathbf{q}(t)$:

$$\dot{\mathbf{q}}(t) = \mathbf{N}^{-1}(\mathbf{q})\dot{\mathbf{x}}(t) \tag{62}$$

From Eq. (62), given the desired rate along the world coordinates, one can easily find the combination of joint motor rates to achieve the desired hand motion. Various methods of computing the inverse Jacobian matrix can be used [23,24]. A resolved motion rate control block diagram is shown in Figure 5.

If $m > n$, then the manipulator is redundant and the inverse Jacobian matrix does not exist. This reduces the problem to finding the generalized

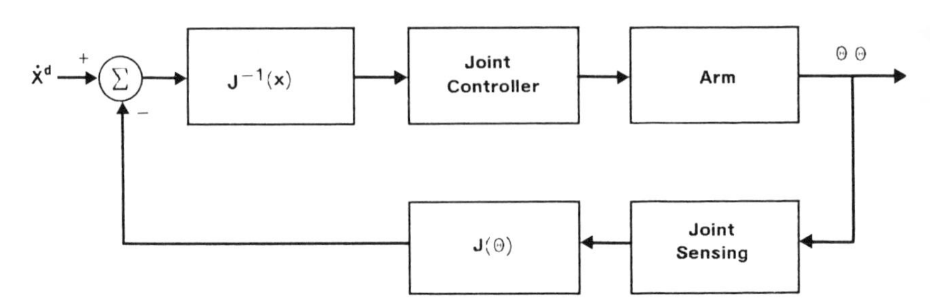

Figure 5. The resolved motion rate control block diagram.

inverse of the Jacobian matrix. In this case, if the rank of $\mathbf{N}(\mathbf{q})$ is n, then $\dot{\mathbf{q}}(t)$ can be found by minimizing an error criterion formed by adjoining Eq. (60) with a Lagrange multiplier to a cost criterion; that is,

$$C = \tfrac{1}{2}\dot{\mathbf{q}}^T\mathbf{A}\dot{\mathbf{q}} + \lambda^T(\dot{\mathbf{x}} - \mathbf{N}(\mathbf{q})\dot{\mathbf{q}}) \tag{63}$$

where λ is a Lagrange multiplier and $\mathbf{A}$ is an $m \times m$ symmetric positive define matrix.

Minimizing the cost criterion C with respect to $\dot{\mathbf{q}}(t)$ and λ, we have, respectively,

$$\dot{\mathbf{q}}(t) = \mathbf{A}^{-1}\mathbf{N}^T(\mathbf{q})\lambda \tag{64}$$

and

$$\dot{\mathbf{x}}(t) = \mathbf{N}(\mathbf{q})\dot{\mathbf{q}}(t) \tag{65}$$

Substituting $\dot{\mathbf{q}}(t)$ from Eq. (64) into Eq. (65), and solving for λ,

$$\lambda = [\mathbf{N}(\mathbf{q})\mathbf{A}^{-1}\mathbf{N}^T(\mathbf{q})]^{-1}\dot{\mathbf{x}}(t) \tag{66}$$

Substituting λ into Eq. (64), we obtain

$$\dot{\mathbf{q}}(t) = \mathbf{A}^{-1}\mathbf{N}^T(\mathbf{q})[\mathbf{N}(\mathbf{q})\mathbf{A}^{-1}\mathbf{N}^T(\mathbf{q})]^{-1}\dot{\mathbf{x}}(t) \tag{67}$$

If the matrix $\mathbf{A}$ is an identity matrix, then Eq. (67) reduces to Eq. (62).

Quite often, one would also like to command the hand motion along the hand coordinate system rather than the world coordinate system (see Figure 4). In this case, the desired hand motion $\dot{\mathbf{h}}(t)$ along the hand coordinate system is related to the world coordinate motion by

$$\dot{\mathbf{x}}(t) = \mathbf{R}_0^h\dot{\mathbf{h}}(t) \tag{68}$$

where $\mathbf{R}_0^h$ is a $n \times 6$ matrix that relates the orientation of the hand coordinate system to the world coordinate system. Given the desired hand rate motion $\dot{\mathbf{h}}(t)$ with respect to the hand coordinate system, and using Eqs. (67) and (68), the joint rate $\dot{\mathbf{q}}(t)$ can be computed by

$$\dot{\mathbf{q}}(t) = \mathbf{A}^{-1}\mathbf{N}^T(\mathbf{q})[\mathbf{N}(\mathbf{q})\mathbf{A}^{-1}\mathbf{N}^T(\mathbf{q})]^{-1}\mathbf{R}_0^h\dot{\mathbf{h}}(t) \tag{69}$$

In Eqs. (67) and (69), the angular position $\mathbf{q}(t)$ depends on time t, so we need to evaluate $\mathbf{N}^{-1}(\mathbf{q})$ at each sampling time t for the calculation of $\dot{\mathbf{q}}(t)$. The added computation in obtaining the inverse Jacobian matrix at each sampling time and the singularity problem associated with the matrix inversion are important issues in using this control method.

4.2 Resolved Motion Accelerated Control

The resolved motion acceleration control (RMAC) [17] extends the concept of resolved motion rate control to include acceleration control. It presents an alternative position control which deals directly with the position and orientation of the hand of a manipulator. All the feedback control is done at the hand level, and it assumes that the desired accelerations of a preplanned hand motion are specified by the user.

The actual and desired position and orientation of the hand of a manipulator can be represented by 4 × 4 homogeneous transformation matrices respectively as

$$\mathbf{H}(t) = \left[\begin{array}{ccc|c} \mathbf{n}(t) & \mathbf{s}(t) & \mathbf{a}(t) & \mathbf{p}(t) \\ \hline 0 & 0 & 0 & 1 \end{array}\right] \quad \text{and} \quad \mathbf{H}^d(t) = \left[\begin{array}{ccc|c} \mathbf{n}^d(t) & \mathbf{s}^d(t) & \mathbf{a}^d(t) & \mathbf{p}^d(t) \\ \hline 0 & 0 & 0 & 1 \end{array}\right] \tag{70}$$

where $\mathbf{n}$, $\mathbf{s}$, $\mathbf{a}$ are the unit vectors along the principal axes $\mathbf{x}$, $\mathbf{y}$, $\mathbf{z}$ of the hand coordinate system respectively and $\mathbf{p}(t)$ is the position vector of the hand with respect to the base coordinate system. The orientation submatrix $[\mathbf{n}, \mathbf{s}, \mathbf{a}]$ can be defined in terms of Euler angles of rotation (α, β, γ) with respect to the base coordinate system as in Eq. (46).

The position error of the hand is defined as the difference between the desired and the actual position of the hand and can be expressed as

$$\mathbf{e}_p(t) = \mathbf{p}^d(t) - \mathbf{p}(t) = \begin{bmatrix} p_x^d(t) - p_x(t) \\ p_y^d(t) - p_y(t) \\ p_z^d(t) - p_z(t) \end{bmatrix} \tag{71}$$

Similarly the orientation error is defined by the discrepancies between the desired and actual orientation axes of the hand and can be represented by

$$\mathbf{e}_0(t) = \tfrac{1}{2}(\mathbf{n}(t) \times \mathbf{n}^d + \mathbf{s}(t) \times \mathbf{s}^d + \mathbf{a}(t) \times \mathbf{a}^d) \tag{72}$$

Thus the control of the manipulator is achieved by reducing these errors of the hand to zero.

Considering a six-link manipulator, we can combine the linear velocities $[\mathbf{v}(t)]$ and the angular velocities $[\omega(t)]$ of the hand into a six-dimensional vector as $\dot{\mathbf{x}}(t)$,

$$\dot{\mathbf{x}}(t) = \begin{bmatrix} \mathbf{v}(t) \\ \omega(t) \end{bmatrix} = \mathbf{N}(\mathbf{q})\dot{\mathbf{q}}(t) \tag{73}$$

where $\mathbf{N}(q)$ is a 6 × 6 matrix as given in Eq. (54). Equation (73) is the basis for resolved motion rate control where joint velocities are solved from the hand velocities. If this idea is extended further to solve for the joint accelerations from the hand acceleration $\ddot{\mathbf{x}}(t)$, then the time derivative of $\dot{\mathbf{x}}(t)$ is the hand acceleration, which is

$$\ddot{\mathbf{x}}(t) = \mathbf{N}(\mathbf{q})\ddot{\mathbf{q}}(t) + \dot{\mathbf{N}}(\mathbf{q}, \dot{\mathbf{q}})\dot{\mathbf{q}}(t) \tag{74}$$

The closed-loop resolved motion acceleration control is based on the idea of reducing the position and orientation errors of the hand to zero. If the

Cartesian path for a manipulator is preplanned, then the desired position $\mathbf{p}^d(t)$, the desired velocity $\mathbf{v}^d(t)$ and the desired acceleration $\dot{\mathbf{v}}^d(t)$ of the hand are known with respect to the base coordinate system. In order to reduce the position error, one may apply joint torques and forces to each joint actuator of the manipulator. This essentially makes the actual linear acceleration of the hand $\dot{\mathbf{v}}(t)$ to satisfy

$$\dot{\mathbf{v}}(t) = \dot{\mathbf{v}}^d(t) + k_1[\mathbf{v}^d(t) - \mathbf{v}(t)] + k_2[\mathbf{p}^d(t) - \mathbf{p}(t)] \tag{75}$$

where k_1 and k_2 are scalar constants. Equation (75) can be rewritten as

$$\ddot{\mathbf{e}}_p(t) + k_1\dot{\mathbf{e}}_p(t) + k_2\mathbf{e}_p(t) = 0 \tag{76}$$

where $\mathbf{e}_p(t) = \mathbf{p}^d(t) - \mathbf{p}(t)$. The input torques and forces must be chosen so as to guarantee the asymptotic convergence of the position error of the hand. This requires k_1 and k_2 be chosen such that the characteristic roots of Eq. (76) have negative real parts.

Similarly, to reduce the orientation error of the hand, one has to choose the input torques and forces to the manipulator so that the angular acceleration of the hand satisfies

$$\dot{\omega}(t) = \dot{\omega}^d(t) + k_1[\dot{\omega}^d(t) - \dot{\omega}(t)] + k_2\mathbf{e}_0 \tag{77}$$

Let us group $\mathbf{v}^d$ and ω^d into a six-dimensional vector and the position and orientation errors into an error vector:

$$\dot{\mathbf{x}}^d(t) = \begin{bmatrix} \mathbf{v}^d(t) \\ \omega^d(t) \end{bmatrix} \quad \text{and} \quad \mathbf{e}(t) = \begin{bmatrix} \mathbf{e}_p(t) \\ \mathbf{e}_0(t) \end{bmatrix} \tag{78}$$

Combining Eqs. (75) and (77), we have

$$\ddot{\mathbf{x}}(t) = \ddot{\mathbf{x}}^d(t) + k_1(\dot{\mathbf{x}}^d(t) - \dot{\mathbf{x}}(t)) + k_2\mathbf{e}(t) \tag{79}$$

Substituting Eqs. (73) and (74) into Eq. (79) and solving for $\ddot{\mathbf{q}}(t)$ gives

$$\begin{aligned} \ddot{\mathbf{q}}(t) &= \mathbf{N}^{-1}(\mathbf{q})[\ddot{\mathbf{x}}^d(t) + k_1(\dot{\mathbf{x}}^d(t) - \dot{\mathbf{x}}(t)) + k_2\mathbf{e}(t) - \dot{\mathbf{N}}(\mathbf{q}, \dot{\mathbf{q}})\dot{\mathbf{q}}(t)] \\ &= -k_1\dot{\mathbf{q}}(t) + \mathbf{N}^{-1}(\mathbf{q})[\ddot{\mathbf{x}}^d(t) + k_1\dot{\mathbf{x}}^d(t) + k_2\mathbf{e}(t) - \dot{\mathbf{N}}(\mathbf{q}, \dot{\mathbf{q}})\dot{\mathbf{q}}(t)] \end{aligned} \tag{80}$$

Equation (80) is the basis for the closed-loop resolved acceleration control for manipulators. In order to compute the applied joint torques and forces to each joint actuator of the manipulator, the recursive Newton–Euler equations of motion are used. The joint position $\mathbf{q}(t)$ and joint velocity $\dot{\mathbf{q}}(t)$ are measured from the potentiometers or optical encoders of the manipulator. Now $\mathbf{v}$, ω, $\mathbf{N}$, $\mathbf{N}^{-1}$, $\dot{\mathbf{N}}$, and $\mathbf{H}(t)$ can be computed from the above appropriate equations easily. These values together with the desired position $\mathbf{p}^d(t)$, desired velocity $\mathbf{v}^d(t)$, and desired acceleration $\dot{\mathbf{v}}^d(t)$ of the hand obtained from a planned trajectory can be used to compute the joint acceleration using Eq. (80). Finally, the applied joint torques and forces can be computed recursively from the Newton–Euler equations of motion.

4.3 Resolved Motion Force Control

Recently a resolved motion force control (RMFC) has been proposed to control the Cartesian position of the end effector of a manipulator [25]. The basic concept of the RMFC is to determine the applied torques to the joint actuators to perform the Cartesian position control of the robot arm. An advantage of the RMFC is that the control is not based on the complicated dynamic equations of motion of the manipulator and still has the ability to compensate for the changing arm configurations, gravity loading forces on the links, and internal friction. The RMFC adapts the force convergent concept (Robbins–Monro stochastic approximation) to control the Cartesian position and force of the end effector instead of the joint servo control. Similar to the RMFC, all the control of RMFC is done at the hand level.

The RMFC is based on the relationship between the resolved force vector **F** obtained from a wrist force sensor and the joint torques at the joint actuators. The control technique consists of the Cartesian position control and the force convergent control. The position control calculates the desired forces and moments to be applied to the end effector in order to track a desired Cartesian trajectory. The force convergent control determines the necessary joint torques to each actuator so that the end effector can maintain the desired forces and moments obtained from the above position control. A control block diagram of the RMFC is shown in Figure 6.

We shall briefly discuss the mathematics that governs this control technique. A more detailed discussion can be found in Ref. [25]. The basic control concept of the RMFC is based on the relationship between the resolved force vector $\mathbf{F} = (F_x, F_y, F_z, M_x, M_y, M_z)^T$ and the joint static torques $\boldsymbol{\tau} = (\tau_1, \tau_2, \ldots, \tau_n)^T$ which are applied to each joint actuator to counterbalance the forces felt at the hand. Now $(F_x, F_y, F_z)^T$ and $(M_x, M_y, M_z)^T$ are the Cartesian forces and moments in the hand coordinate system, respectively.

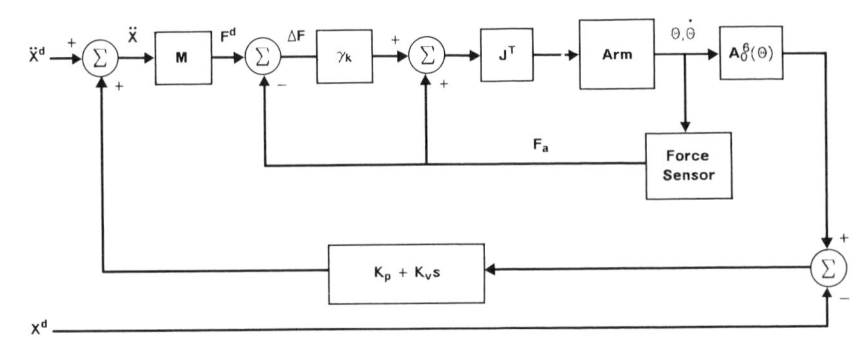

Figure 6. Resolved motion force control.

The underlying relationship between them is

$$\boldsymbol{\tau}(t) = \mathbf{N}^T(\mathrm{q})\mathbf{F}(t) \tag{81}$$

where $\mathbf{N}$ is the Jacobian matrix as in Eq. (54).

Since the objective of the RMFC is to track the Cartesian position of the end effector, an appropriate time-based position trajectory has to be specified as a function of the arm transformation matrix $\mathbf{A}_0^6(t)$, the velocity $(v_x, v_y, v_z)^T$, and the angular velocity $(\omega_x, \omega_y, \omega_z)^T$ about the hand coordinate system That is, the time-varying desired arm transformation matrix $\mathbf{A}_0^6(t+\Delta t)$ can be represented as

$$\mathbf{A}_0^6(t+\Delta t) = \mathbf{A}_0^6(t)\begin{bmatrix} 1 & -\omega_z(t) & \omega_y(t) & v_x(t) \\ \omega_z(t) & 1 & -\omega_x(t) & v_y(t) \\ -\omega_y(t) & \omega_x(t) & 1 & v_z(t) \\ 0 & 0 & 0 & 1 \end{bmatrix}\Delta t \tag{82}$$

Then the desired Cartesian velocity $\mathbf{x}^d(t) = (v_x, v_y, v_z, w_x, w_y, w_z^T)$ can be obtained from the element of the following equations:

$$\begin{bmatrix} 1 & -\omega_z(t) & \omega_y(t) & v_x(t) \\ \omega_z(t) & 1 & -\omega_x(t) & v_y(t) \\ -\omega_y(t) & \omega_x(t) & 1 & v_z(t) \\ 0 & 0 & 0 & 1 \end{bmatrix} = \frac{1}{\Delta t}((\mathbf{A}_0^6)^{-1}(t)\cdot\mathbf{A}_0^6(t+\Delta t)) \tag{83}$$

Using the above equation, the Cartesian velocity error $\dot{\mathbf{x}}^d - \dot{\mathbf{x}}$ can be obtained. The velocity error $\dot{\mathbf{x}}^d - \dot{\mathbf{x}}$ used in Eq. (75) is different from the above velocity error because the above error equation uses the homogeneous transformation matrix method. In Eq. (75), the velocity error is obtained simply by differentiating $\mathbf{p}^d(t) - \mathbf{p}(t)$.

Similarly, the desired Cartesian acceleration $\ddot{\mathbf{x}}^d(t)$ can be obtained as

$$\ddot{\mathbf{x}}^d(t) = \frac{\dot{\mathbf{x}}^d(t+\Delta t) - \dot{\mathbf{x}}^d(t)}{\Delta(t)} \tag{84}$$

Based on the position plus derivative control servomechanism, if there is no error in position and velocity of the hand, then we want the actual Cartesian acceleration $\ddot{\mathbf{x}}(t)$ to track the desired Cartesian acceleration as closely as possible. This can be done by setting the actual Cartesian acceleration as

$$\ddot{\mathbf{x}}(t) = \ddot{\mathbf{x}}^d(t) + K_v(\dot{\mathbf{x}}^d(t) - \dot{\mathbf{x}}(t)) + K_p(\mathbf{x}^d(t) - \mathbf{x}(t)) \tag{85}$$

or

$$\ddot{\mathbf{x}}_e(t) + K_v\dot{\mathbf{x}}_e(t) + K_p\mathbf{x}_e(t) = 0 \tag{86}$$

By choosing the values of K_v and K_p so that the characteristic roots of Eq. (86) have negative real parts, $\mathbf{x}(t)$ will converge to $\mathbf{x}^d(t)$ asymptotically.

Based on the above control technique, the desired Cartesian forces and moments to correct the position errors can be obtained using Newton's second law:

$$\mathbf{F}^d(t) = \mathbf{M}\ddot{\mathbf{x}}(t) \tag{87}$$

where $\mathbf{M}$ is the "mass" matrix with diagonal elements of total mass of the load m and the moments of inertia i_{xx}, i_{yy}, i_{zz} at the principal axes of load.

Then, using the Eq. (81), the desired Cartesian forces $\mathbf{F}^d$ can be resolved into the joint torques:

$$\tau(t) = \mathbf{N}^T(\mathbf{q})\mathbf{F}^d = \mathbf{N}^T(\mathbf{q})\mathbf{M}\ddot{\mathbf{x}}(t) \tag{88}$$

From a computer simulation study, the above RMFC works well when the mass and the load are negligible as compared with the mass of the manipulator. But if the mass and the load are approaching to the mass of the manipulator, the position of the hand does not converge to the desired position. This is due to the fact that some of the joint torques are spent to accelerate the links. In order to compensate for these loading and acceleration effects, a force convergence control is proposed as a second part of the RMFC.

The force convergent control method is based on the Robbins–Monro stochastic approximation method to determine the actual Cartesian force $\mathbf{F}_a$ so that the observed Cartesian force $\mathbf{F}_0$ (measured by a wrist force sensor) at the hand will converge to the desired Cartesian force $\mathbf{F}^d$ obtained from the above position control technique. If the error between the measured force vector $\mathbf{F}_0$ and the desired Cartesian force is greater than a user-designed threshold $\Delta\mathbf{F}(k) = \mathbf{F}^d(k) - \mathbf{F}_0(k)$, then the actual Cartesian force is updated by

$$\mathbf{F}_a(k+1) = \mathbf{F}_a(k) + \gamma_k\, \Delta\mathbf{F}(k) \tag{89}$$

where $\gamma_k = 1/(k+1)$ for $k = 0, 1, \ldots, N$. Theoretically, the value of N must be large. However, in practice, the value of N can be chosen based on the force convergence. Based on a computer simulation study [25], the value of $N = 1$ or 2 gives a fairly good convergence of the force vector.

In summary, the RMFC with force convergent control has the advantage that the control method can be extended to various loading conditions and any-number-of-degree-of-freedom manipulator without increasing the computational difficulties.

5. ADAPTIVE CONTROLS

Most of the existing control schemes discussed in the above sections control the arm at the hand or the joint level and emphasize nonlinear compensations of the interaction forces among the various joints. These control algorithms are inadequate because they require accurate modeling of the arm dynamics and neglect the changes of the load in a task cycle. These changes in the payload of the controlled system are significant enough to render conventional feedback control strategies ineffective. The result is reduced servo response speed and damping, which limits the precision and speed of the end effector. Any significant performance gain in tracking the desired time-based trajectory as closely as possible for all times over a wide range of manipulator motion and payloads requires the consideration of adaptive control techniques.

5.1 Model Referenced Adaptive Control

Among various adaptive control methods, the model referenced adaptive control (MRAC) is most widely used and relatively easy to implement. The concept of model referenced adaptive control is based on selecting an appropriate referenced model and adaptation algorithm which modifies the feedback gains to the actuators of the actual system. The adaptation algorithm is driven by the errors between the referenced model outputs and the actual system outputs. A general control block diagram of the model referenced adaptive control system is shown in Figure 7.

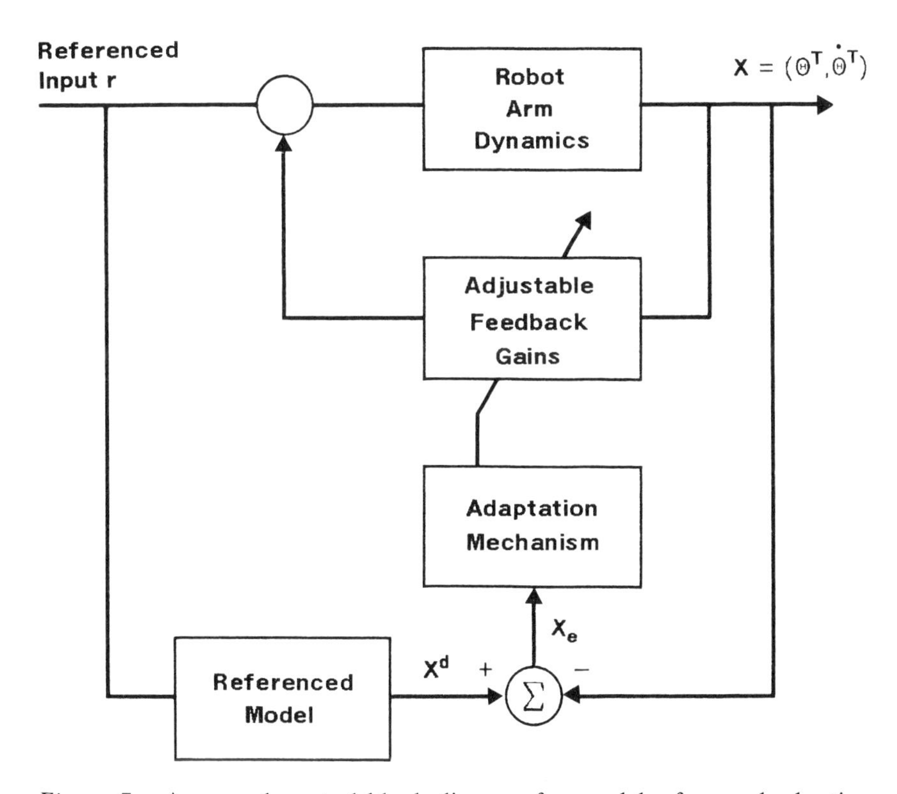

Figure 7. A general control block diagram for model referenced adaptive control.

Dubowsky [3] proposed a simple model referenced adaptive control for the control of mechanical manipulators. In his analysis, the payload is taken into consideration by combining it to the final link and the end effector dimension is assumed to be small compared to the length of other links. Then the selected referenced model provides an effective and flexible means of specifying desired closed loop performance of the controlled system. A linear second-order time-invariant differential equation is selected as the referenced model for each degree of freedom of the robot arm. The manipulator is controlled by adjusting the position and velocity feedback gains to follow the model so that its closed-loop performance characteristics closely match the set of desired performance characteristics in the referenced model. As a result, the above simple model referenced adaptive control only requires moderate computations to compute the control effort; and this can be done relatively easy on a low-cost microprocessor. Such model referenced adaptive control algorithm does not require complex mathematical models of the system dynamics nor the a priori knowledge of the environment (loads etc.). The resulting model referenced adaptive system is capable of maintaining uniformly good performance over a wide range of motions and payloads.

Defining the vector $\mathbf{y}(t)$ to represent the referenced model response and the vector $\mathbf{x}(t)$ to represent the manipulator response, the joint i of the referenced model can be described by

$$a_i \ddot{y}_i(t) + b_i \dot{y}_i(t) + y_i(t) = r_i(t) \tag{90}$$

In terms of natural frequency w_n and damping ratio ξ of a second-order linear system, a_i and b_i correspond to

$$a_i = \frac{1}{w_{ni}^2}, \qquad b_i = \frac{2\xi}{w_{ni}} \tag{91}$$

if it is assumed that the manipulator is controlled by position and velocity feedback gains and that the coupling terms are negligible, then the manipulator dynamic equation for joint i can be written as

$$\alpha_i(t)\ddot{x}_i(t) + \beta_i(t)\dot{x}_i(t) + x_i(t) = r_i(t) \tag{92}$$

where the system parameters $\alpha_i(t)$ and $\beta_i(t)$ are assumed to be slowly time-varying.

Several techniques are available to adjust the feedback gains of the controlled system. Due to its simplicity, a steepest-descent method is used to minimize a quadratic function of the system error which is the difference between the response of the actual system [Eq. (92)] and the response of the referenced model [Eq. (90)]:

$$J_i(e_i) = \tfrac{1}{2}(k_2^i \ddot{e}_i + k_1^i \dot{e}_i + k_0^i e_i)^2, \qquad i = 1, 2, \ldots, n \tag{93}$$

where $e_i = y_i - x_i$, and the values of the weighting factors k_j^i are selected from stability considerations to obtain stable system behavior.

Using a steepest-descent method, the system parameters adjustment mechanism which will minimize the system error is governed by

$$\dot{\alpha}_i(t) = (k_2^1\ddot{e}_i(t) + k_1^i\dot{e}_i(t) + k_0^i e_i(t))(k_2^i\ddot{u}_i(t) + k_1^i\dot{u}_i(t) + k_0^i u_i(t)) \quad (94)$$

$$\dot{\beta}_i(t) = (k_2^i\ddot{e}_i(t) + k_1^i\dot{e}_i(t) + k_0^i e_i(t))(k_2^i\ddot{w}_i(t) + k_1^i\dot{w}_i(t) + k_0^i w_i(t)) \quad (97)$$

where $u_i(t)$ and $w_i(t)$ and their derivatives are obtained from the solutions of the following differential equations:

$$a_i\ddot{u}_i(t) + b_i\dot{u}_i(t) + u_i(t) = -\ddot{y}_i(t) \quad (96)$$

$$a_i\ddot{w}_i(t) + b_i\dot{w}_i(t) + w_i(t) = -\dot{y}_i(y) \quad (97)$$

and $\dot{y}_i(t)$ and $\ddot{y}_i(t)$ are the time response of the referenced model. The closed-loop adaptive system involves solving the referenced model equations for a given desired input; then the differential equations in Eqs. (96) and (97) are solved to yield $u_i(t)$ and $w_i(t)$ and their derivatives for Eqs. (94) and (95). Finally, solving the differential equations in Eqs. (94) and (95), $\alpha_i(t)$ and $\beta_i(t)$ are obtained.

The fact that this adaptive control algorithm is not dependent on complex mathematical models is one of the major advantages. However, the stability problem of the closed adaptive system is very critical. This stability analysis is very difficult, and Ref. [3] presents an investigation of this adaptive system using a linearized model. However, if the interaction forces among the various joints are severe, then the adaptability of the controller is questionable. This requires further investigation of the method.

5.2 Adaptive Control Using an Autoregressive Model

Koivo [8] proposed an adaptive self-tuning controller using an autoregressive model to fit the input–output data from the manipulator. The control algorithm assumes that the interaction forces among the joints are negligible. A control block diagram of the control system is shown in Figure 8. Let the input torque to joint i be u_i and the output of the manipulator be the angular position y_i, the input–output pairs (u_i, y_i) may be described by an autoregressive model which match these pairs as closely as possible:

$$y_i(k) = \sum_{m=1}^{n} \left[a_i^m y_i(k-m) + b_i^m u_i(k-m) \right] + a_i^0 + e_i(k) \quad (98)$$

where a_i^0 is a constant forcing term and $e_i(k)$ is the modeling error which is assumed to be zero mean white Gaussian noise and independent of $y_i(k-m)$ for $m \geq 1$ and u_i. The parameters a_i^m and b_i^m are determined so as to obtain the best least-squares fit of the measured input–output data pairs. These

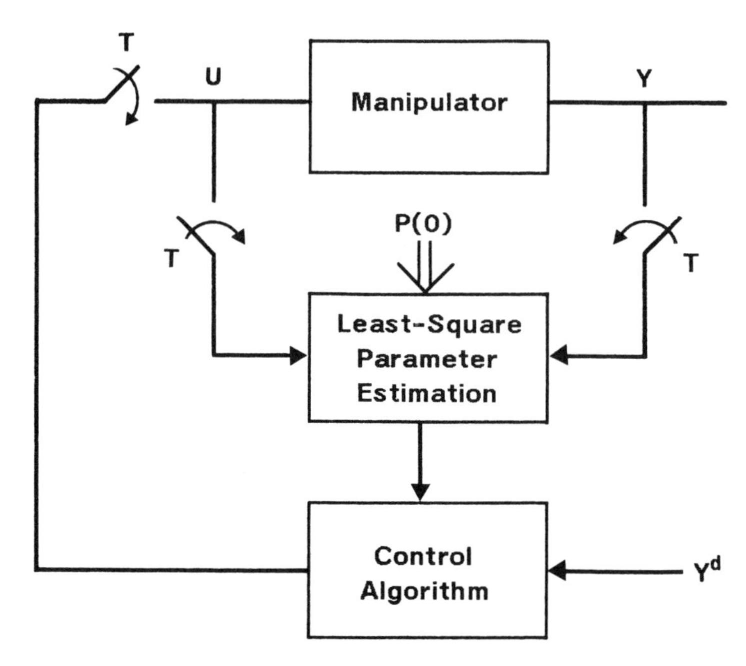

Figure 8. Adaptive control with autoregressive model.

parameters can be obtained by minimizing the following criterion:

$$E_N^i(\alpha) = \frac{1}{N+1} \sum_{k=0}^{N} e_i^2(k) \tag{99}$$

where N is the number of last measurements. Let α be the parameter vector

$$\alpha_i = [a_i^0, a_i^1, \ldots, a_i^n, b_i^0, b_i^1, \ldots, b_i^n]^T \tag{100}$$

and $\psi_i(k-1)$ be the vector of the input–output pairs

$$\psi_i(k-1) = [1, y_i(k-1), \ldots, y_i(k-n), u_i(k-1), \ldots, u_i(k-n)] \tag{101}$$

Then a recursive least-squares estimation of α_i can be found as

$$\hat{\alpha}_i(N) = \hat{\alpha}_i(N-1) + \mathbf{P}_i(N)\psi_i(N-1)[y_i(N) - \hat{\alpha}_i^T(N-1)\psi_i(N-1)]$$
$$\mathbf{P}_i(N) = \frac{1}{\mu_i}\left[\frac{\mathbf{P}_i(N-1)\psi_i(N-1)\psi_i^T(N-1)\mathbf{P}_i(N-1)}{\mu_i + \psi_i^T(N-1)\mathbf{P}_i(N-1)\psi_i(N-1)}\right] \tag{102}$$

where $0 < \mu_i \leq 1$ is a forgetting factor, which provides an exponential weighting of past data in the estimation algorithm and by which the algorithm allows a slow drifting of the parameters; $\mathbf{P}_i$ is a $(2n+1) \times (2n+1)$ symmetric matrix.

Using the above equations to compute the estimates of the autoregressive model, the model can be represented by

$$y_i(k) = \hat{\alpha}_i^T \psi_i(k-1) + e_i(k) \tag{103}$$

In order to track the trajectory set points, a performance criterion for joint i is defined as

$$J_i^k(\mathbf{u}) = E\left[(y_i(k+2) - y_i^d(k+2))^2 + \frac{\gamma_i u_i^2(k+1)}{\psi_i(k)} \right] \tag{104}$$

where $E[\cdot]$ represents an expectation operation conditioned on $\psi_i(k-1)$ and γ_i is a user-defined nonnegative weighting factor.

The optimal control that minimizes the above performance criterion is found to be

$$u_i(k+1) = \frac{-\hat{b}_i^1(k)}{(\hat{b}_i^1(k))^2 + \gamma_i} \left[\hat{a}_i^0(k) + \hat{a}_i^1(k)(\hat{\alpha}/\psi_i(k)) + \sum_{m=2}^{n} \hat{a}_i^m(k) y_i(k+2-m) \right.$$
$$\left. + \sum_{m=2}^{n} \hat{b}_i^m(k) u_i(k+2-m) - y_i^d(k+2) \right] \tag{105}$$

where $\hat{a}_i^m$, $\hat{b}_i^m$, and $\hat{\alpha}_i^m$ are the estimates of the parameters from Eq. (102).

In summary, this adaptive control uses an autoregressive model [Eq. (98)] to fit the input–output data from the manipulator. The recursive least-squares identification scheme [Eq. (102)] is used to estimate the parameters which are used in the optimal control [Eq. (105)] to servo the manipulator.

5.3 Adaptive Perturbation Control

Based on perturbation theory, Lee and Chung [11] proposed an adaptive control strategy which tracks a desired time-based manipulator trajectory as closely as possible for all times over a wide range of manipulator motion and payloads. The adaptive control differs from the above adaptive controls by taking all the interactions among the various joints into consideration. The adaptive control is based on the linearized perturbation equations in the vicinity of a nominal trajectory. The nominal trajectory is specified by an interpolated joint trajectory whose angular position, angular velocity, and angular acceleration are known at every sampling instant. The highly coupled nonlinear dynamic equations of a manipulator are then linearized about the planned manipulator trajectory to obtain the linearized perturbation system. The controlled system is characterized by feedforward and feedback components which can be computed separately and simultaneously. Using the Newton–Euler equations of motion as an inverse dynamics of the manipulator, the feedforward component computes the nominal torques which compensate all the interaction forces among the various joints along the nominal trajectory. The feedback component computes the variational torques which reduce the position and velocity errors of the manipulator to zero along the nominal trajectory. An efficient recursive real-time least-squares identification scheme is used to identify the system parameters in the perturbation equations. A one-step optimal control law is designed to control the linearized perturbation system about the nominal trajectory. The parameters and the feedback gains of the linearized system are updated and adjusted in each sampling period to obtain the necessary control effort. The total torques applied to the joint actuators then consist of the nominal torques computed from the Newton–Euler equations of motion and the variational torques computed from the one-step optimal control law of the linearized system. This adaptive control strategy reduces the manipulator control problem from a nonlinear control to controlling a linear control system about a nominal trajectory.

The adaptive control is based on the linearized perturbation equations about the referenced trajectory. We need to derive an appropriate linearized perturbation equation suitable for developing the feedback controller which computes variational joint torques to reduce position and velocity errors along the nominal trajectory. The L–E equations of motion of an n-link manipulator can be expressed in state space representation as in Eq. (11). With this formulation, the control problem is to find a feedback control law $\mathbf{u}(t) = \mathbf{g}(\mathbf{x}(t))$ such that the closed-loop control system $\dot{\mathbf{x}}(t) = \mathbf{f}(\mathbf{x}(t), \mathbf{g}(x(t)))$ is asymptotically stable and tracks a desired trajectory as closely as possible over a wide range of payloads for all times.

Suppose that the nominal states $\mathbf{x}_n(t)$ of the system [Eq. (11)] are known from the planned trajectory and that the corresponding nominal torques $\mathbf{u}_n(t)$ are also known from the computations of the joint torques using the N–E equations of motion. Then both $\mathbf{x}_n(t)$ and $\mathbf{u}_n(t)$ satisfy Eq. (11),

$$\dot{\mathbf{x}}_n(t) = \mathbf{f}(\mathbf{x}_n(t), \mathbf{u}_n(t)) \tag{106}$$

Using the Taylor series expansion on Eq. (11) about the nominal trajectory and subtracting Eq. (106) from it and assuming that the higher-order terms are negligible, the associated linearized perturbation model for this control system can be obtained:

$$\delta\dot{\mathbf{x}}(t) = \nabla_x \mathbf{f}|_n \, \delta\mathbf{x}(t) + \nabla_u \mathbf{f}|_n \, \delta\mathbf{u}(t) \tag{107}$$

$$= \mathbf{A}(t)\, \delta\mathbf{x}(t) + \mathbf{B}(t)\, \delta\mathbf{u}(t) \tag{108}$$

where $\nabla_x \mathbf{f}|_n$ and $\nabla_u \mathbf{f}|_n$ are the Jacobian matrices of $\mathbf{f}(\mathbf{x}(t), \mathbf{u}(t))$ evaluated at $\mathbf{x}_n(t)$ and $\mathbf{u}_n(t)$, respectively; $\delta\mathbf{x}(t) = \mathbf{x}(t) - \mathbf{x}_n(t)$; and $\delta\mathbf{u}(t) = \mathbf{u}(t) - \mathbf{u}_n(t)$.

The system parameters $\mathbf{A}(t)$ and $\mathbf{B}(t)$ of Eq. (108) depend on the instantaneous manipulator position and velocity along the nominal trajectory and are thus slowly varying in time. Because of the complexity of the manipulator equations of motion, it is extremely difficult to find the elements of $\mathbf{A}(t)$ and $\mathbf{B}(t)$ explicitly. However, the design of a feedback control law for the perturbation equations requires that the system parameters of Eq. (108) be known at all times. Thus parameter identification techniques must be used to identify the unknown elements in $\mathbf{A}(t)$ and $\mathbf{B}(t)$.

As a result of this formulation, the manipulator control problem is reduced to determining $\delta\mathbf{u}(t)$ which drives $\delta\mathbf{x}(t)$ to zero at all times along the nominal trajectory. The overall controlled system is thus characterized by a feedforward component and a feedback component. Given the planned trajectory set points $[\mathbf{q}^d(t), \dot{\mathbf{q}}^d(t), \ddot{\mathbf{q}}^d(t)]$, the feedforward component computes the corresponding nominal torques $\mathbf{u}_n(t)$ from the N–E equations of motion. The feedback component computes the corresponding variational torques $\delta\mathbf{u}(t)$ which provide control efforts to compensate for small deviations from the nominal trajectory. The computation of the variational torques is based on a one-step optimal control law. The main advantages of this formulation are twofold. Firstly, it reduces a nonlinear control problem to a linear control problem about a nominal trajectory, and secondly, the computations of the nominal and variational torques can be performed separately and simultaneously. Because of the parallel computational structure, the proposed adaptive control can be easily implemented using present-day low-cost microprocessors. A control block diagram of the adaptive control is shown in Figure 9.

For implementation on digital computer, Eq. (108) needs to be discretized to obtain an appropriate discrete linear equation for parameter identification

$$\mathbf{x}((k+1)T) = \mathbf{F}(kT)\mathbf{x}(kT) + \mathbf{G}(kT)\mathbf{u}(kT), \qquad k = 0, 1, \ldots \tag{109}$$

where T is the sampling period, $\mathbf{u}(kT)$ is an n-dimensional piecewise constant control input vector of $\mathbf{u}(t)$ over the time interval between any two consecutive sampling instants for $kT \le t < (k+1)T$, and $\mathbf{x}(kT)$ is a $2n$-dimensional

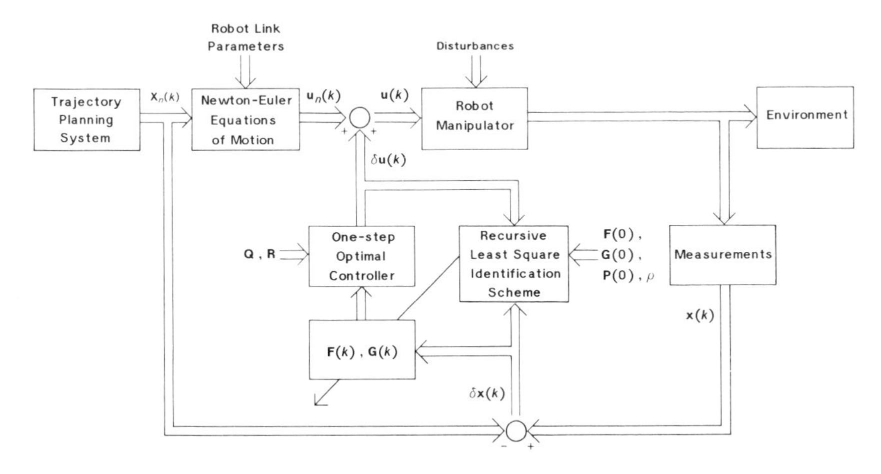

Figure 9. The adaptive perturbation control.

perturbed state vector which is given by

$$\mathbf{x}(kT) = \Gamma(kT, t_0)\mathbf{x}(t_0) + \int_{t_0}^{kT} \Gamma(kT, t)\mathbf{B}(t)\mathbf{u}(t)\, dt \tag{110}$$

and $\Gamma(kT, t_0)$ is the state transition matrix of the system. Now $\mathbf{F}(kT)$ and $\mathbf{G}(kT)$are, respectively, $2n \times 2n$ and $2n \times n$ matrices and are given by

$$\mathbf{F}(kT) = \Gamma((k+1)T, kT) \tag{111}$$

and

$$\mathbf{G}(kT)\mathbf{u}(kT) = \int_{kT}^{(k+1)T} \Gamma((k+1)T, t)\mathbf{B}(t)\mathbf{u}(t)\, dt \tag{112}$$

With this model, a total of $6n^2$ parameters in the $\mathbf{F}(kT)$ and $\mathbf{G}(kT)$ matrices need to be identified. Without confusion, we shall drop the sampling period T from the rest of the equations for clarity and simplicity.

Various identification algorithms, such as the methods of least squares, maximum likelihood, instrumental variable, cross-correlation, and stochastic approximation, have been applied successfully to the parameter identification problem [1]. Due to its simplicity and ease of applying, a recursive real-time least-squares parameter identification scheme is selected for identifying the system parameters in $\mathbf{F}(k)$ and $\mathbf{G}(k)$. In the parameter identification scheme, we make the following assumptions: (a) the parameters of the system are slowly time-varying but the variation speed is slower than the adaptation speed; (b) measurement noise is negligible; and (c) the state variables $\mathbf{x}(k)$ of Eq. (109) are measurable.

In order to apply the recursive least-squares identification algorithm to Eq. (109), we need to rearrange the system equations in a form that is suitable for parameter identification. Defining and putting the ith row of the unknown parameters of the system at the kth instant of time in a $3n$-dimensional vector, we have

$$\vartheta_i^T(k) = (f_{i1}(k), \ldots, f_{ip}(k), g_{i1}(k), \ldots, g_{in}(k)), \qquad i = 1, 2, \ldots, p \tag{113}$$

or expressed in matrix form as

$$\Theta(k) = \begin{bmatrix} f_{11}(k) & \cdots & f_{p1}(k) \\ \cdot & \cdots & \cdot \\ \cdot & \cdots & \cdot \\ \cdot & \cdots & \cdot \\ f_{1p}(k) & \cdots & f_{pp}(k) \\ g_{11}(k) & \cdots & g_{p1}(k) \\ \cdot & \cdots & \cdot \\ \cdot & \cdots & \cdot \\ \cdot & \cdots & \cdot \\ g_{1n}(k) & \cdots & g_{pn}(k) \end{bmatrix} = [\vartheta_1(k), \vartheta_2(k), \ldots, \vartheta_p(k)] \tag{114}$$

where $p = 2n$. Similarly, defining the outputs and inputs of the perturbation system [Eq. (109)] at the kth instant of time in a $3n$-dimensional vector as

$$\mathbf{z}^T(k) = (x_1(k), x_2(k), \ldots, x_p(k), u_1(k), u_2(k), \ldots, u_n(k)) \tag{115}$$

and the states at the kth instant of time in a $2n$-dimensional vector as

$$\mathbf{x}^T(k) = (x_1(k), x_2(k), \ldots, x_p(k)) \tag{116}$$

the corresponding system equation expressed in Eq. (109) can be written as

$$x_i(k+1) = \mathbf{z}^T(k)\vartheta_i(k), \qquad i = 1, 2, \ldots, p \tag{117}$$

With this formulation, we would like to identify the parameters in each column of $\Theta(k)$ based on the measurement vector $\mathbf{z}(k)$. In order to examine the "goodness" of the least-squares estimation algorithm, a $2n$-dimensional error vector $\mathbf{e}(k)$, often called residual, is included to account for the modeling error and noise in Eq. (109):

$$\mathbf{e}_i(k) = x_i(k+1) - \mathbf{z}^T(k)\hat{\vartheta}_i(k), \qquad i = 1, 2, \ldots, p \tag{118}$$

The basic least-squares parameter estimation assumes that the unknown parameters are constant values and the solution is based on batch-processing N sets of measurement data, which are weighted equally, to estimate the unknown parameters. Unfortunately, this algorithm cannot be applied to the time-varying parameters. Furthermore, the solution requires matrix inversion which is computational-intensive. In order to reduce the number of numerical computations and to track the time-varying parameters $\Theta(k)$ at each sampling period, a sequential least-squares identification scheme which updates the unknown parameters at each sampling period based on the new set of measurements at each sampling interval provides an efficient algorithmic solution to the identification problem. Such a recursive real-time least-squares parameter identification algorithm can be found by minimizing an exponentially weighted error criterion which has an effect of placing more weights on the squared errors of the more recent measurements [1],

$$J_N = \sum_{j=1}^{N} \rho^{N-j} e_i^2(j) \tag{119}$$

where the error vector is weighted as

$$\mathbf{e}_i^T(N) = (\sqrt{\rho^{N-1}}\, e_i(1), \sqrt{\rho^{N-2}}\, e_i(2), \ldots, e_i(N)) \tag{120}$$

and $N > 3n$ is the number of measurements used to estimate the parameters $\vartheta_i(N)$. Minimizing the error criterion in Eq. (119) with respect to the unknown parameters vector ϑ_i and utilizing the matrix inverse lemma, a recursive real-time least-squares identification scheme can be obtained for $\vartheta_i(k)$ after simple algebraic manipulations [1],

$$\hat{\vartheta}_i(k+1) = \hat{\vartheta}_i(k) + \gamma(k)\mathbf{P}(k)\mathbf{z}(k)[\mathbf{x}_i(k+1) - \mathbf{z}^T(k)\hat{\vartheta}_i(k)] \tag{121}$$

$$\mathbf{P}(k+1) = \mathbf{P}(k) - \gamma(k)\mathbf{P}(k)\mathbf{z}(k)\mathbf{z}^T(k)\mathbf{P}(k) \tag{122}$$

$$\gamma(k) = [\mathbf{z}^T(k)\mathbf{P}(k)\mathbf{z}(k) + \rho]^{-1} \tag{123}$$

where $0 < \rho < 1$ and the "hat" $(\hat{\ })$ is used to indicate the estimate of the parameters $\vartheta_i(k)$ and $\mathbf{P}(k) = \rho[\mathbf{Z}(k)\mathbf{Z}^T(k)]^{-1}$ is a $3n \times 3n$ symmetric positive definite matrix, where $\mathbf{Z}(k) = [\mathbf{z}(1), \mathbf{z}(2), \ldots, \mathbf{z}(k)]$ is the measurement matrix up to the kth sampling instant. If the errors $e_i(k)$ are identically distributed and independent with zero mean and variance σ^2, then $\mathbf{P}(k)$ can be interpreted as the covariance matrix of the estimate if ρ is chosen as σ^2.

The above recursive equations indicate that the estimate of the parameters $\hat{\vartheta}_i(k+1)$ at the $(k+1)$th sampling period is equal to the previous estimate $\hat{\vartheta}_i(k)$ corrected by the term proportional to $[x_i(k+1) - \mathbf{z}^T(k)\hat{\vartheta}_i(k)]$. The $\mathbf{z}^T(k)\hat{\vartheta}_i(k)$ is the prediction of the value $x_i(k+1)$ based on the estimate of the parameters $\vartheta_i(k)$ and the measurement vector $\mathbf{z}(k)$. The components of the vector $\gamma(k)\mathbf{P}(k)\mathbf{z}(k)$ are weighting factors which indicate how the corrections and the previous estimate should be weighted to obtain the new estimate $\hat{\vartheta}_i(k+1)$. The parameter ρ is a weighting factor and is commonly used for tracking slowly time-varying parameters by exponentially forgetting the "aged" measurements. If $\rho \ll 1$, a large weighting factor is placed on the more recent sampled data by rapidly weighing out previous samples. If $\rho \approx 1$, accuracy in tracking the time-varying parameters will be lost due to the truncation of measured data sequence. We can compromise between fast adaptation capabilities and loss of accuracy in parameter identification by adjusting the weighting factor ρ. In most applications for tracking slowly time-varying parameters, ρ is usually chosen to be $0.90 \leq \rho < 1.0$. The estimate of the parameters will be used to determine the optimal control solution for the perturbation system in the next section.

Finally, the above identification scheme [Eqs. (121)–(123)] can be started by choosing the initial values of $\mathbf{P}(0)$ to be

$$\mathbf{P}(0) = \sigma\mathbf{I}_{3n} \tag{124}$$

where σ is a very large positive scalar and $\mathbf{I}_{3n}$ is a $3n \times 3n$ identity matrix. The initial estimate of the unknown parameters $\mathbf{F}(k)$ and $\mathbf{G}(k)$ can be approximated by the following equations:

$$\mathbf{F}(0) \approx \mathbf{I}_{2n} + \left[\frac{\partial \mathbf{f}}{\partial \mathbf{x}}(\mathbf{x}_n(0), \mathbf{u}_n(0))\right]T + \left[\frac{\partial \mathbf{f}}{\partial \mathbf{x}}(\mathbf{x}_n(0), \mathbf{u}_n(0))\right]^2 \frac{T^2}{2} \tag{125}$$

$$\begin{aligned}\mathbf{G}(0) \approx &\left[\frac{\partial \mathbf{f}}{\partial \mathbf{u}}(\mathbf{x}_n(0), \mathbf{u}_n(0))\right]T + \left[\frac{\partial \mathbf{f}}{\partial \mathbf{x}}(\mathbf{x}_n(0), \mathbf{u}_n(0))\right]\left[\frac{\partial \mathbf{f}}{\partial \mathbf{u}}(\mathbf{x}_n(0), \mathbf{u}_n(0))\right]T^2 \\ &+ \left[\frac{\partial \mathbf{f}}{\partial \mathbf{x}}(\mathbf{x}_n(0), \mathbf{u}_n(0))\right]^2\left[\frac{\partial \mathbf{f}}{\partial \mathbf{u}}(\mathbf{x}_n(0), \mathbf{u}_n(0))\right]\frac{T^3}{2}\end{aligned} \tag{126}$$

where T is the sampling period.

With the determination of the parameters in $\mathbf{F}(k)$ and $\mathbf{G}(k)$, proper control laws can be designed to obtain the required correction torques to reduce the position and velocity errors of the manipulator along a nominal trajectory.

This can be done by finding an optimal control $\mathbf{u}^*(k)$ which minimizes the performance index $J(k)$ while satisfying the constraints of Eq. (109):

$$J(k) = \tfrac{1}{2}[\mathbf{x}^T(k+1)\mathbf{Q}\mathbf{x}(k+1) + \mathbf{u}^T(k)\mathbf{R}\mathbf{u}(k)] \qquad (127)$$

where $\mathbf{Q}$ is a $p \times p$ semipositive definite weighting matrix and $\mathbf{R}$ is an $n \times n$ positive definite weighting matrix. The one-step performance index in Eq. (127) indicates that the objective of the optimal control is to drive the posiion and velocity errors of the manipulator to zero along the nominal trajectory in a coordinated position and rate control per interval step, while at the same time a cost is attached to the use of control efforts. The optimal control solution which minimizes the functional in Eq. (127) subject to the constraints of Eq. (109) is well known and is found to be [22]

$$\mathbf{u}^*(k) = -[\mathbf{R} + \hat{\mathbf{G}}^T(k)\mathbf{Q}\hat{\mathbf{G}}(k)]^{-1}\hat{\mathbf{G}}^T(k)\mathbf{Q}\hat{\mathbf{F}}(k)\mathbf{x}(k) \qquad (128)$$

where $\hat{\mathbf{F}}(k)$ and $\hat{\mathbf{G}}(k)$ are the system parameters obtained from the identification algorithm [Eqs. (121)–(123)] at the kth sampling instant.

The above identification and control algorithms in Eqs. (121)–(123) and Eq. (128) do not require complex computations. In Eq. (123), $(\mathbf{z}^T(k)\mathbf{P}(k)\mathbf{z}(k) + \rho)$ gives a scalar value which simplifies its inversion. Although the weighting factor ρ can be adjusted for each ith parameter vector $\vartheta_i(k)$ as desired, this requires excessive computations in the $\mathbf{P}(k+1)$ matrix. For real-time robot arm control, such adjustment is not desirable. Here $\mathbf{P}(k+1)$ is computed only once at each sampling time using the same weighting factor ρ. Moreover, since $\mathbf{P}(k)$ is a symmetric positive definite matrix, only the upper diagonal matrix of $\mathbf{P}(k)$ needs to be computed. The combined identification and control algorithm can be computed in $O(n^3)$ time. The computational requirements of the proposed adaptive control are tabulated in Table 1. Based on a PDP 11/45 computer and its manufacturer's specification sheet, an ADDF (floating-point addition) instruction requires 5.17 μs and a MULF (floating-point multiply) instruction requires 7.17 μs. If we assume that for

Table 1. Computations of the Adaptive Controller

Adaptive Controller	*Multiplications*	*Additions*
Newton–Euler Equations of Motion	$117n - 24$	$103n - 21$
Least-Squares Identification Algorithm	$30n^2 + 5n + 1$	$30n^2 + 3n - 1$
Control Algorithm	$8n^3 + 2n^2 + 39$	$8n^3 - n^2 - n + 18$
Total Mathematical Operations of Adaptive Controller	$8n^3 + 32n^2 + 5n + 40$	$8n^3 + 29n^2 + 2n + 17$

Table 2. Comparisons of the PD and Adaptive Controllers

		PD Controller			*Adaptive Controller*		
Various Loading Conditions	*Joint*	*Max. Error (degree)*	*Trajectory Tracking Max. Error (mm)*	*Final Position Error (degree)*	*Max. Error (degree)*	*Trajectory Tracking Max. Error (mm)*	*Final Position Error (degree)*
No-load and 10% error In inertia tensor	1	0.089	1.55	0.025	0.020	0.34	0.000
	2	0.098	1.71	0.039	0.020	0.36	0.004
	3	0.328	2.86	0.121	0.032	0.28	0.002
$\frac{1}{2}$ max. load and 10% error In inertia tensor	1	0.121	2.11	0.054	0.045	0.78	0.014
	2	0.147	2.57	0.078	0.065	1.14	0.050
	3	0.480	4.19	0.245	0.096	0.83	0.077
Max. load and 10% error In inertia tensor	1	0.145	2.53	0.082	0.069	1.20	0.023
	2	0.185	3.23	0.113	0.069	1.22	0.041
	3	0.607	5.30	0.360	0.066	0.58	0.019

each ADDF and MULF instruction we need to fetch data from the core memory twice and the memory cycle time is 450 ns, then the proposed adaptive control requires approximately 7.5 ms to compute the necessary joint torques to servo the first three joints of a PUMA robot arm for a trajectory set point.

A computer simulation study of a three-joint PUMA manipulator was conducted to evaluate and compare the performance of the adaptive controller with the controller [Eq. (18)] obtained from the computed torque technique which is basically a proportional plus derivative control (PD controller) for various loading conditions along a given trajectory. The performances of the PD and adaptive controllers are compared and evaluated for three different loading conditions and the results are tabulated in Table 2: (a) no-load and 10% error in inertia tensor matrix; (b) half of maximum load

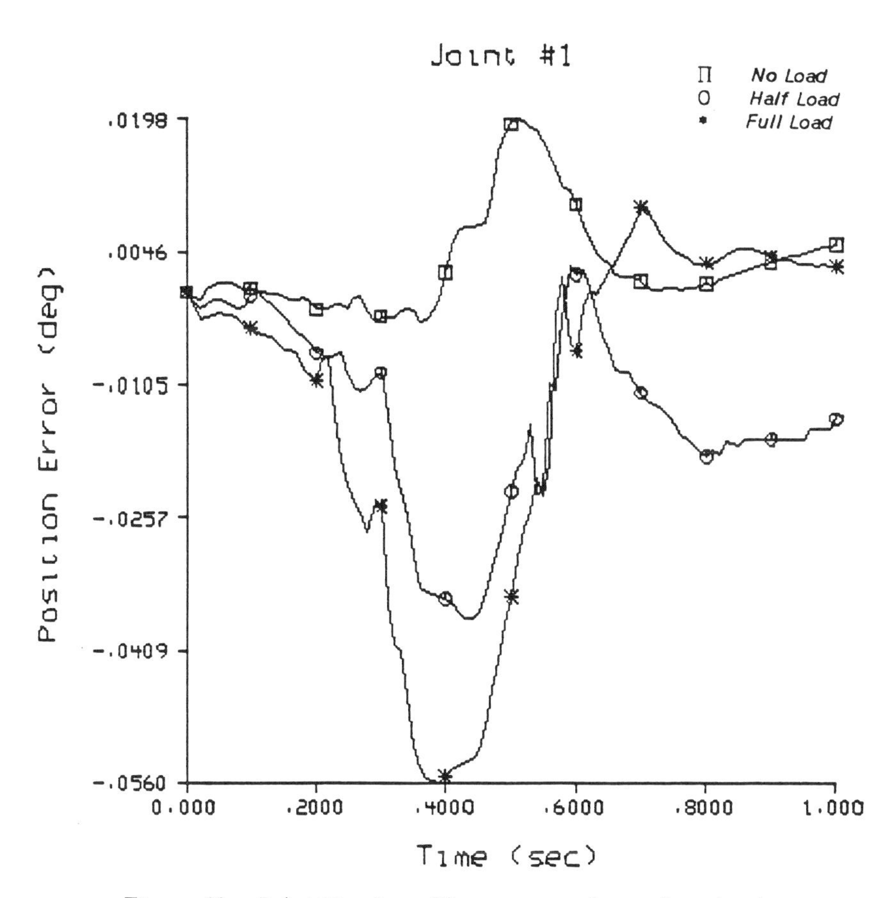

Figure 10. Joint No. 1 position error under various loads.

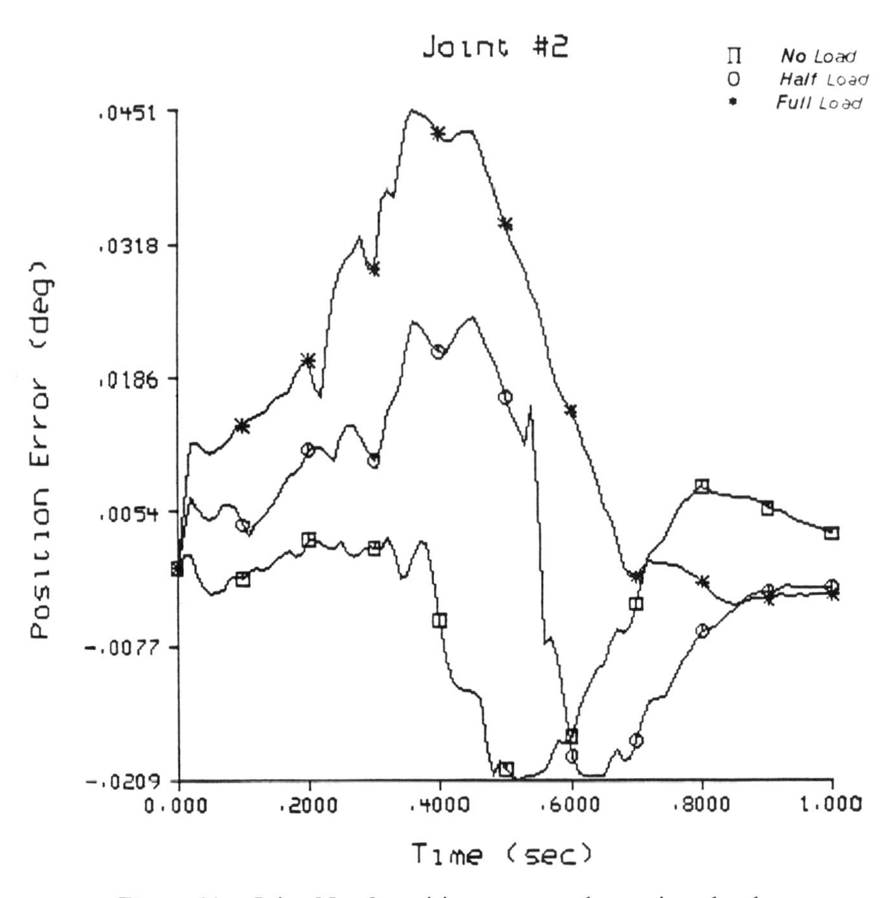

Figure 11. Joint No. 2 position error under various loads.

and 10% error in inertia tensor matrix; and (c) maximum load (5 lb) and 10% error in inertia tensor matrix. In each case, a 10% error in inertia matrices means $= 10\%$error about its measured inertial values. For all the above cases, the adaptive controller shows better performance than the PD controller with constant feedback gains both in trajectory tracking and the final position errors. Plots of angular position errors for the above cases for the adaptive control are shown in Figures 10–12. Details of the simulation result can be found in Ref. [11].

5.4 Resolved Motion Adaptive Control

The above adaptive control strategy in the joint variable space can be extended to control the manipulator in Cartesian coordinates under various loading conditions by adopting the ideas of resolved motion rate and accel-

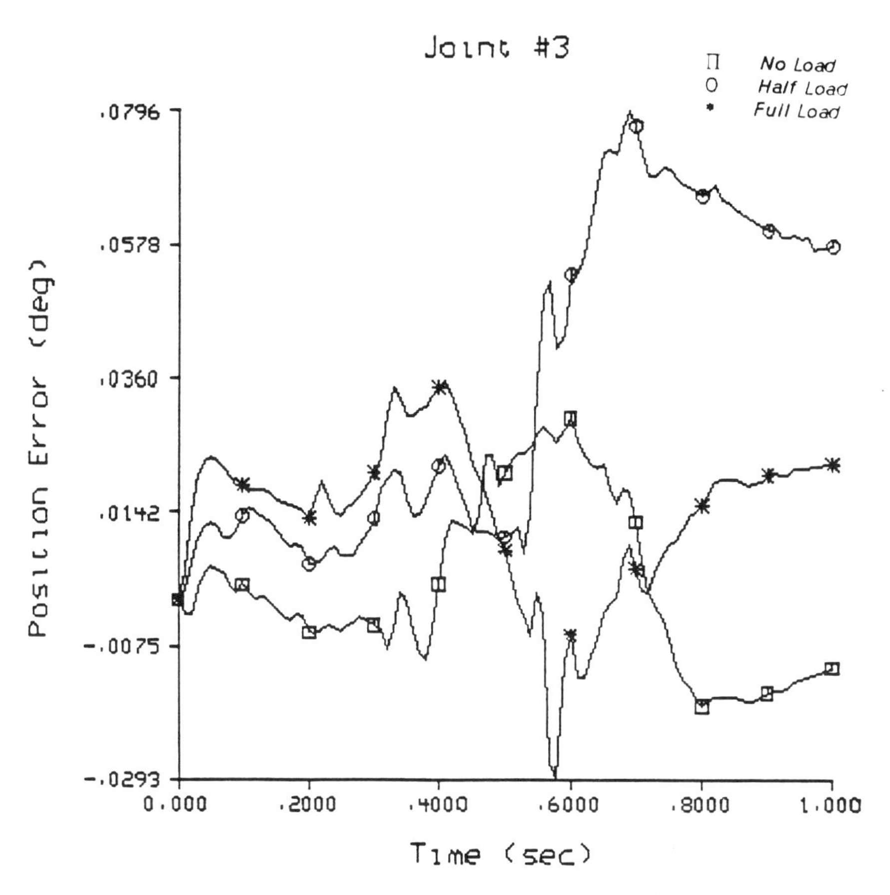

Figure 12. Joint No. 3 position error under various loads.

eration controls [17, 23, 24]. The resolved motion adaptive control [14] is performed at the hand level and is based on the linearized perturbation system along a desired time-based hand trajectory. The resolved motion adaptive control differs from the resolved motion acceleration control by minimizing the position/orientation and angular and linear velocities of the manipulator hand along the hand coordinate axes instead of position error and orientation vector error. Similar to the previous adaptive control, the controlled system is characterized by feedforward and feedback components which can be computed separately and simultaneously. The feedforward component resolves the specified positions, velocities, and accelerations of the hand into a set of values of joint positions, velocities, and accelerations from which the nominal joint torques are computed using the Newton–Euler equations of motion to compensate all the interaction forces among the various joints.

The feedback component computes the variational joint torques which reduce the manipulator hand position and velocity errors along the nominal hand trajectory. A recursive least-squares identification scheme is again used to perform on-line parameter identification of the linearized system.

Using the kinematic relationship between the joint coordinates and the Cartesian coordinates derived previously in Eqs. (45)–(57), the equations of motion of the manipulator in Cartesian coordinates can be easily obtained. The acceleration of the manipulator has been obtained previously in Eq. (56) and is repeated here for convenience:

$$\begin{bmatrix} \dot{\mathbf{v}}(t) \\ \dot{\Omega}(t) \end{bmatrix} = [\dot{\mathbf{N}}(\mathbf{q}, \dot{\mathbf{q}})][\mathbf{N}^{-1}(\mathbf{q})] \begin{bmatrix} \mathbf{v}(t) \\ \Omega(t) \end{bmatrix} + [\mathbf{N}(\mathbf{q})]\ddot{\mathbf{q}}(t) \tag{129}$$

In order to include the dynamics of the manipulator into the above kinematics equation [Eq. (129)], we need to use the L–E equations of motion as in Eq. (8). Since $\mathbf{D}(\mathbf{q})$ is always nonsingular, $\ddot{\mathbf{q}}(t)$ can be obtained from Eq. (8) and substituted into Eq. (129) to obtain the accelerations of the manipulator hand:

$$\begin{bmatrix} \dot{\mathbf{v}}(t) \\ \Omega(t) \end{bmatrix} = [\dot{\mathbf{N}}(\mathbf{q}, \dot{\mathbf{q}})][\mathbf{N}^{-1}(\mathbf{q})] \begin{bmatrix} \mathbf{v}(t) \\ \Omega(t) \end{bmatrix} + [\mathbf{N}(\mathbf{q})][\mathbf{D}^{-1}(\mathbf{q})][\tau(t) - \mathbf{h}(\mathbf{q}, \dot{\mathbf{q}}) - \mathbf{c}(\mathbf{q})] \tag{130}$$

For convenience, let us partition $\mathbf{N}(\mathbf{q})$, $\mathbf{N}^{-1}(\mathbf{q})$, and $\mathbf{D}^{-1}(\mathbf{q})$ into 3×3 submatrices and $\mathbf{h}(\mathbf{q}, \dot{\mathbf{q}})$, $\mathbf{c}(\mathbf{q})$, and $\tau(t)$ into 3×1 submatrices:

$$[\mathbf{N}(\mathbf{q})] \triangleq \left[\begin{array}{c|c} \mathbf{N}_{11}(\mathbf{q}) & \mathbf{N}_{12}(\mathbf{q}) \\ \hline \mathbf{N}_{21}(\mathbf{q}) & \mathbf{N}_{22}(\mathbf{q}) \end{array}\right], \qquad [\mathbf{N}^{-1}(\mathbf{q})] \triangleq [\mathbf{K}(\mathbf{q})] \triangleq \left[\begin{array}{c|c} \mathbf{K}_{11}(\mathbf{q}) & \mathbf{K}_{12}(\mathbf{q}) \\ \hline \mathbf{K}_{21}(\mathbf{q}) & \mathbf{K}_{22}(\mathbf{q}) \end{array}\right] \tag{131}$$

$$[\mathbf{D}^{-1}(\mathbf{q})] \triangleq [\mathbf{E}(\mathbf{q})] \triangleq \left[\begin{array}{c|c} \mathbf{E}_{11}(\mathbf{q}) & \mathbf{E}_{12}(\mathbf{q}) \\ \hline \mathbf{E}_{21}(\mathbf{q}) & \mathbf{E}_{22}(\mathbf{q}) \end{array}\right], \qquad [\mathbf{h}(\mathbf{q}, \dot{\mathbf{q}})] \triangleq \left[\begin{array}{c} \mathbf{h}_1(\mathbf{q}, \dot{\mathbf{q}}) \\ \hline \mathbf{h}_2(\mathbf{q}, \dot{\mathbf{q}}) \end{array}\right] \tag{132}$$

$$[\mathbf{c}(\mathbf{q})] \triangleq \left[\begin{array}{cc} \mathbf{c}_1 & \mathbf{c}_1(\mathbf{q}) \\ \hline \mathbf{c}_2 & \mathbf{c}_2(\mathbf{q}) \end{array}\right], \qquad [\tau(t)] \triangleq \left[\begin{array}{c} \tau_1(t) \\ \hline \tau_2(t) \end{array}\right] \tag{133}$$

Combining Eqs. (48), (52), and Eq. (130) and using Eqs. (131)–(133), we can obtain the state equations of the manipulator in Cartesian coordinates:

$$\left[\begin{array}{c} \dot{\mathbf{p}}(t) \\ \Phi(t) \\ \hline \dot{\mathbf{v}}(t) \\ \Omega(t) \end{array}\right] = \left[\begin{array}{ccc} 0 & 0 & \mathbf{I}_3 \\ 0 & 0 & 0 \\ \hline 0 & 0 & \dot{\mathbf{N}}_{11}(\mathbf{q}, \dot{\mathbf{q}})\mathbf{K}_{11}(\mathbf{q}) + \dot{\mathbf{N}}_{12}(\mathbf{q}, \dot{\mathbf{q}})\mathbf{K}_{21}(\mathbf{q}) \\ 0 & 0 & \dot{\mathbf{N}}_{21}(\mathbf{q}, \dot{\mathbf{q}})\mathbf{K}_{11}(\mathbf{q}) + \dot{\mathbf{N}}_{22}(\mathbf{q}, \dot{\mathbf{q}})\mathbf{K}_{21}(\mathbf{q}) \end{array}\right.$$

(equation continues)

$$
\left.\begin{bmatrix} 0 \\ \mathbf{S}(\Phi) \\ \hline \dot{\mathbf{N}}_{11}(\mathbf{q},\dot{\mathbf{q}})\mathbf{K}_{12}(\mathbf{q}) + \dot{\mathbf{N}}_{12}(\mathbf{q},\dot{\mathbf{q}})\mathbf{K}_{22}(\mathbf{q}) \\ \dot{\mathbf{N}}_{21}(\mathbf{q},\dot{\mathbf{q}})\mathbf{K}_{12}(\mathbf{q}) + \dot{\mathbf{N}}_{22}(\mathbf{q},\dot{\mathbf{q}})\mathbf{K}_{22}(\mathbf{q}) \end{bmatrix}\begin{bmatrix} \mathbf{p}(t) \\ \Phi(t) \\ \hline \mathbf{v}(t) \\ \Omega(t) \end{bmatrix}\right.
$$

$$
+ \left[\begin{array}{c|c} 0 & 0 \\ 0 & 0 \\ \hline \mathbf{N}_{11}(\mathbf{q})\mathbf{E}_{11}(\mathbf{q}) + \mathbf{N}_{12}(\mathbf{q})\mathbf{E}_{21}(\mathbf{q}) & \mathbf{N}_{11}(\mathbf{q})\mathbf{E}_{12}(\mathbf{q}) + \mathbf{N}_{12}(\mathbf{q})\mathbf{E}_{22}(\mathbf{q}) \\ \mathbf{N}_{21}(\mathbf{q})\mathbf{E}_{11}(\mathbf{q}) + \mathbf{N}_{22}(\mathbf{q})\mathbf{E}_{21}(\mathbf{q}) & \mathbf{N}_{21}(\mathbf{q})\mathbf{E}_{12}(\mathbf{q}) + \mathbf{N}_{22}(\mathbf{q})\mathbf{E}_{22}(\mathbf{q}) \end{array}\right]
$$

$$
\times \begin{bmatrix} -\mathbf{h}_1(\mathbf{q},\dot{\mathbf{q}}) - \mathbf{c}_1(\mathbf{q}) + \tau_1(t) \\ \hline -\mathbf{h}_2(\mathbf{q},\dot{\mathbf{q}}) - \mathbf{c}_2(\mathbf{q}) + \tau_2(t) \end{bmatrix} \tag{134}
$$

Equation (134) represents the state equations of the manipulator and will be used to derive an adaptive control scheme in Cartesian coordinates.

Defining the state vector for the manipulator hand as

$$
\begin{aligned}
\mathbf{x}(t) &\triangleq (\mathbf{x}_1, \mathbf{x}_2, \ldots, \mathbf{x}_{12})^T \triangleq (p_x, p_y, p_z, \alpha, \beta, \gamma, v_x, v_y, v_z, \omega_x, \omega_y, \omega_z)^T \\
&\triangleq (p^T, \Phi^T, \mathbf{v}^T, \Omega^T)
\end{aligned} \tag{135}
$$

and the input torque vector as

$$
\mathbf{u}(t) \triangleq (\tau_1, \ldots, \tau_6)^T \triangleq (u_1, \ldots, u_6)^T \tag{136}
$$

Eq. (134) can be expressed in state space representation as

$$
\dot{\mathbf{x}}(t) = \mathbf{f}(\mathbf{x}(t), \mathbf{u}(t)) \tag{137}
$$

where $\mathbf{x}(t) \in R^{2n}$, $\mathbf{u}(t) \in R^n$, $\mathbf{f}(\cdot)$ is a $2n \times 1$ nonlinear vector-valued function and continuously differentiable, and $n = 6$ is the number of degree of freedom of the manipulator. Equation (5.137) can be expressed explicitly as

$$
\begin{aligned}
\dot{x}_1(t) &= f_1(\mathbf{x}, \mathbf{u}) = x_7(t) \\
\dot{x}_2(t) &= f_2(\mathbf{x}, \mathbf{u}) = x_8(t) \\
\dot{x}_3(t) &= f_3(\mathbf{x}, \mathbf{u}) = x_9(t) \\
\dot{x}_4(t) &= f_4(\mathbf{x}, \mathbf{u}) = -\sec x_5(x_{10} \cos x_6 + x_{11} \sin x_6) \\
\dot{x}_5(t) &= f_5(\mathbf{x}, \mathbf{u}) = \sec x_5(x_{10} \cos x_5 \sin x_6 - x_{11} \cos x_5 \cos x_5) \\
\dot{x}_6(t) &= f_6(\mathbf{x}, \mathbf{u}) = -\sec x_5(x_{10} \sin x_5 \cos x_6 + x_{11} \sin x_5 \sin x_6 + x_{12} \cos x_5)
\end{aligned} \tag{138}
$$

$$
\begin{aligned}
\dot{x}_{i+6}(t) &= f_{i+6}(\mathbf{x}, \mathbf{u}) \\
&= g_{i+6}(\mathbf{q}, \dot{\mathbf{q}})\mathbf{x}(t) + b_{i+6}(\mathbf{q})\lambda(\mathbf{q}, \dot{\mathbf{q}}) + b_{i+6}(\mathbf{q})\mathbf{u}(t), \quad i = 1, \ldots, 6
\end{aligned}
$$

where $g_{i+6}(\mathbf{q}, \dot{\mathbf{q}})$ is the $(i + 6)$th row of the matrix:

$$
\left[\begin{array}{cc|cc}
0 & 0 & \mathbf{I}_3 & 0 \\
0 & 0 & 0 & \mathbf{S}(\Phi) \\
\hline
0 & 0 & \dot{\mathbf{N}}_{11}(\mathbf{q}, \dot{\mathbf{q}})\mathbf{K}_{11}(\mathbf{q}) + \dot{\mathbf{N}}_{12}(\mathbf{q}, \dot{\mathbf{q}})\mathbf{K}_{21}(\mathbf{q}) & \dot{\mathbf{N}}_{11}(\mathbf{q}, \dot{\mathbf{q}})\mathbf{K}_{12}(\mathbf{q}) + \dot{\mathbf{N}}_{12}(\mathbf{q}, \dot{\mathbf{q}})\mathbf{K}_{22}(\mathbf{q}) \\
0 & 0 & \dot{\mathbf{N}}_{21}(\mathbf{q}, \dot{\mathbf{q}})\mathbf{K}_{11}(\mathbf{q}) + \dot{\mathbf{N}}_{22}(\mathbf{q}, \dot{\mathbf{q}})\mathbf{K}_{21}(\mathbf{q}) & \dot{\mathbf{N}}_{21}(\mathbf{q}, \dot{\mathbf{q}})\mathbf{K}_{12}(\mathbf{q}) + \dot{\mathbf{N}}_{22}(\mathbf{q}, \dot{\mathbf{q}})\mathbf{K}_{22}(\mathbf{q})
\end{array}\right]
$$

and $b_{i+6}(\mathbf{q})$ is the $(i + 6)$th row of the matrix:

$$
\left[\begin{array}{c|c}
0 & 0 \\
0 & 0 \\
\hline
\mathbf{N}_{11}(\mathbf{q})\mathbf{E}_{11}(\mathbf{q}) + \mathbf{N}_{12}(\mathbf{q})\mathbf{E}_{21}(\mathbf{q}) & \mathbf{N}_{11}(\mathbf{q})\mathbf{E}_{12}(\mathbf{q}) + \mathbf{N}_{12}(\mathbf{q})\mathbf{E}_{22}(\mathbf{q}) \\
\mathbf{N}_{21}(\mathbf{q})\mathbf{E}_{11}(\mathbf{q}) + \mathbf{N}_{22}(\mathbf{q})\mathbf{E}_{21}(\mathbf{q}) & \mathbf{N}_{21}(\mathbf{q})\mathbf{E}_{12}(\mathbf{q}) + \mathbf{N}_{22}(\mathbf{q})\mathbf{E}_{22}(\mathbf{q})
\end{array}\right]
$$

and

$$
\lambda(\mathbf{q}, \dot{\mathbf{q}}) = \begin{bmatrix} -\mathbf{h}_1(\mathbf{q}, \dot{\mathbf{q}}) - \mathbf{c}_1(\mathbf{q}) \\ -\mathbf{h}_2(\mathbf{q}, \dot{\mathbf{q}}) - \mathbf{c}_2(\mathbf{q}) \end{bmatrix}
$$

Equation (138) describes the complete manipulator dynamics in Cartesian coordinates. The control problem is to find a feedback control law $\mathbf{u}(t) = \mathbf{g}(\mathbf{x})t)$) to minimize the manipulator hand error along the desired hand trajectory over a wide range of payloads. Again perturbation theory is used, and Taylor series expansion is applied to Eq. (138) to obtain the associated linearized system and to design a feedback control law about the desired hand trajectory. The determination of the feedback control law for the linearized system is identical to the one in the joint coordinates. [Eqs. (109)–(128)]. The resolved motion adaptive control block diagram is shown in Figure 13.

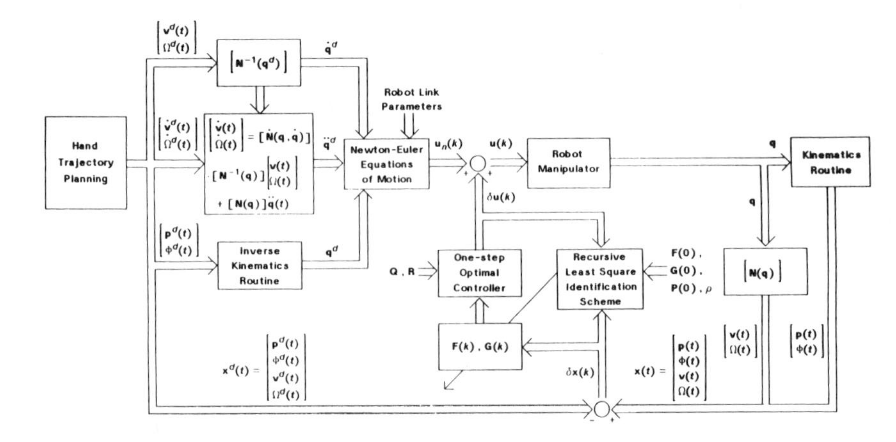

Figure 13. The resolved motion adaptive control.

The overall resolved motion adaptive control system is again characterized by a feedforward component and a feedback component. Such formulation has the advantage of employing parallel computation schemes in computing these components. The feedforward component computes the desired joint torques as follows: (i) the hand trajectory set points $[\mathbf{p}^d(t), \Phi^d(t), \mathbf{v}^d(t), \dot{\Omega}^d(t)]$ are resolved into a set of values of desired joint positions, velocities, and accelerations; (ii) the desired joint torques along the hand trajectory are computed form the Newton–Euler equations of motion using the computed sets of values of joint positions, velocities, and accelerations. These computed torques constitute the nominal torque values $\mathbf{u}_n(t)$. The feedback component computes the variational joint torques $\delta\mathbf{u}(t)$ in the same way as in Eq. (128) with a recursive least-squares identification scheme in Eqs. (121)–(123).

A feasibility study of implementing the adaptive controller using present-day low-cost microprocessors is conducted by looking at the computational requirements in terms of mathematical multiplication and addition operations [14]. The study assumes that multiprocessors are available for parallel computation of the controller. The feedforward component which computes the nominal joint torques along a desired hand trajectory can be computed serially in four separate stages. It requires a total of 1386 multiplications and 988 additions for a six-joint manipulator. The feedback control component which computes the variational joint torques can be conveniently computed serially in three separate stages. It requires about 3348 multiplications and 3118 additions for a six-joint manipulator. Since the feedforward and feedback components can be computed in parallel, the resolved motion adaptive control requires a total of 3348 multiplications and 3118 additions in each sampling period. Computational requirements in term of multiplications and additions for the adaptive controller for a n-joint manipulator are tabulated in Table 3.

Based on the specification sheet of an INTEL 8087 microprocessor, an interger multiply requires 19 μs, an addition requires 17 μs, and a memory fetch or store requires 9 μs, and assuming that each multiplication and addition operation requires two memory fetches, then the proposed controller can be computed in about 233 ms which is not fast enough for closing the servo loop. Similarly, looking at the specification sheet of a Motorola MC68000 microprocessor, an integer multiply requires 5.6 μs, an addition requires 0.96 μs, and a memory fetch or store requires 0.32 μs, and so the proposed controller can be computed in about 26.24 ms, which is still not fast enough for closing the servo loop. Finally, looking at the specification sheet of a PDP 11/45 computer, an integer multiply requires 3.3 μs, an addition requires 300 ns, and a memory fetch or store requires 450 ns, so the proposed controller can be computed in about 18 ms, which translates to approximately a sampling frequency of 55 Hz. However, the PDP 11/45 is a uniprocessor machine and parallel computation assumption is not valid. But it does give us an

Table 3. Computations of the Resolved Motion Adaptive Control*

Adaptive Controller		*Number of Multiplications*	*Number of Additions*
Stage 1	Compute q^d (Inverse Kinematics)	(39)	(32)
Stage 2	Compute $\dot{q}^d$	$n^2 + 27n + 327$ (525)	$n^2 + 18n + 89$ (233)
Stage 3	Compute $\ddot{q}^d$	$4n^2$ (144)	$4n^2 - 3n$ (126)
Stage 4	Compute τ (see [16])	$117n - 24$ (678)	$103n - 21$ (597)
Total Feedforward Computations		$5n^2 + 144n + 342$ (1386)	$5n^2 + 118n + 100$ (988)
Stage 1	Compute $(\mathbf{p}^T, \mathbf{\Phi}^T)^T$	(48)	(22)
	Compute $(\mathbf{v}^T, \mathbf{\Omega}^T)^T$	$n^2 + 27n - 21$ (177)	$n^2 + 18n - 15$ (129)
Stage 2	Compute Hand Errors $(\mathbf{x}(k) - \mathbf{x}_n(k))$ and Identification Scheme	0 (0) + $33n^2 + 9n + 2$ (1244)	$2n$ (12) + $34\frac{1}{2}n^2 - 1\frac{1}{2}n$ (1233)
Stage 3	Compute Adaptive Controller	$8n^3 + 4n^2 + n + 1$ (1879)	$8n^3 - n$ (1722)
Total Feedback Computations		$8n^3 + 38n^2 + 37n + 30$ (3348)	$8n^3 + 35\frac{1}{2}n^2 + 17\frac{1}{2}n + 7$ (3118)
Total Mathematical Operations		$8n^2 - 38n^2 - 37n + 30$ (3348)	$8n^3 + 35\frac{1}{2}n^2 - 17\frac{1}{2}n + 7$ (3118)

* Number inside the parentheses indicates computations for $n = 6$.

indication of the required processing speed of the microprocessors. We anticipate faster microprocessors just around the corner that will be able to compute the proposed resolved motion adaptive controller within 10 ms.

6. SUMMARY AND CONCLUSIONS

We have reviewed various robot manipulator control methods. They vary from simple servomechanism to advanced control such as adaptive control with identification algorithm. The control techniques are discussed in joint motion control, resolved motion control, and adaptive control. Most of the joint motion and resolved motion control methods discussed servo the arm at the hand or the joint level and emphasize nonlinear compensations of the coupling forces among the various joints. We have also discussed various adaptive control strategies. The model referenced adaptive control is easy to implement, but suitable referenced models are difficult to choose and it is difficult to establish any stability analysis of the controlled system. Self-tuning adaptive control fits the imput–output data of the system with an auto-regressive model. Both methods neglect the coupling forces between the joints which may be served for manipulators with rotary joints. The adaptive control using perturbation theory may be more appropriate for various manipulators because it takes all the interaction forces between the joints into consideration. The adaptive control strategy was found suitable for controlling the manipulator in both the joint coordinates and Cartesian coordinates. The adaptive control system is characterized by a feedforward component and a feedback component which can be computed separately and simultaneously in parallel. The computations of the adaptive control for a six-link robot arm may be implemented in low-cost microprocessors for controlling in the joint variable space, while the resolved motion adaptive control cannot be implemented in present-day low-cost microprocessors because they still do not have the required speed to compute the proposed controller within 10 ms. This may be changing in the next year or two.

All the control analyses discussed above presented an ideal system study because they all neglected such nonlinear effects as gear friction and backlash. The physical implementation of the above controls may require further investigation on the effects of gear friction, backlash, control device dynamics, and flexible link structure to the controllers.

ACKNOWLEDGMENT

This work was supported in part by the National Science Foundation under Grant ECS-81-06954. Any opinions, findings, and conclusions or recommendations expressed in this publication are those of the author and do not necessarily reflect the views of the funding agency.

NOTES

1. Here, matrices and vectors are represented in boldface upper- and lowercase alphabets, respectively.
2. PUMA is a trademark of Unimation, Inc.
3. VAL is a software form Unimation, Inc. for the control of PUMA robot arms.

REFERENCES

[1] Astrom, K. J., Eykhoff, P., System identification—A survey. *Automatica, 7*, 123–162, 1971.

[2] Bejczy, A. K., Robot arm dynamics and control. Technical Memorandum 33–669, Jet Propulsion Laboratory, Feb. 1974.

[3] Dubowsky, S., and DesForges, D. T., The application of model referenced adaptive control to robotic manipulators. *Transaction of the ASME, Journal of Dynamic Systems, Measurement and Control, 101* (Sep.), 193–200, 1979.

[4] Hemani, H., and Camana, P. C., Nonlinear feedback in simple locomotion system. *IEEE Transactions on Automatic Control, AC-19* (Dec.), 855–860, 1976.

[5] Hollerbach, J. M., A recursive lagrangian formulation of manipulator dynamics and a comparative study of dynamics formulation complexity. *IEEE Trans. on Systems, Man, and Cybernetics, SMC-10*(11), 730–736, 1980.

[6] Kahn, M. E., and Roth, B., The near-minimum-time control of open-loop articulated kinematic chains. *Transaction of the ASME, Journal of Dynamic Systems, Measurement and Control, 93*(Sep), 164–172, 1971.

[7] Kirk, D. E. *Optimal Control Theory, An Introduction* Englewood Cliffs, NJ: Prentice-Hall, 1970, pp. 227–258.

[8] Koivo, A. J., and Guo, T. H., Adaptive linear controller for robotic manipulators, *IEEE Transactions on Automatic Control, AC-28*(1) 162–171, 1983.

[9] Lee, C. S. G, Robot arm kinematics, dynamics, and control. *IEEE Computer*, 15(12), 62–80,1982.

[10] Lee, C. S. G., and Chen, M. H., A suboptimal control design for mechanical manipulators. *Proceedings of the 1983 American Control Conference*, June 22–24, 1983, San Francisco, California, pp. 1056–1061.

[11] Lee, C. S. G., and Chung, M. J., "An Adaptive Control Strategy for Mechanical Manipulators," *IEEE Transactions on Automatic Control, AC-29*(9):837–840, 1984.

[12] Lee, C. S. G., Chung, M. J., Turney, J. L., and Mudge, T. N., On the control of mechanical manipulators. *Proceedings of the Sixth IFAC Conference in Estimation and Parameter identification*, Washington D.C., June 1982, pp. 1454–1459.

[13] Lee, C. S. G., Gonzalez, R. C., and Fu, K. S., *Tutorial on Robotics*. IEEE Computer Society Press, November 1983.

[14] Lee, C. S. G., and Lee, B. H., "Resolved Motion Adaptive Control for Mechanical Manipulators." *Transactions of ASME, Journal of Dynamic Systems, Measurement and Control, 106*(2):134–142, 1984.

[15] Lee, C. S. G., Mudge, T. N., and Turney, J. L., Hierarchial control structure using special purpose processors for the control of robot arms. *Proceedings of 1982 Pattern Recognition and image Processing Conference*, Las Vegas, Nevada, June 14–17, 1982, pp. 634–640.

[16] Luh, J. Y. S., Walker, M. W., and Paul, R. P. C., On-line computational scheme for mechanical manipulators. *ASME Journal of Dynamics Systems, Measurement and Control, 102*(June), 69–76, 1980.

[17] Luh, J. Y. S., Walker, M. W., and Paul, R. P., Resolved-acceleration control of mechanical manipulators. *IEEE Transactions on Automatic Control, AC*-25(3), 468–474, 1980.

[18] Markiewicz, B. R., Analysis of the computed: torque drive method and comparison with conventional position servo for a computer-controlled manipulator, Technical Memo 33-601, Jet Propulsion Lab, March 1973.

[19] Paul, R. P., *Robot Manipulators–Mathematics, Programming, and Control*. Cambridge, MA: MIT Press, 1981.

[20] Paul, R. P., Modeling, trajectory calculation and servoing of a computer controlled arm. Stanford Artificial Intelligence Laboratory, A.l. Memo 177, Sep. 1972.

[21] Saridis, G. N., and Lee, C. S. G., An approximation theory of optimal control for train-

able manipulators. *IEEE Transaction on Systems, Man and Cybernetics, SMC-9*(3), 152–159, 1979.

[22] Saridis, G. N., and Lobbia, R. N., Parameter identification and control of linear discrete-time systems. *IEEE Transactions on Automatic Control, AC-17*(1), 52–60, 1972.

[23] Whitney, D. E., Resolved motion rate control of manipulators and human prostheses. *IEEE Transactions on Man-Machine System, MMS-10*(2) 47–53, 1969.

[24] Whitney, D. E., The mathematics of coordinated control of prosthetic arms and manipulators. *Journal of Dynamic Systems, Measurement, and Control*, Dec., 303–309, 1972.

[25] Wu, C. H., and Paul, R. P., Resolved motion force control of robot manipulator. *IEEE Trans. on Systems, Man and Cybernetics, SMC-12*(3), 266–275, 1982.

[26] Young, K. K. D., Controller design for a manipulator using theory of variable structure systems. *IEEE Transactions on Systems, Man, and Cybernetics, SMC-8*(2), 101–109, 1978.

Chapter 4

THE EVOLUTION OF ROBOT MANIPULATOR PROGRAMMING

Richard P. Paul

1. INTRODUCTION

The major problem of robot manipulator programming is the tremendous amount of real-time computation necessary to intelligently control the robot. This was solved initially by employing a simulation phase in which the motion would be simulated and data revelant to the actual performance of the robot would be computed. There were no time constraints in the simulation phase. Sensor processing, such as vision, required a great amount of time and fitted in well with this strategy. After the simulation and data reduction phase was completed the robot was actually run with only the minimum necessary computation performed in real time, which was taxing enough to the early computers. The early robot programming languages, while appearing to be

Advances in Automation and Robotics, Volume 1, page 117–139.

ISBN: 0-89232-399-X

real-time control languages, were in fact simulation languages with the side effect of providing a data file from which an actual manipulator could be run, in much the same way that one might drive a computer graphics display. In fact, the basic data type of robotics, the homogeneous transformation, comes from computer graphics. With the further development of computers and of high-level languages and the simplification of robot manipulator control, it has finally become possible to embed robot manipulator control into a high-level language and to exercise real-time control over the manipulator. We describe the embedding of the manipulator control into PASCAL, but many other languages would do as well. Current problems of robot programming lie in the area of operating systems necessary to support the parallel operation of the many sensor and control processes related to robotics.

2. THE INDUSTRIAL ROBOTS

The industrial robot is based on both the teleoperator and the numerically controlled machine tool. The teleoperator was developed during the Second World War to handle radioactive materials [1]; it was a substitute for the operator's hands and consisted of a pair of tongs (the slave) in the radioactive enclosure and two handles on the outside (the master). Both tongs and handles were connected by a linkage to provide for the arbitrary positioning and orientation of the master and slave. The mechanical linkage caused the slave to replicate the motion of the master.

In 1947, the first servoed electric-powered teleoperator was developed. The slave was servo-controlled to follow the position of the master. Force information was, however, no longer available to the operator, and tasks requiring parts to be brought into contact were difficult to perform. In 1948, a servo system was introduced in which the force exerted by the tongs could be relayed to the operator by back-driving the master; the operator could once again feel what was going on.

In 1948, the Air Force sponsored research in the development of a numerically controlled (NC) milling machine [2]. This research was to combine sophisticated servo system expertise with the new, developing digital computer techniques. The pattern to be cut was stored in digital form on a punched tape and then a servo-controlled milling machine, using the tape as input, would cut the metal. The MIT Radiation Laboratory was awarded a subcontract and demonstrated such a machine in 1953.

In the 1960s, George Devol demonstrated what was to become the first Unimate industrial robot, a device combining the articulated linkage of the teleoperator with the servoed axes of the numerically controlled milling machine. The industrial robot could be taught to perform any simple job by guiding it by hand through a sequence of task positions which were recorded in digital memory. Task execution consisted in replaying these positions by servoing the individual joint axes. Task interaction was limited to opening and closing the end effector and to either signaling external equipment or waiting for a synchronizing signal. The industrial robot was ideal for pick and place jobs such as unloading a diecasting machine. The part would appear in a precise position, defined with respect to the robot; it would be grasped, moved out of the die, and dropped on a conveyor. The success of the industrial robot, like the NC milling machine, relied on precise, repeatable digital servo loops. There was no interaction between the robot and its work.

3. THE SENSOR-CONTROLLED ROBOTS

Simultaneously with the development of the industrial robot, an attempt was made to automate the teleoperator, an attempt made possible only by the development of digital computers. In 1961, Ernst [3] developed a robot with touch sensors located in the hand which could be programmed to perform tasks such as locating and picking up blocks and putting them in a box (see Figure 1). Programming was in the form of instructions such as "*move* in direction x with speed v *until* sense element s indicates a "or" if sense element s indicates *1*, *go to* the next instruction, otherwise continue the same action." A program of 600 lines of code, made up of instructions of this form was required for the block program. The lack of any global idea of the position of objects limited this robot as much as the complete lack of task information limited the position-controlled industrial robot.

4. VISION

In 1963, Roberts demonstrated the feasibility of processing a digitized halftone picture of a scene to obtain a mathematical description of the blocklike objects in the scene, expressing their location and orientation by homogeneous transformations [4] (see the Appendix). This work was important for two reasons: It demonstrated that objects could be identified and located in a digitized halftone image, and it introduced homogeneous transformations as a suitable data structure for the description of the relative position and orientation between objects. If the relative position and orientation between objects is represented by homogeneous transformations, the operation of matrix multiplication of homogeneous transformations can establish the overall relationship between any two objects [5,6].

Figure 1. The first sensor-controlled robot.

5. HAND–EYE SYSTEM

Touch feedback, because of its slow, groping nature, was dropped in favor of vision as an input mechanism. By 1967, a computer equipped with a television camera as an input mechanism could, in real time, identify objects and their location [7].

By 1970, a camera- and arm-equipped computer could play real-world games and the "instant insanity" puzzle was successfully solved at Stanford University [8]. In this puzzle, four cubes with different-colored faces must be stacked so that no two similar colors appear on any side. At MIT, a block structure could be observed and copied. In Japan, research led to a hand-eye system which could assemble block structures when presented with an assembly drawing [9]. In the hand–eye system a world model of fixed objects and of objects located by the vision system was represented as instances of prototype objects whose orientation and position were described by homogeneous transformations [10]. The manipulator system was told what object to move and the position to which it should be moved. The manipulator system then determined a stable grasping position. It also determined whether the object needed to be set down and regrasped in order to be moved to the specified destination [11]. A collision avoider (never implemented) was to determine a safe path for the arm through the world model, describing the path as a sequence of homogeneous transformations. In order to avoid stopping at every point making up such a path or trajectory, a continuous curve was fitted through the sequence of joint angles corresponding to the sequence of transformations making up the path.

A manipulator trajectory was specified as six sequences of joint angles through which the six joints were to pass in a time coordinated manner. At the first and last points of each of the six trajectories, zero velocity and acceleration constraints were imposed. At all the intermediate points, continuity of velocity and acceleration was required. A sequence of polynomial splines was calculated to meet these requirements. Unfortunately, the spline fit was a lengthy procedure and could not be performed as the manipulator moved. However, in the blocks world system, as all the positions were known before the manipulator was moved, the lengthy spline fit could be computed before the motion began. This introduced the concept of a planning phase and a run-time phase. By dividing the task in this manner, the planning could be performed off-line, with no time constraints. Additional calculations relating to dynamics, servo gains, and offsets could also be calculated. At runtime, no delays were incurred while the solutions for the next positions were obtained, and the resulting motion was smooth and continuous.

While this system was excellent for generating graphics displays of an ideal arm moving ideal objects, problems occurred in a real environment. No interaction was specified between the arm and the environment; the arm simply moved through space, opening and closing its gripper. At every interaction with the environment, the forces and torques generated were ignored. Consider, for example, the task of placing an object on a surface. The arm is commanded to move the object to zero height and the gripper opened. The position tolerance of this move is zero, for if the arm stops above the surface and opens its gripper, then the object is dropped, not placed. If the arm tries to move below the surface, it is stopped by the object while infinite forces are exerted on the object. The placement task should be specified to move the object toward the surface until an appropriate contact force is detected and the gripper then opened. Similarly, in grasping an object, the hand should be centered over the object using touch feedback; the fingers should not simply be closed, possibly displacing the object.

Force and touch feedback were added to the arm, in a rather crude manner, to perform the above functions. With the addition of this feedback came a great deal of sensory information from the environment. While the position of objects could be modeled using homogeneous transformations and problem solvers could function with such a model, there was no model for the interactive forces, torques, and touches. This information was not used by the problem solver, other than to call vision to locate an unknown object. The manipulator programming became ad hoc and experimental. In order to meet the needs of this type of programming, the WAVE system was developed.

6. THE WAVE SYSTEM

In the WAVE system [12] the procedure calls previously embedded in a high-level language could be typed in directly by name with parameters or read from a file. A macro facility was added to build simple sequences of manipulator primitives to make up higher-level commands. An on-line macro editor provided a quick interactive way of developing these macros. Motion, force, and touch commands became the primitives of the language. The WAVE system functioned in two modes: a planning mode in which instructions read in were assembled into a file for later execution, and a direct mode in which each instruction was executed as it was typed in.

The force and touch primitives in WAVE were modifiers of motion statements; for example, the WAVE instruction to insert a pin along the z axis of station coordinates is as follows:

```
FREE 2 X,Y   ; COMPLY WITH FORCES IN X AND Y
SPIN 2 X,Y   ; COMPLY WITH TORQUES IN X AND Y
STOP Z 100   ; STOP WHEN THE Z FORCE EXCEEDS 100
MOVE IN      ; WHILE MOVING TO POSITION 'IN'
SKIPN 23     ; IF THE STOPPING CONDITION WAS MET
             ; SKIP THE NEXT INSTRUCTION
JUMP EROC    ; JUMP TO ERROR RECOVERY ROUTINE
```

7. CARTESIAN MOTION

The WAVE system specified positions and orientations in Cartesian coordinates but moved the manipulator in joint coordinates. In order to provide a system capable of working on moving objects of known position and velocity one must also move the manipulator in Cartesian coordinates. The position and orientation of the object, described by a homogeneous transformation, are simply postmultiplied by a second transformation representing the desired position and orientation of the gripper with respect to the object. The product yields a transformation representing the required base coordinate position and orientation of the manipulator. From this transformation the joint coordinates can be obtained. These transformations must be performed at a rate sufficiently high to provide for continuous tracking motion of the manipulator. A system designed to function in this manner was developed at SRI [13] and performed these transformations at a 20-Hz rate in order to control a Unimate manipulator. The preplanned spline fit trajectories of WAVE were discarded in favor of a simple on-line method which calculated trajectories segment by segment. The resultant Cartesian motion, although elegant, was simply a result of the necessity of combining Cartesian positions of objects when one object was in motion.

The programming style was also changed in that a move through a series of positions was programmed as a move to each individual position. During execution, the moves through a series of positions were turned into a continuous motion by the run-time trajectory calculator. During each move segment a series of functions could be performed, such as opening and closing the gripper. This form of programming was very similar to the original Unimate style of programming. The main differences were the following: (1) positions were specified in terms of homogeneous transformation products, one of which could represent a moving coordinate system such as a conveyor; (2) motions were made in a coordinated, well-controlled manner; (3) motions did not stop at each intermediate point but transitioned smoothly through intermediate points. This system eliminated the need for a planning phase.

8. INTEGRATING ROBOT CONTROL INTO HIGH-LEVEL LANGUAGES

With the elimination of the need for a planning phase, robot manipulator control could be represented as a sequence of program statements to move the manipulator from one position to another. A manipulator control process, much like an input–output device driver, would actually move the manipulator. The manipulator control process would also provide for smooth path motion transitions and for bringing the manipulator to rest when no further move statements were pending. The move statement would provide for synchronization between program execution and manipulator motion.

With the development of new high-level programming languages such as Algol 60, c, and PASCAL, it became possible to represent the data structures necessary for manipulation directly in the high-level language. The integration of sensors with a manipulator could then be achieved by embedding the manipulator control directly into one of the above languages. Sensors would be treated as input devices and all the well-understood control and data types of the language would be available to the robot programmer. We will describe the embedding of manipulator control into PASCAL [14,15] but any of the other languages would do as well. It is first necessary to describe the data structures to define transformations and transform expressions. We will then show how our data representation can be used to solve the necessary transform equations. Motion primitives will then be introduced which correspond to joint motion, Cartesian motion, and a new form of functional motion. Program execution will be considered in terms of a coprocessor structure: one processor for the program and one for the manipulator. A synchronization structure will be developed to coordinate the two processors. These processors may, of course, be implemented in one physical processor or in a multiprocessor configuration.

8.1 Task Description

Consider the following task. The task consists in picking up pins and inserting them into holes in a subassembly. By defining a series of manipulator end effector positions pn (see Figure 2), we can describe the task as a sequence

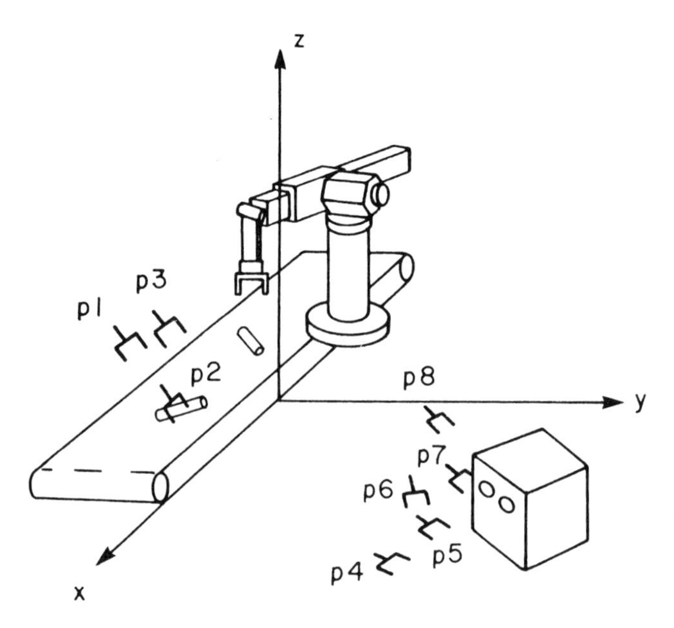

Figure 2. Task positions.

of manipulator moves and actions referring to these numbered positions:

```
MOVE      p1   Approach pin
MOVE      p2   Move over pin
GRASP          Grasp the pin
MOVE      p3   Lift it vertically
MOVE      p4   Approach hole at an angle
MOVE      p5   Stop on contact with hole
MOVE      p6   Stand the pin up
MOVE      p7   Insert the pin
RELEASE
MOVE      p8   Move away
```

Such a program could be executed by any one of a number of industrial robots, and it also exhibits all the limitations of such robots. There is no provision for compliance, such as is required during the pin insertion or the contact between the pin and the hole. No provision is made for storing information related to the actual position of any objects. After the pin is inserted in the first hole, information relating to the position of the hole needs to be retained in order to simplify the insertion of the second pin. If the manipulator is moved, the entire program must be retaught. To insert the second pin, the entire program must be repeated, but with slightly different positions relating to the second hole. What is missing is the structure of the task.

Let us begin to define the structure of the task by defining the structure of the manipulator. We will describe the manipulator by the product of three transformations such that the positions in the task description are replaced by

$$\text{MOVE pn} = \text{MOVE } Z\ T_6\ E$$

where Z represents the position of the manipulator with respect to the base coordinate system;
T_6 represents the end of the manipulator with respect to its base (T_6 is a computable function of joint coordinates);
E represents a tool or end effector at the end of the manipulator.

With such a description, the calibration of a manipulator to the work station is represented by Z. If the task is to be performed with a change of tool, only E must be changed.

We will now represent the structure of the task in terms of the following transforms:

P the position of the pin in base coordinates;
H the position of the block with the two holes;

$^{H}HR_i$	the position of the *i*th hole in the block with respect to the H coordinate system;
^{P}PG	the position of the gripper holding the pin with respect to the pin;
^{P}PA	the gripper approaching the pin;
^{P}PD	the gripper departing with the pin;
^{HR}PHA	the pin approaching the *i*th hole;
^{HR}PCH	the pin at contact with the hole;
^{HR}PAL	the pin at the beginning of insertion;
^{HR}PN	the pin inserted.

The task can now be represented as a series of transform equations solvable for T_6, the manipulator control input, as follows:

p1: $Z\ T_6\ E = P\ PA$
p2: $Z\ T_6\ E = P\ PG$
GRASP
p3: $Z\ T_6\ E = P\ PD\ PG$
p4: $Z\ T_6\ E = H\ HR_i\ PHA\ PG$
p5: $Z\ T_6\ E = H\ HR_i\ PCH\ PG$
p6: $Z\ T_6\ E = H\ HR_i\ PAL\ PG$
p7: $Z\ T_6\ E = H\ HR_i\ PN\ PG$
RELEASE
p8: $Z\ T_6\ E = H\ HR_i\ PN\ PA$

While this may appear complicated, it represents the essential structure of the task, and each transformation represents a separate piece of information.

8.2 Declarations and Data Structures

We will begin this section by defining two data types: vectors and transforms. A vector presents no real challenge and is represented in PASCAL simply as

```
type vector = record x,y,z: real end;
```

A transform is then defined as four vectors (see the Appendix):

```
type transform = record n,o,a,p: vector end;
```

Transforms also have inverses, which we will store in another record and link together

```
type transpointer = ↑transform;
     transform = record n,o,a,p: vector;
          inverse: transpointer;
end;
```

The field inverse is a pointer to another record of type transform. Some transforms are functionally defined, a transform describing the instantaneous position of a moving conveyor, for example. We will enumerate all current functions by a scalar-type functionname and include it in the transform description

```
type transform = record n,o,a,p: vector;
      inverse: transpointer;
      fn: functionname;
end; {transform}
```

The scalar-type functionname is a list of all transform functions. For each function included in functionname a function must be declared.

8.3 Transform Equation Data Structure

We are now ready to represent the transform equations defining a manipulator task. We will do this in two stages. Initially, we will represent the left-hand side of an equation, then the right-hand side, and finally we will link them together to represent the equation. We will represent each term of a transform equation by a term record which links the transform into the equation.

```
type transform = ↑term;
   term = record
      nxt, inv:termpointer;
      trans:transpointer
end; {term}
```

The first (rightmost) element of either the left- or right-hand side of the equation is formed by the function atom. Additional elements of the equation (reading right to left) are added to the front of the data structure by the function cons. In order to construct the data structure corresponding to p1 of the task described above,

$$\text{p1: } Z\ T_6\ E = P\ PA$$

we would first define a series of functions which could be used to create data structure lists corresponding to equations which contain 2, 3, . . . , n transformations: list2, list3, . . . , listn, respectively. For example,

```
function list3 (a,b,c:transpointer):termpointer;
   begin {list3}
      list3:=cons(a,cons(b,atom(c)))
   end; {list3}
```

and then make the following declarations and execute the following calls:

```
var z, t6, e, p, pa; transpointer;
  l hs, rhs:termpointer;

l hs:=list3(z,t6,e);
rhs:=list2(p,pa);
```

The resulting data structure is shown in Figure 3.

The full transform equation is represented by forming two circular data structures to represent the closed-link chain in both directions (see Figure 4).

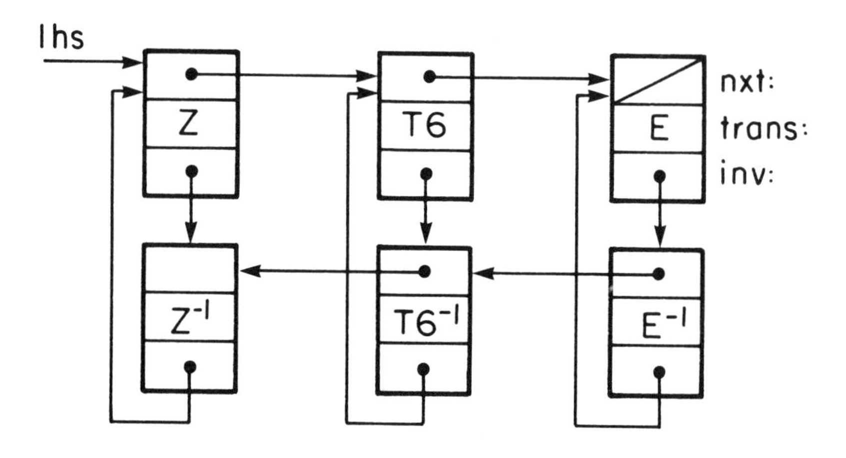

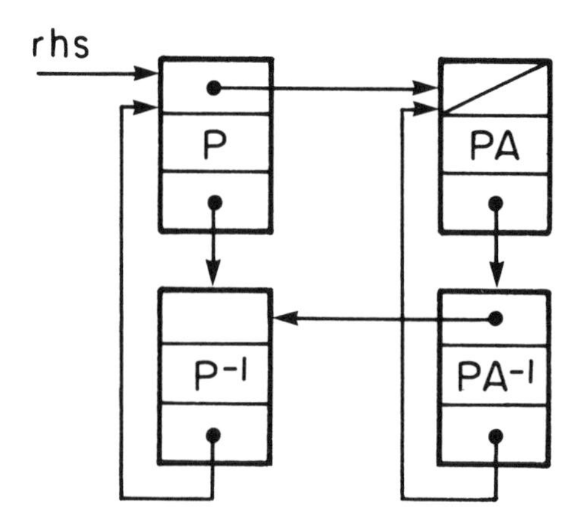

Figure 3. Transform equation data structure.

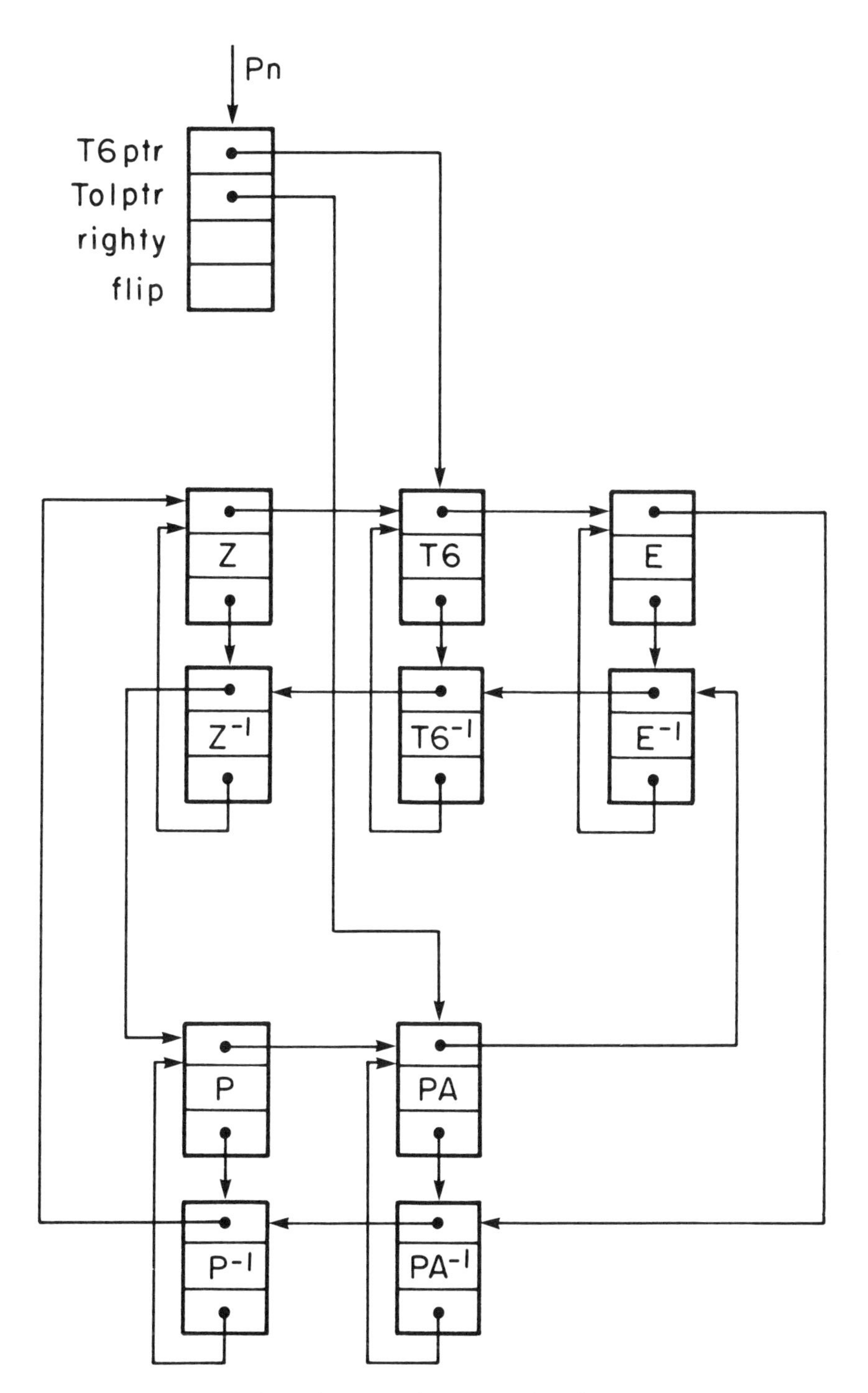

Figure 4. Complete transform equation data structure.

This is accomplished by a position function makeposition, which returns a position record. This record points to the equation data structure in terms of T_6 and to the tool coordinate frame. Information relating to the manipulator configuration might also be stored in this record, thus defining position as follows:

```
type positionpointer = ↑position;
  position = record
    t6ptr, tolptr:termpointer;
    rightly, flip:boolean
end
```

We are now in a position to define all of the manipulator positions in a task and to solve the resulting transform equations for all the information necessary to execute the task. How is this to be done? We will define a motion procedure move with two arguments: a position equation and a mode record. The position equation describes the next position to which the manipulator is to be moved. The mode record describes how the manipulator is to be moved and is defined partially as follows:

```
type modepointer = ↑mode;
  mode record
  typeofmotion:(joint,Cartesian);
  {joint motion or Cartesian}
tacc, tsegment:integer;
  {acceleration time and segment time}
mass:real;{mass of load}
end;
```

Times are specified in milliseconds. The procedure move is aware of the minimum possible acceleration and segment times and will use these if smaller times are specified. Thus times of 0 inform move to compute the times.

Let us define the positions and modes for the task we have described:

```
p1: = make position (list3(z,t6,e),
  list3(conv,p,pa),e,true,true);
with m1 ↑ do begin {set up mode}
 typeofmotion: = joint;
  tacc: = 0; tsegment: = 0; mass: = 0 end;
p2: = makeposition(list3(z,t6,e),
  list3(conv,p,pg),e,true,true);
m2 ↑: = m1 ↑; with m2 ↑ do typeofmotion: = Cartesian;
p3: = makeposition(list3(z,t6,e),
  list4(conv,p,pd,pg),pg,true,true);
```

```
m3 ↑: = m2 ↑; with m3 ↑ do begin
 tsegment: = 500; mass: = 10 end;
p4: = makeposition(list3(z,t6,e),
  list4(h,ht,pha,pg),pg,true,false);
m4 ↑: = m3 ↑; with m4 ↑ do begin
 typeofmotion: = joint; tsegment: = 0 end;
p5; = makeposition(list3(z,t6,e),
  list4(h,ht,pch,pg),pg,true,false);
m5 ↑: = m4 ↑; with m5 ↑ do begin
 typeofmotion: = Cartesian; tsegment: = 1000 end;
p6: = makeposition(list3(z,t6,e),
  list4(h,ht,pac,pg),pg,true,false);
m6 ↑: = m5 ↑; with m6 ↑ do tsegment: = 0;
p7: = makeposition(list3(z,t6,e),
  list3(h,ht,pn),pg,true,false);
m7 ↑: = m6 ↑; with m7 ↑ do tsegment: = 500;
p8: = makeposition(list3(z,t6,e),
  list4(h,ht,pn,pa)e,true,false);
m8 ↑: = m2 ↑ {an unloaded move}
```

The program then becomes

```
for i: = 1,2 do begin
  read(camera, pc);{Read in position of pin}
  matrixmultiply(p,cam,pc);{Set p}
  move(p1,m1);{approach pin}
  move(p2,m2);{over pin}
  move wait;
  close;
  move (p3,m3);{departure position}
  ht ↑: = hr[i] ↑;{Temp position of hole}
  move(p4,m4);{Hole approach}
  move(p5,m5);{Contact hole}
  move(p6,m6);{Stand up}
  move(p7,m7);{Insert pin}
  move wait;
  open(10)
  move(p8,m8){Depart from pin}
end
```

8.4 Software Organization

The procedure move communicates with another process which, driven by a real-time interrupt, actually runs the manipulator. The move procedure simply passes to the interrupt process two pointers defining the position and mode. If, at the time of the call, the manipulator is already in motion, then program execution waits until the manipulator approaches the currently specified position. At this time, a transition to the next position commences without the manipulator stopping. Program execution then continues. If the manipulator reaches the transition point without a move statement pending, then the manipulator is brought to rest at the specified position. Sometimes it is necessary to bring the manipulator to rest, as in the case at positions p2 and p7, when the object is grasped and released. This is accomplished by a procedure movewait, which holds program execution until the manipulator is brought to rest at its current final position.

The camera and end effector are also asynchronous processes, and we will define some procedures by which to communicate requests for image processing:

```
camerafindpin;
{scan for a pin and record conveyor position}
```

and for the communication of results:

```
readcamera(p:transpointer; conveyorposition:integer);
  {set the transform pointed to by transpointer equal
    to the image transformation at the time the
    image was processed.
    set conveyorposition to the position of the
    conveyor when the image was processed}
```

If no image has been found, then readcamera will hold program execution until a valid image is processed. Finally, we define the next procedure to hold program execution while the end effector is operating:

```
operatewait;
  {wait unit end effector operation finishes}
```

Armed with these new procedures, we can rewrite the program in terms of three processes—the manipulator process, the end effector process, and the camera process:

```
camerafindpin;{find first pin};
for i: = 1,2 do begin
  ht ↑:-hr[i] ↑;{Temp position of hole}
  readcamera(pc,sc);{wait here until pin found}
  matrixmultiply(p,cam,pc);{Set p}
  move(p1,m1);{approach pin}
  move(p2,m2);{over pin}
  movewait;
  close;
  operatewait;{wait until grasped}
```

```
    move(p3,m3);{departure position}
    move(p4,m4);{Hole approach}
    camerafindpin;{start looking for next pin}
    move(p5,m5);{Contact hole}
    move(p6,m6);{Stand up}
    move(p7,m7);{Insert pin}
    movewait;
    open(10);{no need to wait here}
    move(p8,m8){Depart from pin}
  end
```

8.5 Specifying Compliance

We will specify the compliance of the manipulator in terms of a servo mode for each of the six Cartesian coordinates. The servo mode is itself a record, which is defined as follows:

```
type servomode = record
   case servo:(position,force,stopforce,goforce) of
   position:(tolerance:real);
   {the position tolerance if the manipulator
   is to stop}
force:(value:real);
   {the force to be exerted in a compliance mode}
stopforce:(limit,distance:real);
   {monitor the force along this axis and change to
   a force command when the force condition is met
   terminate the current motion when this condition
   is met or distance is exceeded}
goforce:(value,limit,distance:real)
   {exert this force until position error in this
   direction exceeds limit or the motion exceeds
   distance, then change to a position servo mode.
   Terminate the current motion when this condition
   is met}
end;{servomode}
```

The manipulator process is aware of the minimum values of tolerance, force, stopping force, and goforce displacement. If these are specified as zero, then the minimum values are employed instead. Mode is redefined to be

```
type mode = record
   typeofmotion:(joint,Cartesian);
   tacc,tsegment:integer;
   mass:real;{mass of load}
   dx,dy,dz,rx,ry,rz:servomode
   {dx,dy, and dz refer to translations or forces
```

```
    along the principal axes of the TOOL frame.
    rx,ry, and rz refer to rotations or torques about
    the axes}
  end;
```

For example, the mode for the pin insertion m7 is expanded to

```
  with m7↑do begin
    tsegment:=500;
    with dx do begin servo:=force; value:=0 end;
    with dy do begin servo:=force; value:=0 end;
    with dz do begin servo:=stopforce; limit:=50;
    distance:=5 end;
    with rx do begin servo:=force; value:=0 end;
    with ry do begin servo:=force; value:=0 end;
  end;
```

8.6 Functionally Defined Motion

Whenever it is waiting for a move statement, the manipulator enters the Cartesian servo mode and continues to evaluate the current transform expression set point. This feature enables us to provide functional defined motions. For example, we might wish to describe a circle with the end effector or trace out some curve defined by a polynomial. But before we discuss functionally defined motions we first need to discuss the time variable.

The time variable, time, simply increments at a millisecond rate. It is, however, assignable, so it may be reset to zero at any time. After it is assigned, it simply continues to increment at a millisecond rate.

We will now employ the Cartesian servo mode and the time variable to illustrate functionally defined motion. Consider the crank shown in Figure 5. The transform equation describing the manipulator holding the crank handle is

$$Z\ T_6\ E = \text{PIVOT ROTPZ CRANK ROTMZ G}$$

where ROTPZ is a function transform representing a positive rotation about the z axis of theta degrees. ROTMZ is another function transform, representing a negative rotation about the z axis of theta degrees. The following program will move the manipulator to the crank handle and then turn the crank around twice:

```
  p1:=makeposition(list3(z,t6,e),
    list5(pivot,rotpz,crank,rotmz,g),
    rotmz,pivot,true,true);
  theta:=0;
```

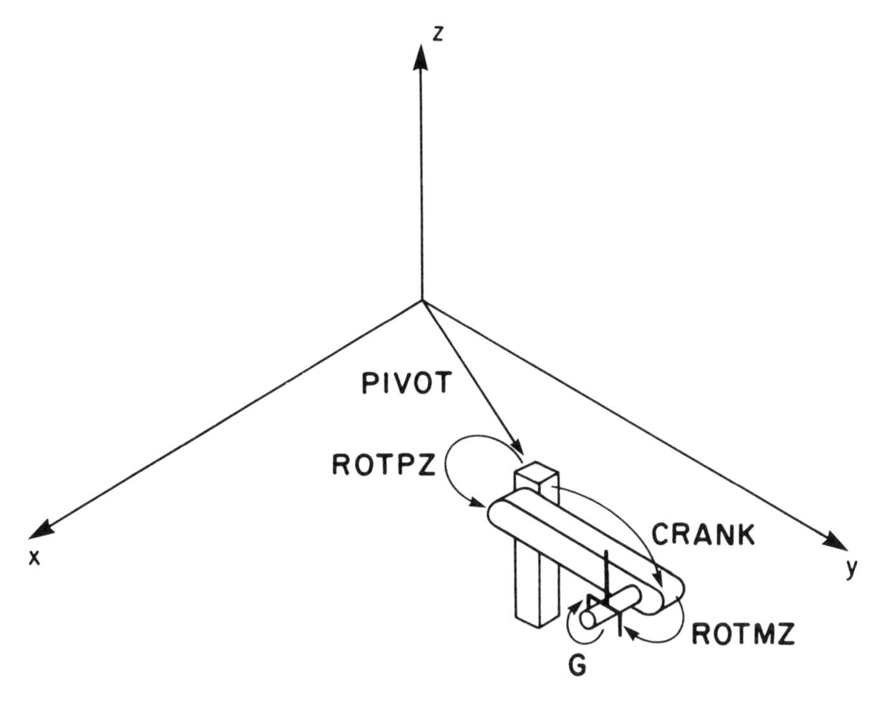

Figure 5. Crank-turning example.

```
move(p1,m0);{move to the crank handle}
movewait;{and stay there}
close;
time:=0;
{reset time}
while time<5000 do theta:=720 * (time / 5000);
```

The manipulator moves to the crank handle and grasps it. The while statement takes 5 sec to execute, and the manipulator remains in Cartesian servo mode, evaluating the transform equation which is functionally dependent on theta. This causes the crank to be rotated twice.

9. CONCLUSIONS

Initially, manipulator control was by means of procedures embedded in a high-level language. This evolved into geometric-based planning and execution systems where planning was time-independent and execution was time-efficient. With the addition of sensor feedback and the need to develop the ad hoc procedures involved in assembly, an interpretative system evolved. Finally, we have come full circle with manipulator control once again embedded in a high-level language. This has become possible because of the development of high-level languages, the simplification of manipulator control eliminating the need to plan trajectories and to precalculate dynamics, and the increasing computing power available to provide for the economic real-time control of the manipulator.

APPENDIX

Homogeneous transformations are used to express the relative position and orientation between two coordinate systems. The components of a vector, from the origin of the first coordinate frame to the origin of the second frame, describe the relative position. The components of each of three unit vectors along the principal axes of the second coordinate frame describe the relative orientation. The components of all vectors are in terms of the first frame (see Figure 6). If we have two objects described in terms of coordinate systems fixed to each object, then, given a homogeneous transformation expressing the relationship between coordinate frames, we can fully describe the position and orientation of one object relative to the other. At the end of the position vector we draw the three unit vectors in terms of their components in the

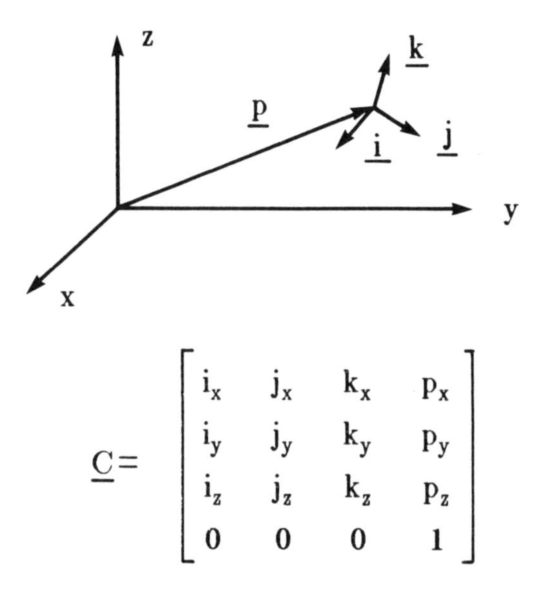

$$\underline{C} = \begin{bmatrix} i_x & j_x & k_x & p_x \\ i_y & j_y & k_y & p_y \\ i_z & j_z & k_z & p_z \\ 0 & 0 & 0 & 1 \end{bmatrix}$$

Figure 6. Coordinate systems.

reference object coordinate frame. If we know the relationship of the described object to its coordinate frame, we can then reconstruct both objects.

We will arrange the components of the three unit vectors as the three columns of a rotation matrix. The x vector is the first column, the y vector is the second column, and the z vector is the third. The position of any vector in the described object, in terms of the reference object, can then be obtained by considering the vector as a column matrix and premultiplying by the rotation matrix. Why does this work? Consider vectors along each of the principal directions of the described object; these vectors consist of two zeros and a number representing the length of the vector. When multiplied by the rotation matrix, this vector will simply scale the appropriate column of the rotation matrix representing the vector to which it is parallel. We must, however, add the vector describing the position of the second object in terms of the first object to the result of the matrix multiplication. We may perform the addition by adding the position vector as a fourth column to the rotation matrix and by adding a "one" as a fourth element to the vector we wish to transform from the second object to the first. In terms of matrix multiplication, this has the effect of adding the components of the position vector as required. Finally, by adding a fourth row to the 3×4 matrix, consisting of three zeros and a "one," we can produce a transformed vector consisting of four elements, the x, y, and z components and another "one" in the last row. This 4×4 matrix is known as a homogeneous transform. If we had a third object described in terms of the second by a homogeneous transformation $\mathbf{T}_2$ and a vector in terms of the third object $\mathbf{v}_3$, then we would have

$$\mathbf{v}_2 = \mathbf{T}_2 \mathbf{v}_3 \tag{1}$$

Further, if we had described the second object in terms of the first by a transform $\mathbf{T}_1$ then we would have the vector $\mathbf{v}_1$ given by

$$\mathbf{v}_1 = \mathbf{T}_1 \mathbf{v}_2 \tag{2}$$

and, combining Eqs. (1) and (2),

$$\mathbf{v}_1 = \mathbf{T}_1 \mathbf{T}_2 \mathbf{v}_3 \tag{3}$$

But if we had described the third object in terms of the first by a transform $\mathbf{T}_3$, we would have

$$\mathbf{v}_1 = \mathbf{T}_3 \mathbf{v}_3 \tag{4}$$

As this is true for all vectors $\mathbf{v}_1$ and $\mathbf{v}_3$, we have, by combining Eqs. (3) and (4),

$$\mathbf{T}_3 = \mathbf{T}_1 \mathbf{T}_2 \tag{5}$$

This means that if we have a description of one object in terms of a second, and a description of the second object in terms of a third, then the matrix product of the two transformations is the transformation which describes the first object in terms of the third. Further, if we have a description $\mathbf{T}_1$ of one object in terms of a second object, and also a description of the second object in terms of the first, $\mathbf{T}_2$, then we must have

$$\mathbf{T}_1\mathbf{T}_2 = \mathbf{I} \tag{6}$$

where $\mathbf{I}$ is the identity transform, and thus

$$\mathbf{T}_2 = \mathbf{T}_1^{-1} \tag{7}$$

where the superscript -1 refers to the operation of matrix inversion, a simple operation for homogeneous transformation.

ACKNOWLEDGMENT

This material is based on work supported by the National Science Foundation under Grant No. MEA-8119884. Any opinions, findings, conclusions, or recommendations expressed in this publication are those of the authors and do not necessarily reflect the views of the National Science Foundation.

REFERENCES

[1] Goertz, R. C., Manipulators used for handling radioactive materials. Chap. 27 of *Human Factors in Technology*, E. M. Bennett, et al., eds., McGraw-Hill, New York, 1963, pp. 425–443.

[2] J. Rosenberg, *A History of Numerical Control 1949–1972: The Technical Development, Transfer to Industry, and Assimilation*, U.S.C. Information Sciences Institute, Marina del Rey, Calif., ISI Report ISI-RR-72-3, 1972.

[3] H. A. Ernst, *A Computer-Operated Mechanical Hand*, Sc.D. Thesis, M.I.T., Cambridge, Mass., 1961.

[4] L. G. Roberts, *Machine Perception of Three-Dimensional Solids*, M.I.T. Lincoln Lab., Cambridge, Mass., Report No. 315, 1963.

[5] L. G. Roberts, *Homogeneous Matrix Representation and Manipulation of N-Dimensional* Constructs, M.I.T. Lincoln Lab., Cambridge, Mass., Document MS1045, 1965.

[6] R. L. Paul, "The Mathematics of Computer Controlled Manipulators," *Proc. 1977 Joint Automatic Control Conference*, San Francisco, June 1977, pp. 124–131.

[7] M. W. Wichman, *Use of Optical Feedback in Computer Control of an Arm*, Stanford Artificial Intelligence Lab., Stanford, Calif., Memo No. 56, 1967.

[8] J. Feldman, et al., "The Use of Vision and Manipulation to Solve the Puzzle," *Proc. Second Int'l. Joint Conf. on Artificial Intelligence*, London, Sept. 1971, pp. 359–364.

[9] M. Ejiri, T. Uno, H. Yoda, T. Goto, and K. Takeyasu, "A Prototype Intelligent Robot That Assembles Objects From Plane Drawings," *IEEE Trans. Computers*, Vol. C-21, No. 2, Feb. 1972, pp. 161–170.

[10] Paul, R. P., "The Computer Representation of Simply Described Scenes," in *Pertinent Concepts in Computer Graphics*, ed. by M. Faiman and J. Nievergelt, University of Illinois Press, pp. 87–103, 1969.

[11] Paul, R. L., *Modeling, Trajectory Calculation and Servoing of a Computer Controlled Arm*, Stanford Artificial Intelligence Laboratory Memo AIM-177, Stanford University, 1972.

[12] Paul, R. P. C., "Wave: A Model-Based Language for Manipulator Control," The Industrial Robot, March 1977.

[13] Paul, R. P., "Manipulator Cartesian Path Control," *IEEE Trans. on Systems, Man and Cybernetics*, SMC-9, 11, 702–711, Nov. 1979.

[14] Jensen, K. and Wirth, N., *PASCAL User Manual and Report*, Springer-Verlag, 1974.

[15] Paul, Richard P., "Robot Manipulators: Mathematics, Programming, and Control," MIT Press, 1981.

Chapter 5

ROBOTIC VISION

J. L. Mundy

1. INTRODUCTION

1.1 The Use of Vision in Robotics

The use of robots in industrial applications is expanding rapidly. In most industrial applications, the task workspace environment must be highly constrained, which implies that extensive fixturing is required to maintain the location of the tools and components being handled by the robot. Extensive engineering effort is necessary to design the workspace and fixturing, which detracts from the economic advantage to be gained from the automation of the task. Also, the time to install a robot and bring the application into production use is greatly extended by these rigid requirements.

The major problem for the near term is thus the intolerance of robots to uncertainty in the size, shape, and location of objects in the work environment. Another issue is the effort needed to program the robot position sequences to perform the application task. This programming effort is com-

Advances in Automation and Robotics, Volume 1, pages 141–208.

ISBN: 0-89232-399-X

plicated by the need to avoid collisions and interferences. Every contingency has to be anticipated and included in the robot program. The result of these limitations is that, once a robot application has been installed, it is usually not reprogrammed to a new task. In effect, the robot is a component imbedded in a fixed automation implementation.

Some of these problems can be solved by extending the flexibility of the robot control. This can take the form of force feedback so that the force applied by the robot is limited in situations where the planned position cannot be reached. The concept of avoiding obstacles by force programming rather then position programming is influencing the design of next-generation robot systems [1].

Passive compliance in fixtures and robot end effectors also allows some further adaptive capability. The remote center compliance principle is a good example of this approach [2]. In this method the wrist of the robot has flexible members which provide a self-correcting restoring force when the arm is performing insertion tasks. This allows the insertion of cylinders into openings with extremely fine tolerances.

While these methods expand the domain of uncertainty within which robots can operate, their ability to adapt is still extremely limited. The evolution to flexible automation with rapid turnover in robot task assignments and the ability to deal with unexpected configurations of the workspace will depend heavily on the use of sensors which can provide detailed descriptions of the environment. The basic information which must be available is the location, orientation, and classification of objects in the work environment. This description must be updated rapidly to allow on-line planning of robot motions. It also must be accurate enough to permit grasping and manipulation of the tools and components in the task environment.

These requirements have motivated extensive research and development in the area of machine vision for the purpose of achieving a sensing system which can describe the work environment rapidly and accurately. The goal of these efforts is a vision system which can recognize objects and their position and orientation. Such information provides the basis for determining an appropriate response to uncertainty and lack of structure in the work environment. This flexibility makes it possible to prepare generic plans which can apply to a family of tasks without requiring detailed reference to specify object locations and orientations. All this implies that for the next generation of industrial robots the emphasis will shift from mechanical and control issues to vision and adaptive on-line planning.

1.2 Approaches to Machine Vision

The use of vision with robots has always been an interesting area for practical research. The approach to vision has evolved from simple binary image analysis methods to current efforts in direct three-dimensional image sensors. The binary case has received the most attention, mainly due to the early pioneering efforts at SRI [3]. This group developed a system of algorithms for the analysis of binary (two-intensity-level) images. These algorithms provide features of objects which allow simple classification, measurement, and location functions.

In order to obtain images that can be successfully interpreted by these algorithms it is necessary to form high-contrast scenes such as silhouettes so that the intensity edges in the image correspond to physical boundaries of the objects. Figure 1(a) shows a typical part, a motor end casing which has been placed on a light table by a robot assembly manipulator [4]. The binary image of this part is shown in Figure 1(b). Typical features which can be extracted and used for recognition are number or holes, distance between holes, and total area of material. The location of the hole pattern relative to the reference coordinate system can be used to orient the part for subsequent use by the robot to add the part to an assembly.

The binary image methods can succeed only if the correspondence between the image and object boundaries is reliable. This correspondence can easily be disturbed unless the illumination is carefully controlled so that a single threshold can separate the object from the background. In addition, it is not

Figure 1(a). A motor end casting.

Figure 1(b). A binary image of the casting formed by backlighting.

often possible to find applications where a set of silhouttes of an object can define all of the information needed in the execution of tasks. For example, many parts can only be distinguished by the configuration of surface features or markings which cannot be defined by a silhouette view.

These limitations motivate interest in the next class of vision methods, which are called *gray-scale* algorithms since they deal with a full range of image intensities rather than just two levels. Typically, a gray-scale image processing system operates on 256 levels of intensity. By detecting the presence of edges and regions of intensity within the gray-scale image considerably more tolerance to illumination and object surface properties is achieved. For example, Figure 2 shows the detection of holes in a metal casting using boundary detection in a rather noisy gray-scale image. These holes could not have been detected using binary methods because of the low and unreliable contrast in the images of scattered light from the casting surface.

The power of existing feature extraction algorithms is not sufficient to allow general approaches to object description for uncontrolled illumination and object orientation. The image intensity can easily vary due to object reflectance variations in a manner which is difficult to distinguish from valid object features. For example, a dark mark or pit can be confused with a hole without additional information such as another view or an additional direction

Figure 2. Boundary detection in a gray-level image. The image is formed by directional illumination of the metal surface.

of illumination. The result of these limitations is that each application must be studied so that the best illumination scheme is selected and the boundary detection algorithms are suitable for the range of conditions which are encountered.

This application dependence gets back to the same problem which is limiting the flexibility of existing robot installations. On the other hand, the consideration of families of applications can constrain the problem enough to allow the design and implementation of suitable algorithms and optical systems to which perform well over the range of cases within the family. For example, the approach to the measurement of holes in the casting shown earlier is adequate to handle a wide range of similar castings as long as the surface finish does not vary over the family and the hole pattern can be described parametrically.

The most powerful level of vision technique is to extract the three-dimensional configuration of objects in the workspace directly by optical triangulation or time-of-flight ranging. This general approach removes many of the restrictions mentioned earlier. In fact, the information which is provided ultimately by binary or gray-scale vision is usually in terms of the three-dimensional properties of the objects in the scene. It is therefore desirable

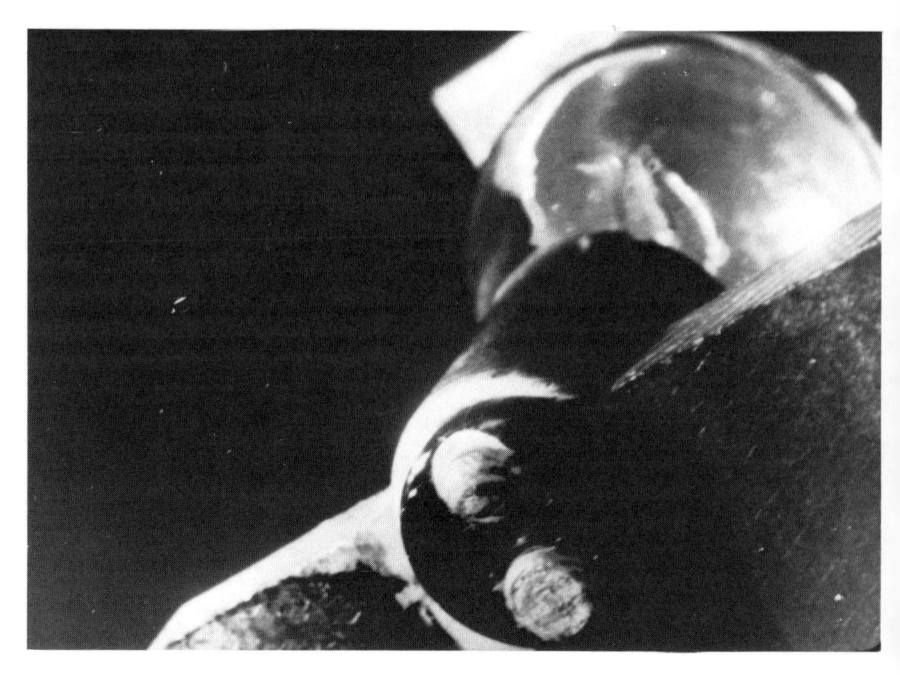

Figure 3. A lamp base illuminated with laser stripes. The faint gray stripes on the lamp contacts are the laser illumination. Note that the stripes are parallel lines on the lower contact and concentric circles on the upper contact.

to provide this information directly without going through the process of extracting it from indirect features of the silhouette or scattered optical image.

An example of this idea is shown in Figure 3. Here the base of a taillight lamp has been illuminated with a series of light stripes. The angle of view is different from the angle of observation so that variations in height of the object surface causes a shift in the observed position of the stripes. It can be seen that the stripe pattern forms concentric circles over the taillight lamp contact, which is spherical. The other, defective, contact is flat so that the stripes are seen as parallel lines. Thus there is a direct correspondence between simple features in the image and the shape of the object surface. The stripes can be detected in the image using simple binary methods, and their location provides the three-dimensional coordinates of the object surface. Since the illumination is controlled, this process does not depend on ambient conditions or is the method paritcularly sensitive to object orientation. Naturally, if the object surface is too shiny, as in the case of a mirror, the method will break down.

In applications where it is not practical to use a structured light source, three-dimensional information can still be acquired through triangulation by acquiring multiple views of the object. Corresponding image features in two views of an object can be matched to determine the coordinates of the feature. This approach is most relevant in situations where the range must be determined at large distances so that the power required from an active source is impractically large.

The remainder of the chapter will treat these methods in more detail. The goal is to present the techniques in enough detail to allow the selection of a suitable approach for robot applications. It is also hoped that this discussion will allow the reader to become quickly aquainted with the current state of the art in robot vision techniques.

2. BINARY VISION METHODS

2.1 Introduction

The classification of image intensity into two classes, light and dark, is the most widely used and oldest form of image analysis. If the images are of sufficiently high contrast, then this method is an efficient way to segment objects from the background. The usual concept is that, through careful lighting, the range of intensities for the objects of interest are widely separated from the intensity range for background or objects not of interest. A classic example of this case is the silhouette, which an object is illuminated from behind so that there is full contrast between the object and the background.

The image samples (pixels) corresponding to the object are separated from the background pixels by testing the image intensity against a threshold. Ideally, the threshold lies about halfway between the intensity mean for the background and the intensity mean for the object. In this way a new image is created, a binary image, which has only two states; the pixels that lie above the threshold and those that lie below. This binary image can then be analyzed to perform object classification and measurement. The boundary of the object is easy to extract from the binary image since the boundary points correspond to the light-to-dark and dark-to-light transitions. These transitions correspond to the simple exclusive-or logic function performed on neighboring pixels. The extraction of object features is usually performed on a connected set of pixels of the intensity state of the objects. The connectivity analysis is carried out in a raster fashion over the image, and various accumulative properties can be calculated as the connected set is expanded. For example, area, perimeter, center of area, maximum and minimum extent, and various moments can all be computed incrementally as the connected set is formed. These features are usually sufficient to classify objects. Critical object measurements can be derived from an analysis of the boundary of the object or from the coordinates of some of the features just mentioned. For example, the distance between two holes is determined from the location of the center of area of the connected sets corresponding to each hole.

The operations on binary images are easy to implement in special-purpose hardware. This has allowed the implementation of inspection and robot vision systems which can analyze binary scenes at hundreds of thousands of pixels per second. Binary image-processing hardware which can easily keep up with standard video rates (8 million pixels/sec) is commonly available.

The inherent simplicity of these binary algorithms is the main factor in the widespread interest in their application. If it is possible to provide the necessary high contrast for reliable object segmentation, then these systems can be implemented with the throughput needed for robot control applications. This section will review a range of techniques for obtaining high-contrast images as well as the basic binary analysis methods.

2.2 Optical Configurations for the Generation of High-Contrast Images

2.2.1 The Silhouette

The success of binary image processing techniques is completely dependent on the availability of high-contrast images. The most common approach to producing such images is the silhouette. The optical setup for generating a silhouette is shown in Figure 4. Here the objects are placed on a transluscent or transparent screen and illuminated from behind, this produces an image field which is clear except for the area subtended by the object. A example of such a silhouette image was given in Figure 1.

The most critical aspect of the silhouette is the region in the vicinity of the object boundary. The accuracy of the definition of the boundary is dependent on the sharpness of the intensity transition at the boundary. The interval of the transition is related to the optical arrangement of the image sensor and the illumination source. The most common limitation on sharpness is the lack of perfect collimation in the illumination source.

Consider Figure 5, which shows a geometrical construction of the light ray paths in a collimated illumination source. The divergence of the rays is due to the finite size of the source. If the rays diverge, then the shadow boundary around the object will not be perfectly sharp. Instead, the interval of intensity change will be smeared out by the effective convolution of the object boundary with the finite image of the source.

This exposes the trade-off between source size and silhouette sharpness. In order to increase sharpness it is necessary to reduce the effective source size by introducing an aperture between the source and the collimating lens. Of course, this reduces the amount of available light and thus requires a

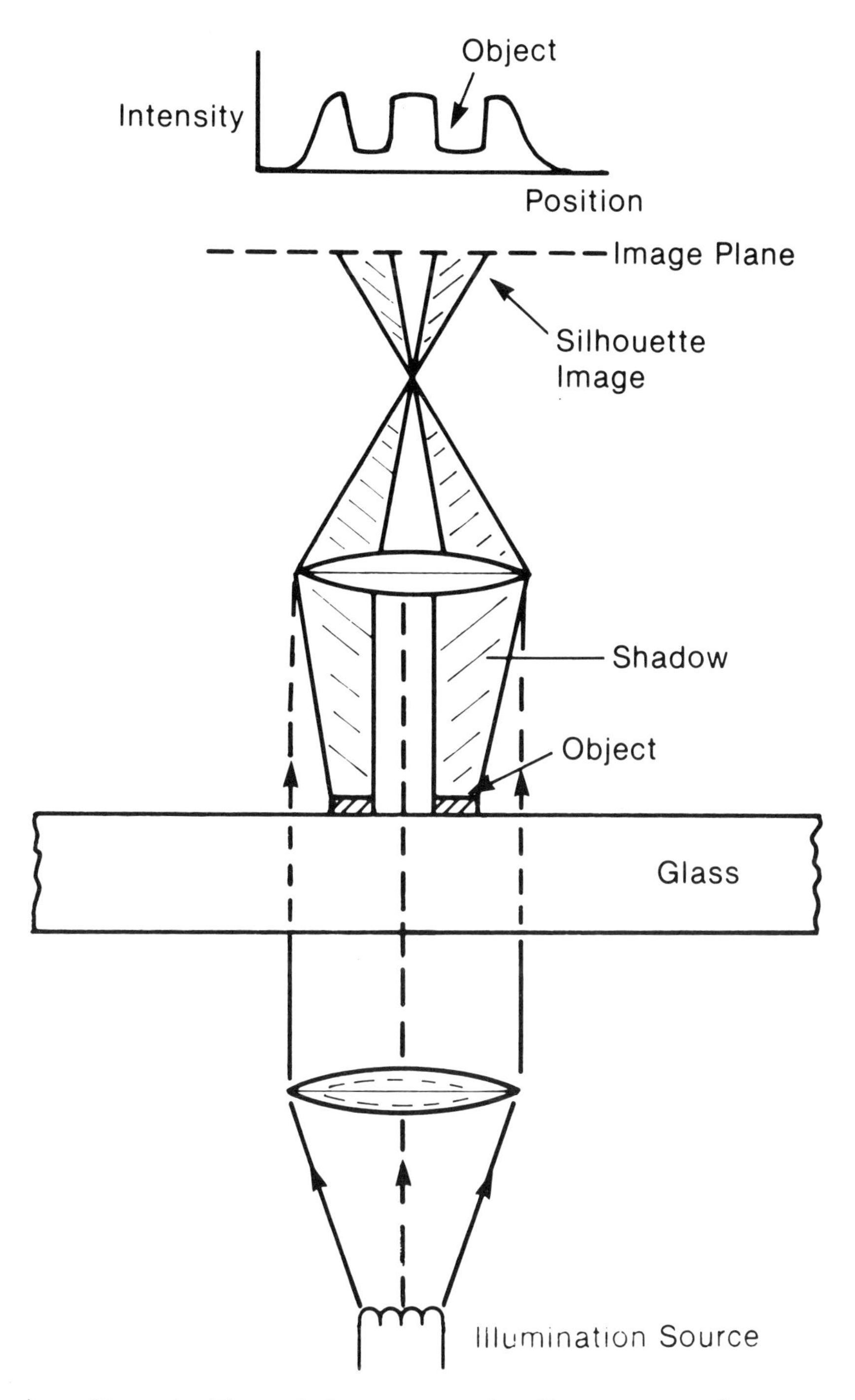

Figure 4. The optical arrangement for silhouette generation.

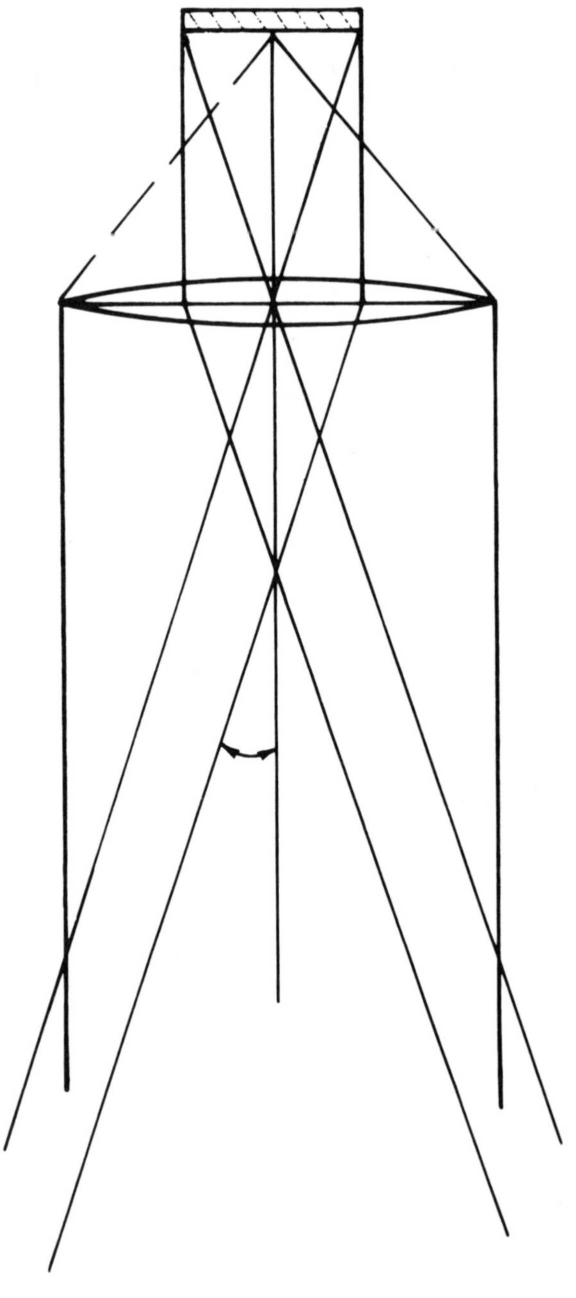

Figure 5. The ray paths in a collimated light source. The finite size of the source produces an angular spread to the rays. The angle shown is directly proportional to the source height (shown as a shaded rectangle).

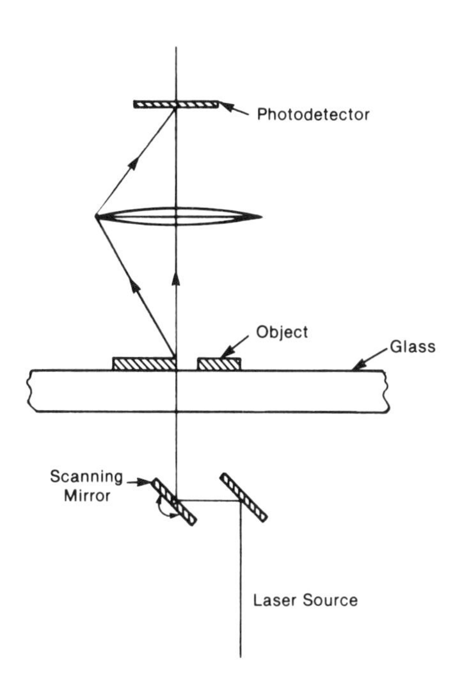

Figure 6. The optical arrangement for silhouette formation using laser illumination. The photodetector collects the rays scattered by the object boundaries.

longer integration time for the sensor detector to reach the same signal-to-noise ratio.

The problem can be overcome through the use of laser illuminators. The source size for a laser is on the order of the wavelength, which is several orders of magnitude smaller then that available from an incandescent or arc source. With this source it is possible to define object boundaries down to the order of the wavelength of the laser. The main drawback to this source is the interference phenomena which occur due to the coherent nature of laser illumination.

These effects are reduced if the arrangement of Figure 6 is used, where the image is formed by a mechanical scan of the laser spot. In this case the light is collected over many path lengths so that the self-interference effects cancel. A similar approach is to introduce a pizeoelectric mirror element into the optical path which changes the effective path length at a rate higher than the image sensor integration time. This also cancels the interference fringes and speckle.

2.2.2 Wavelength Discrimination

In many applications, it is not possible to control the illumination to the extent necessary to form an acceptable silhouette. Another approach is to provide contrast by discriminating on other optical parameters. It is often the case that contrast can be achieved by selecting wavelength intervals for illumination or detection in which the reflectance of the object or background is distinct. This can be thought of as color discrimination, although the wavelength bands may not be in the range of the human visual system. For example, the use of near-infrared illumination can produce contrast for some materials which does not occur in the normal visual range. Also included in this concept is the use of fluorescence, where a material is illuminated with a short wavelength which causes emission of light at longer wavlengths. This method is quite powerful since the fluorescent material can be added or applied selectively to object surfaces which must be discriminated and not to other elements of the environment.

The use of fluorescence is illustrated in Figure 7. Here an object has been coated with a fluorescent dye. In Figure 7(a) the scene is illuminated with a conventional incandescent source. There are defects which are not visible under normal illumination. In Figure 7(b) the illumination has been changed to ultraviolet which has wavelengths on the order of 200 nm (nanometers). The defective locations are clearly visible. This short-wavelength illumination

(a)

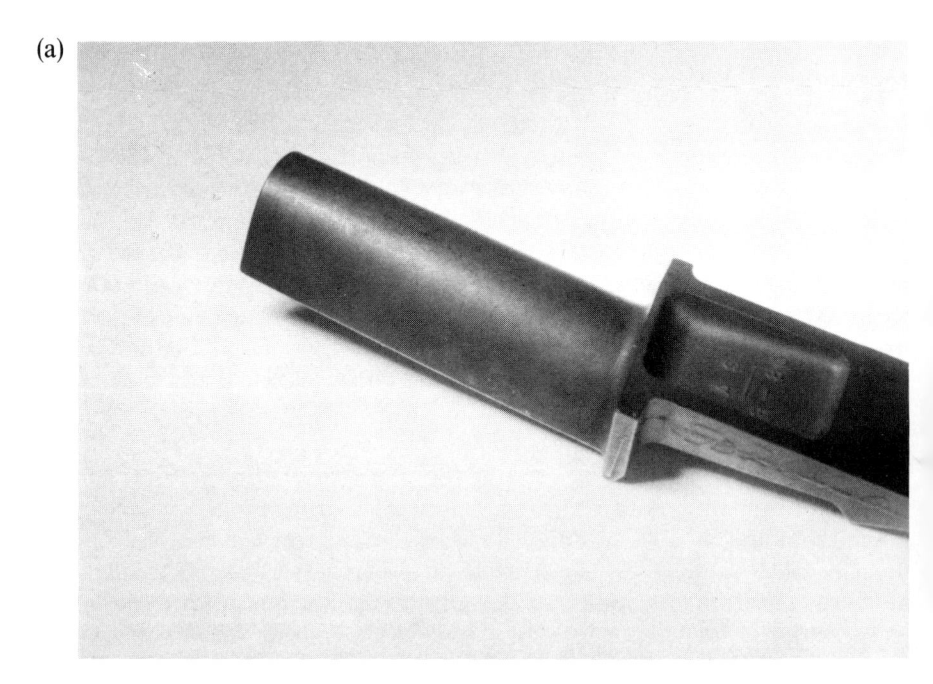

Figure 7. The use of ultraviolet illumination: (a) a metal casting under visible wavelength illumination;

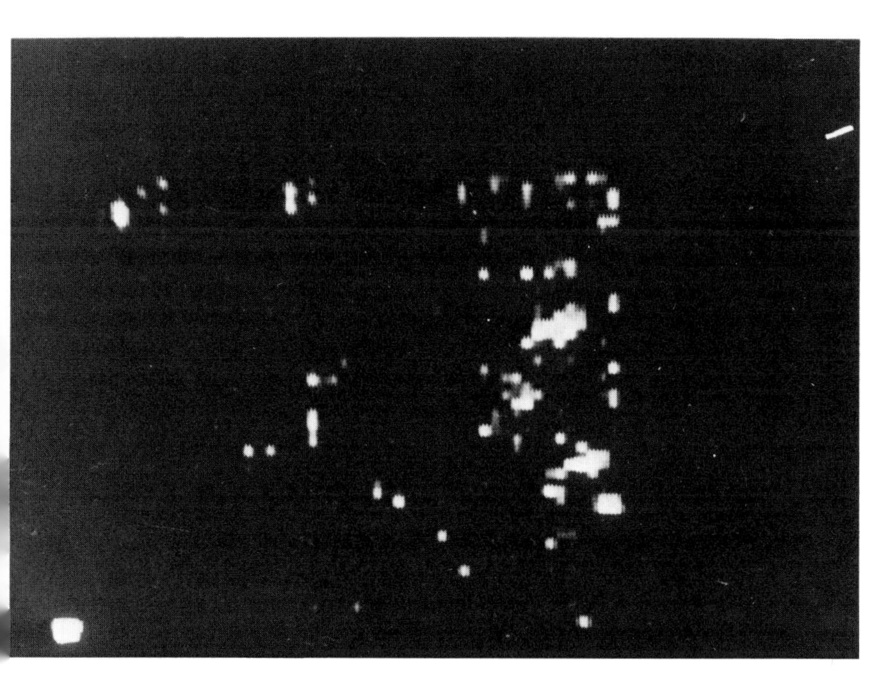

Figure 7 (continued) (b) a magnified version of 7(a) with ultraviolet illumination.

caüses the emmission of light in the 550-nm range, which can be easily separated from the ultraviolet.

Most image sensors are not very sensitive for wavelengths shorter that 300 nm, so a natural filtering occurs. The resulting contrast is very high and forms a very reliable binary representation of the object. The main drawback to the use of fluorescent dye is the need to maintain a well-controlled environment so that the only source of illumination is ultraviolet. The range of wavelengths produced by the dye fluorescence is too broad to be effectively filtered with respect to ambient illumination.

2.2.3 Directional Illumination

It is possible to make use of illumination direction to achieve contrast. This approach is the most effective for surfaces which are characterized by a highly directional distribution of reflected power. The relationship describing the reflection properties of metal surfaces is a good example of such a directional distribution. The relation becomes relatively simple for the case of equal angles of incidence and observation. The scattering theory will be discussed in Section 3.4, which shows that the reflected power distribution is centered about the angle of observation. An optical arrangement

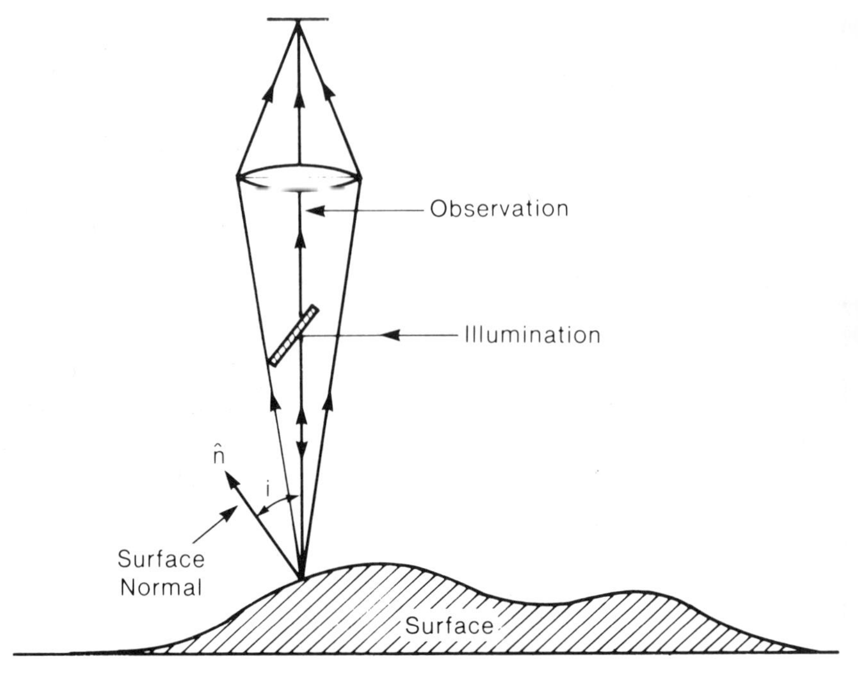

Figure 8. An imaging configuration that provides illumination along the direction of observation. The illumination is introduced with a half-silvered mirror shown as the shaded rectangle. The scattered power is a function of the surface-normal direction and the surface reflectance.

to achieve illumination along the direction of observation is shown in Figure 8. The angular width of the scattered power distribution decreases with decreasing surface roughness. In the extreme limit, where the surface is essentially a mirror, this relation becomes an impulse.

This behavior suggests a method for obtaining contrast between regions of surface that have distinct surface normals. For example, consider the bevel shape in Figure 9. The illumination is directed in such a fashion as to provide high contrast between the middle facet and the other surfaces. In general, the reflected power is a measure of the direction of the surface normal. If the basic reflectivity of the surface is uniform. then the variation in the surface normal can be directly related to the image intensity [5]. This relation can be integrated to derive the surface shape itself. From this more general context it is clear that directional illumination can be used to define regions of distinct surface orientation. The contrast between these regions is dependent on the sharpness of the reflectance function and the difference in orientation.

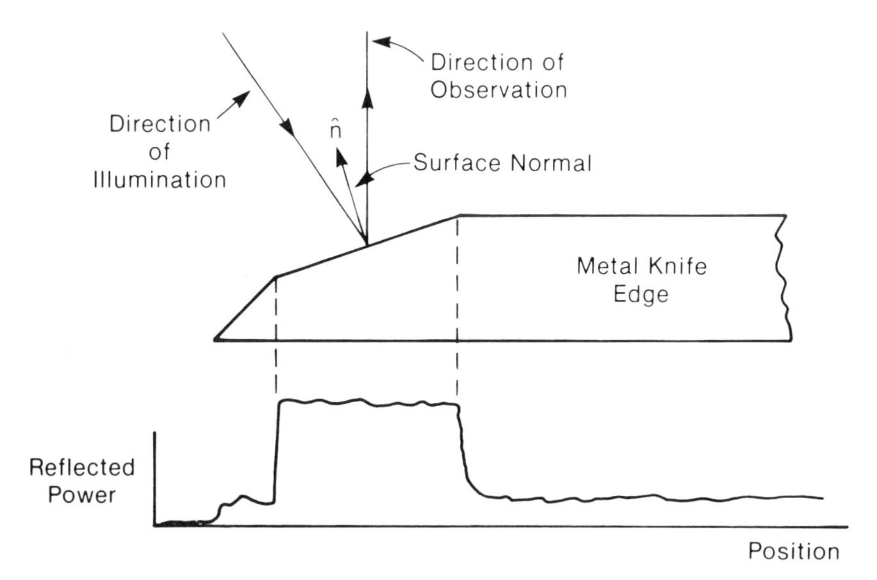

Figure 9. An example of the application of directional illumination. The surface roughness is low so that the scattered power is concentrated in a direction corresponding to the specular condition. The image intensity is shown in correspondence to surface position.

2.3 Threshold Methods

2.3.1 Introduction

The next step in obtaining a binary image from a high-contrast image is to determine a threshold which separates the intensity distribution into two classes. The selection of a threshold has most often been approached using a histogram, since this is a revealing representation of the intensity distribution. The histogram gives the number of pixels at each intensity level. In most cases, the histogram is distributed over 256 levels, corresponding to eight-bit digitization of the image intensities.

Under ideal circumstances, the histogram is bimodal, with one distribution corresponding to the object intensity and the other corresponding to background. An example is shown in Figure 10, which shows a high-contrast image and the histogram. The peak at low intensity is the distribution of pixel belonging to the object, while the peak at high intensity is the background. It is clear that the ideal position for the threhold is at the minimum of the valley between the two distributions. The result of thresholding the image

(a)

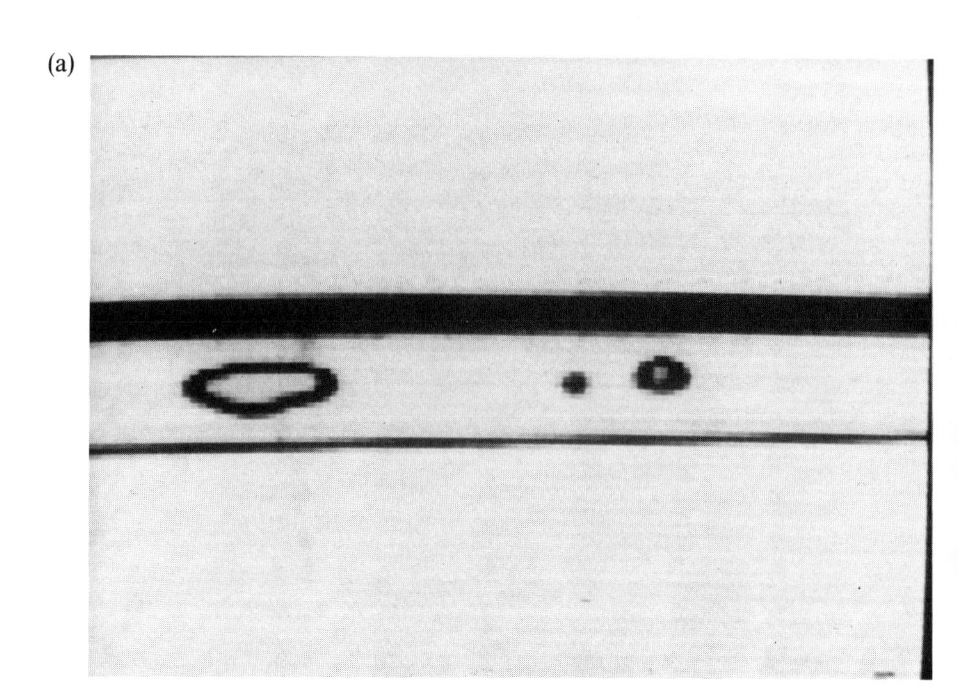

(b)

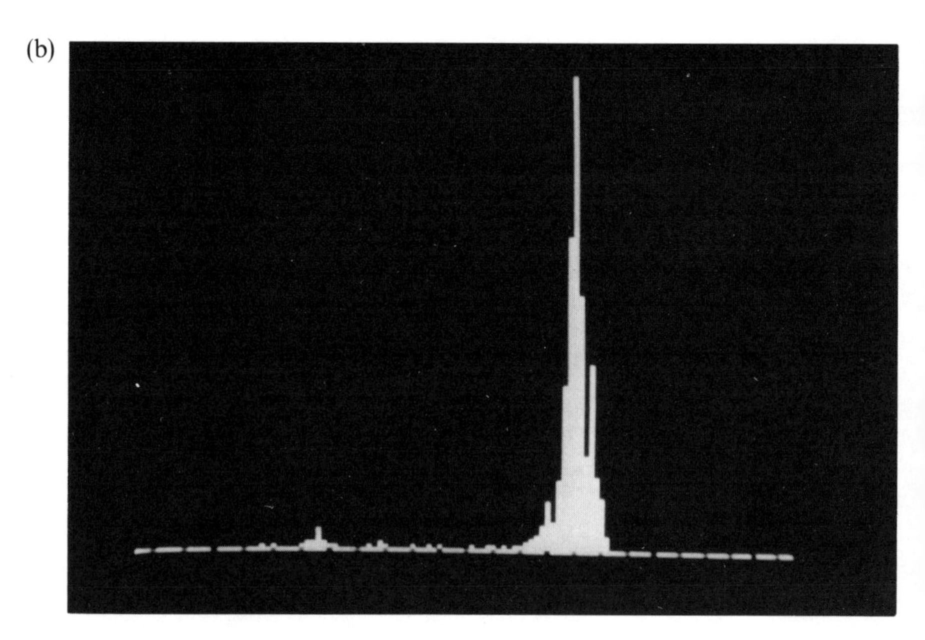

Figure 10. An example of binary image formation. From top to bottom: (a) a high-contrast; (b) the histogram;

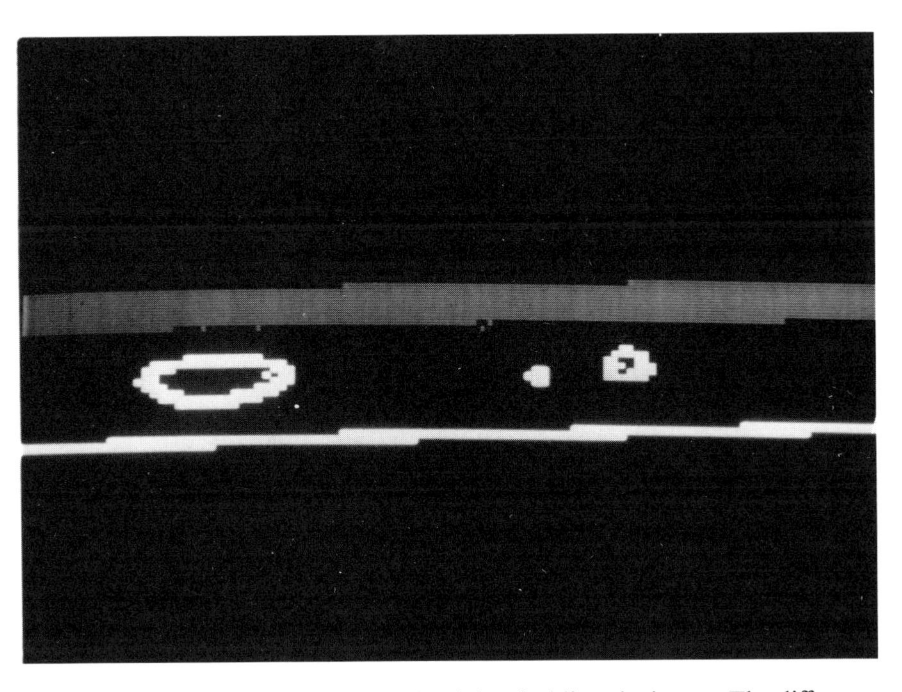

Figure 10 (continued) (c) the result of thresholding the image. The different gray levels in the bottom image represent separate connected regions.

at this location is also shown in Figure 10. The main problem is to define a procedure which provides the best threshold for a given image which incurs the minimum number of pixel misclassifications. This can be difficult for images with graded intensity distributions such as that shown in Figure 11. As seen, the histogram has no distinct bimodality and thus no minimum to determine the threshold location. The image corresponding to the threshold defined by the method in the next section is also shown in Figure 11. The threshold was selected by a procedure based on first-order statistics, described in the next section.

2.3.2 Threshold Selection from First-Order Statistics

The optimum threshold setting is one which minimizes the classification errors between elements of two intensity classes. A method is available which maximizes the class separation [6]. The method also produces a binary image which is an optimum fit in the least-squares sense to the original gray-level image. Suppose we have an image defined by L intensity levels and N pixels. Let the number of pixels at intensity level i be denoted as $n(i)$. With these

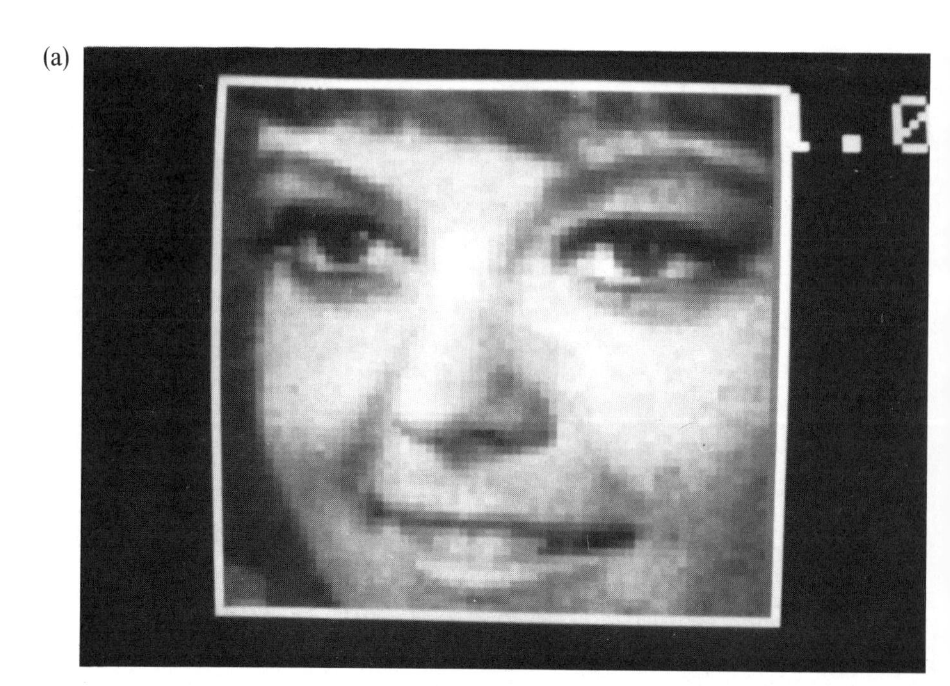

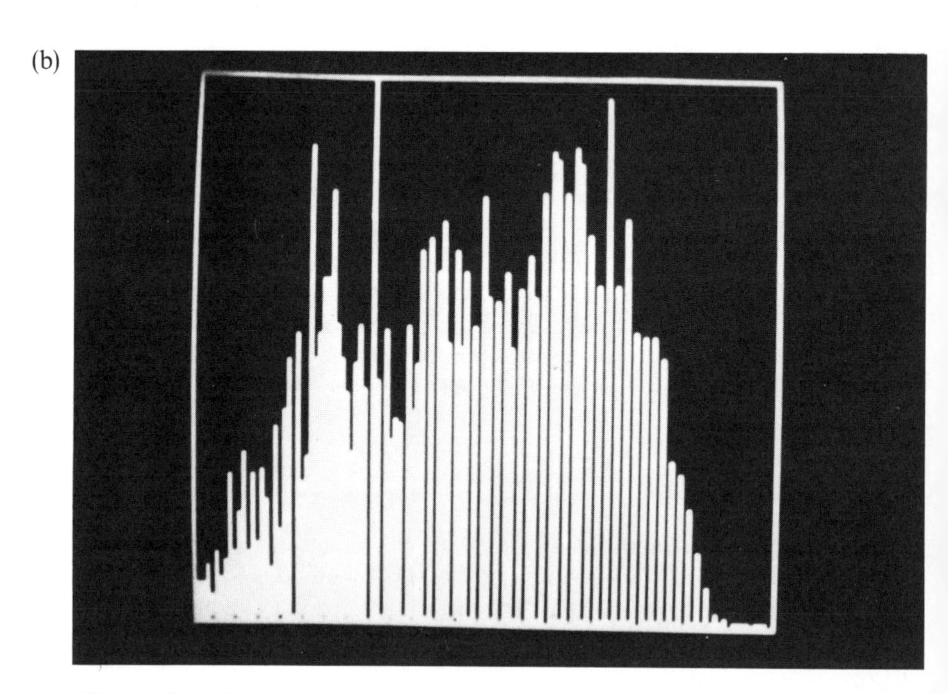

Figure 11. An image with graded contrast. From top to bottom: (a) the image; (b) the corresponding histogram;

Figure 11 (continued) (c) a binary image corresponding to the best-intensity threshold.

definitions the standard statisitical quantities are given by

$$u_T = \frac{1}{N} \sum_{i=1}^{L} [i \cdot n(i)]$$

$$u(k) = \frac{1}{N} \sum_{i=1}^{k} [i \cdot n(i)]$$

$$w(k) = \frac{1}{N} \sum_{i=1}^{k} n(i)$$

One reasonable measure of the separation of the classes of intensity representing the background and object is the variance between classes. This is given by

$$\sigma^2 = \frac{[u_T w(k) - u(k)]^2}{w(k)[1 - w(k)]}$$

The best threshold is that which maximizes the variance between classes where the two classes consist of pixels which lie above and below the threshold, respectively. This definition is expressed in terms of first-order statistics which are easy to compute. It is necessary to maintain the two cumulative sums u and w. At each intensity level k, the sums are evaluated and used to compute σ. The optimum threshold is where σ passes through an absolute maximum. The advantage of this method is that it does not require a feature analysis of the histogram to determine the valley point. In fact, such a valley may not exist or may be very subtle compared to random fluctuations in the histogram distribution. A typical problem is encountered when the number of pixels in the object or background is vastly different. In this case, the minimum is obscure and not easily dectected by direct analysis of the histogram such as searching for a minimum. The point where σ passes through a maximum is unique for a bimodal histogram even with a highly disparate population in the two intensity classes.

2.3.3 *Local Threshold*

It is impossible to define a single threshold for images where the general intensity level varies over the image. This can be due to nonuniform illumination or variable object reflectance. In this case the histogram of image intensities is a continuous distribution with no prominent maxima or minima. With this situation, even the optimal threshold selection procedure discussed above cannot provide a satisfactory segmentation of the image intensities.

The simplest way to avoid this problem is to consider smaller regions of the image where the intensity distribution does satisfy the bimodal assumption. Instead of computing a histogram for the entire image, individual histograms are calculated on small image blocks. In order to obtain a meaningful characterization of the image data it is necessary to include several hundred pixels in each region. This imposes a lower limit on the size of the region and thus on the rate at which the background intensity can vary over the image.

A typical case is to compute a histogram over a 16×16 neighborhood of the image surrounding each pixel. The threshold selection scheme of Section 2.3.2. is then used to decide the class of the center pixel. This concept is best illustrated by the one-dimensional curve in Figure 12(a) which shows a feature superimposed on a gradual baseline shift. In Figure 12(b) the data have been separated into two classes based on the histogram of a 16-sample interval about each point on the curve. As can be seen, the effects of the baseline shift are removed and the local features are segmented from the background. This is a significant improvement over the single-threshold approach which would have not classified the curve into meeningful segments.

2.4 Region Analysis

The previous discussion has centered on the problem of obtaining a high-contrast image in which the objects of interest present a distribution of intensity levels widely separated from the background. Then the image intensities

(a)

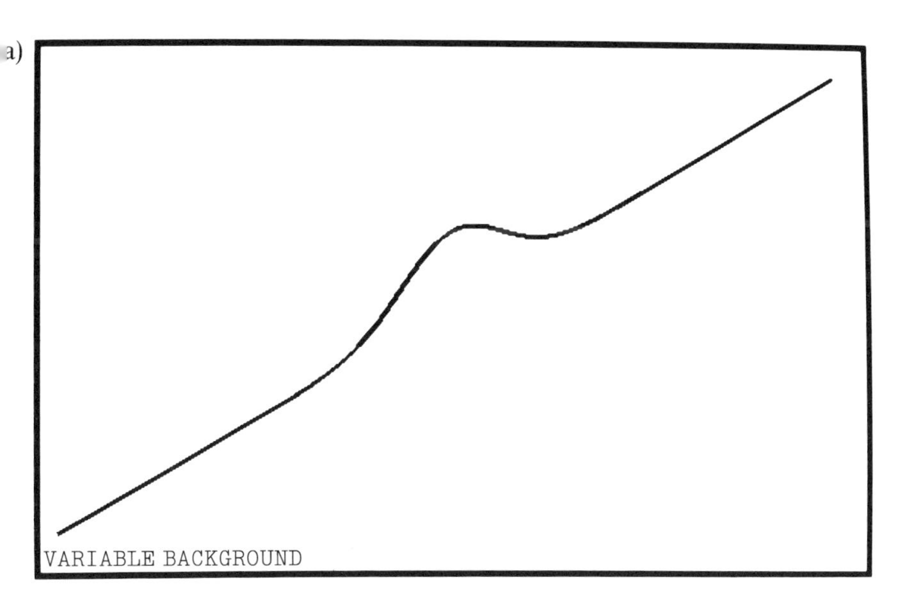

(b)

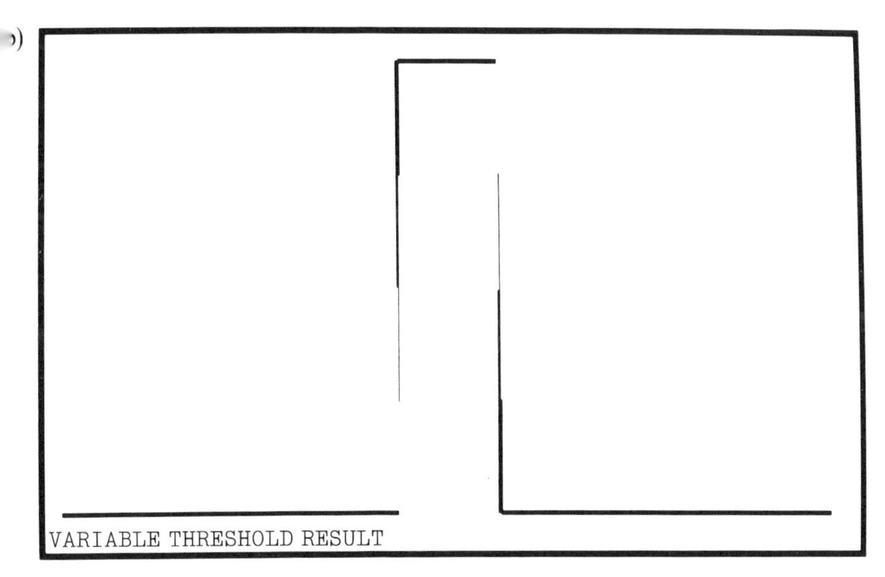

Figure 12. An illustration of adaptive thresholding. (a) A feature superimposed on a variable background. (b) The feature segmented using a threshold determined by averaging over a 16-point interval.

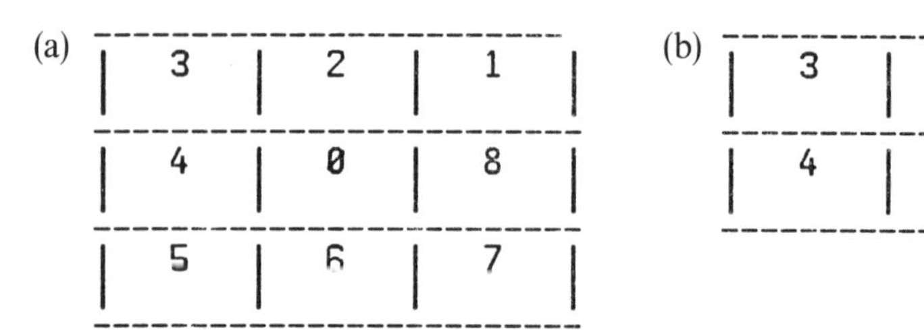

Figure 13. (a) The connected neighborhood for pixel 0. (b) The neighborhood for connectivity analysis during a raster scan of the image.

are classified by comparing the intensity a each pixel against a threshold. To the extent that this classification strategy is successful, new binary image is formed where the object pixels are in one class and the background in another. The next step is to obtain information about the spatial characteristics of the regions corresponding to the objects in the scene. This process first involves the creation of connected sets of pixels which belong to the object class. The identification of these connected sets is known as region growing [7].

The formation of connected sets of pixels is based on the neighborhood shown in Figure 13(a), which is used in a left-to-right top-to-bottom, or raster, scan of the image. The connectivity shown is eight-way in the sense that it is possible for a pixel to have eight connected neighbors. Since the neighborhood is processed in a raster fashion, it is necessary to consider only the pixel shown in Figure 13(b). The connectivity of the center pixel can be determined by linking it upward and to the left as the scan progresses. There are only a small number of cases to consider depending on the region assignments of the pixels in the neighborhood and the state of the center pixel.

Table 1 shows the distinct combinations and the resulting action in each case.

Table 1. Neighborhood Merging Rules

Pixel 0	*Neighborhood*	*Action*
Background ("off")	Don't care	None
Object ("on")	All "off"	Start new region
	2 "on" and don't care for others in the neighborhood	Connect 0 to region at 2
	2 "off" and	
	1 or 4 "on"; 3 "off"	Connect 0 to region at 1 or 4
	1 and 4 "off"; 3 "on"	Connect 0 to region at 3
	1 or 4 "on" and 3 "on"	Merge regions at 1 or 4 and 3; add 0 to merged region

In the case where the image is a fixed-size array, such as 512 × 512 or 256 × 256, the size and number of regions will be finite. This limits the amount of memory required to store the region data. Also, a limit is imposed on the complexity of region connectivity. It is possible to form images of indefinite size so that such assumptions become invalid. For example, high-resolution applications may use a linear array of photodetectors where only the image width is limited. The other dimensions of the image is scanned mechanically and can be unbounded. In order to bound the size of regions it is necessary to arbitrarily truncate the region as shown in Figure 14.

The most difficult aspect of the connectivity analysis is the merging of regions as their connection is discovered by the algorithm just presented. The merging problem is illustrated in Figure 15.

In Figure 15(a), a configuration of regions is illustrated by the letters A through D. It is assumed that the letter assignment is made in the order in which the region is formed. It is also assumed that the regions are connected

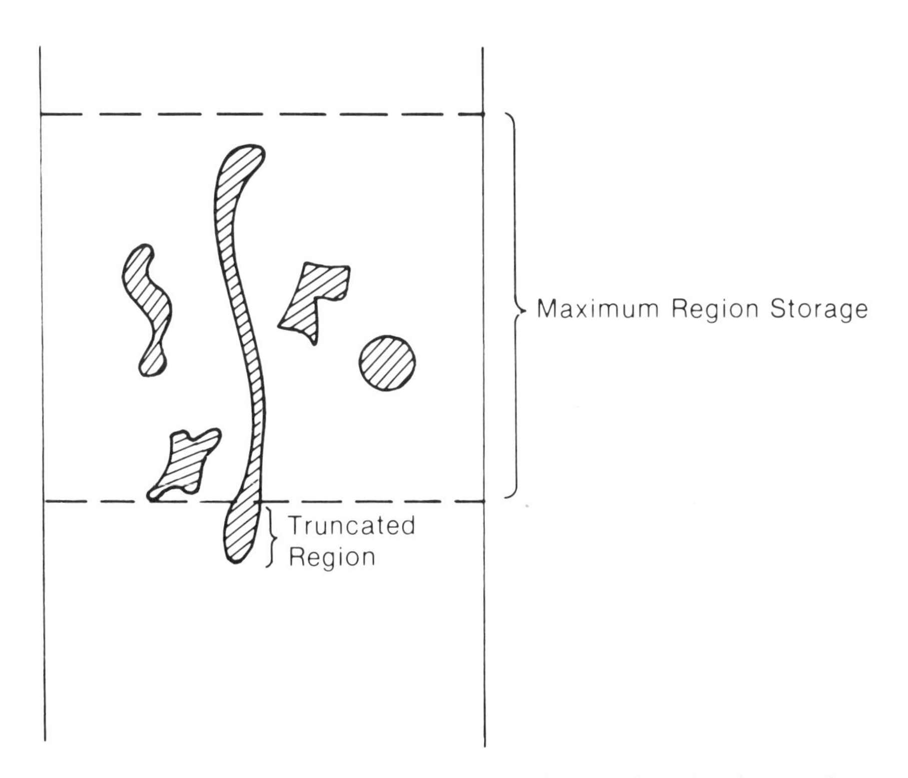

Figure 14. Finite storage of region information requires that long regions are truncated.

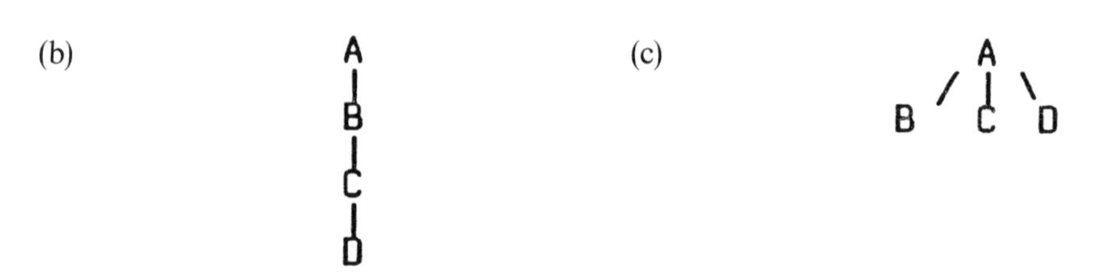

Figure 15. An illustration of label merging: (a) A set of regions; (b) Initial label equivalences; (c) Label equivalences after reverse pass.

in a pass of analysis which proceeds in raster fashion, namely left to right and top to bottom. At the right end of the row which contains region D, the connectivity of the regions in the image will be as shown in Figure 15(b). The tree shown has region labels as nodes, and an edge is present in the tree if two regions are connected.

The depth of the tree in 15(b) is bounded only by the finite width of the image. This is not a convienient situation for subsequent stages of analysis. In order to establish a regular pipiline of region analysis it is desirable to have a fixed number of processing steps for each label equivalence. This is the case for the connection tree in Figure 15(c). Fortunately, the tree in 15(b) can be converted to that in 15(c) by a fixed number of operations for each pixel in a reverse scan of each image line. The table corresponding to Figure 15(b) is modified for each region encountered in a reverse scan by looking up the previous connected region index. The table entry for that region is replaced by the previous index. As the sweep proceeds from right to left, the connection tree will be transformed from Figure 15(b) to Figure 15(c). Notice that each step in the transformation is accomplished by a fixed number of table operations, namely a label lookup and an entry replacement. The algorithm is successful because the order of region label formation is monotonic in the depth of the tree in Figure 15(b).

2.4.2 Region Features

Each connected set of pixels represents an object in the scene or a portion of an object which is made up of a complex configuration of regions. For example, the region shown in Figure 16(a) has a number of levels of containment as shown in Figure 16(b). The containment tree along with a description of the geometric properties of each region forms an adequate description of the object.

The geometric description of a region is embodied in the set of connected pixels, but this is not a convienient format for carrying out recognition and measurement tasks. The most common approach is to summarize the region by a set of geometric features which can be computed incrementally as the region is formed. This also implies that the features can be combined when

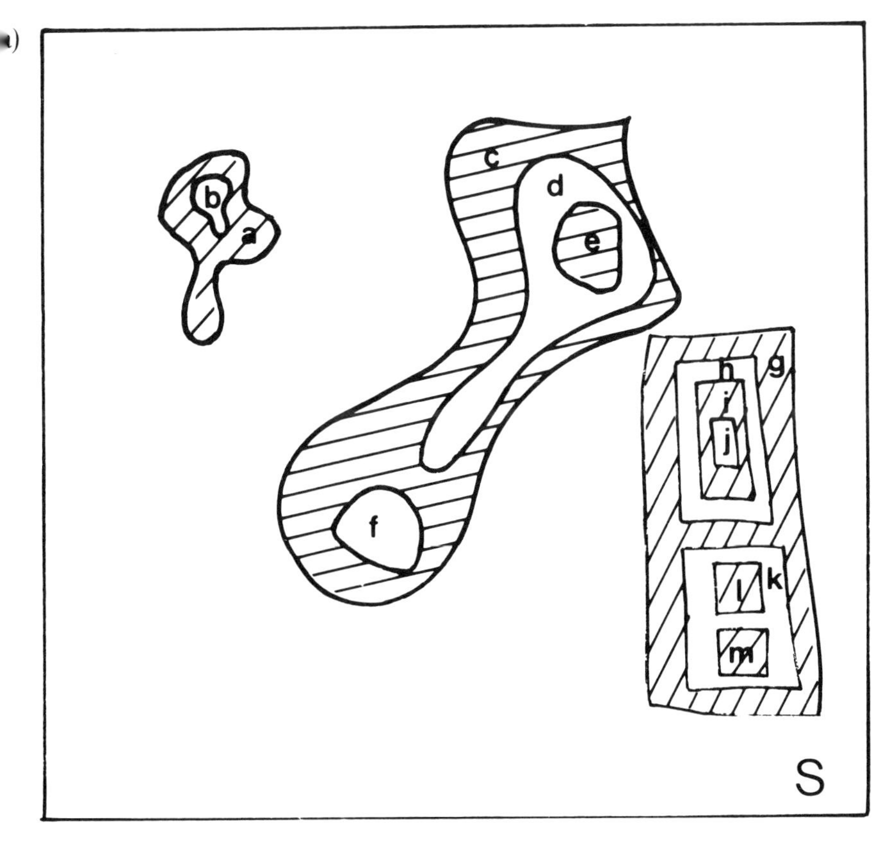

Figure 16. The containment tree for nested regions: (a) a set of regions;

(b)

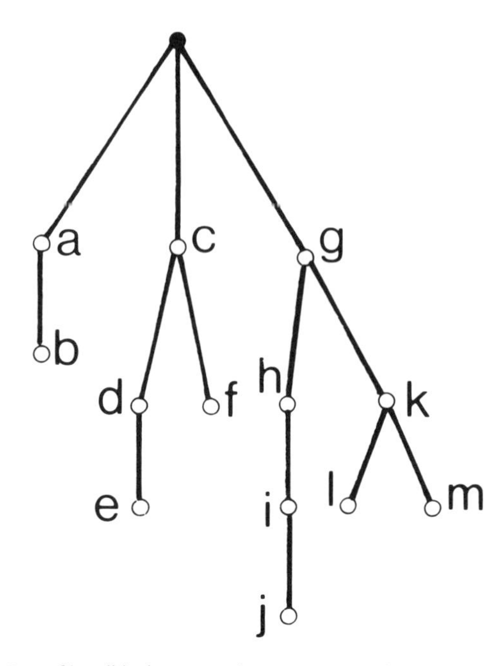

Figure 16 (continued) (b) the containment tree where nodes are regions and edges indicate that the subordinate node is inside the parent node.

the connection between two components of a region is discovered. These properties fall into two major classes, area features and boundary features. A representative list of these is given below.

Area features:
- Area
- Center of area
- Area moments

Boundary features:
- Perimeter
- Bounding rectangle
- Corner locations

The geometery of these features are illustrated in Figure 17. The boundary features are typically computed from a chain code for the boundary which is extracted during region growth [8]. The boundary code is expressed in terms of the eight directions defined in Figure 17. The length of the code elements is taken to be unity for the horizontal and vertical vectors and the square root of 2 for the elements at 45°. It is also assumed that the area of

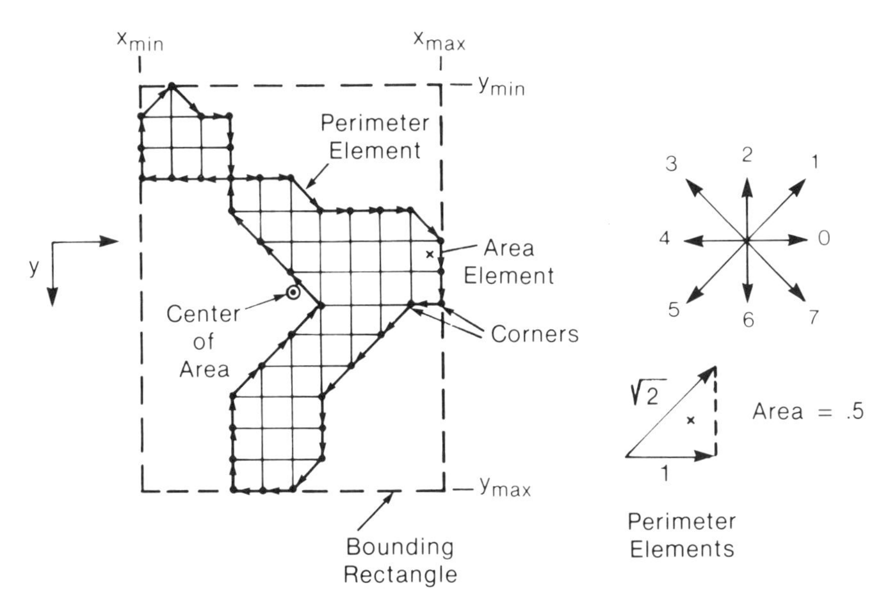

Figure 17. The definition of region features. A region is defined by area elements and perimeter elements. A chain code is shown along the boundary using the eight direction vectors in the upper right corner of the figure. The bounding rectangle is shown as dashed lines.

one resolution element (pixel) is unity. Each of the features will now be defined in terms of the incremental rules for generating the feature as the region is grown as well as the rule for combining features values for region components. In these rules the state of the region feature before and after adding a pixel is denoted by i and $i + 1$, respectively. It is also assumed that region components i and j are being combined to form an updated region, i. The coordinates of a pixel are denoted as x, y. Some of the area features, such as center of area, are represented as a running summation which must be normalized by dividing the sum by the area of the region. (See Table 2.)

Other useful features can be extracted from this basic set with simple computations. For example, the orientation of the region can be determined by finding a direction which diagonalizes the area moment matrix. First the normalized moment matrix is computed.

$$M = \frac{\begin{bmatrix} XX_n & XY_n \\ XY_n & YY_n \end{bmatrix}}{A_n} = \begin{bmatrix} xx & xy \\ xy & yy \end{bmatrix}$$

where n is the number of pixels in the completed region.

Table 2. Region and Perimeter Incremental Features

Feature	*Pixel Increment*	*Component Merge*
Area	$A_{i+1} = A_i + 1$	$A_i = A_i + A_j$
Center of area	$X_{i+1} = X_i + x$	$X_i = X_i + X_j$
	$Y_{i+1} = Y_i + y$	$Y_i = Y_i + Y_j$
Second-order area moments	$XX_{i+1} = XX_i + x^2$	$XX_i = XX_i + XX_j$
	$YY_{i+1} = YY_i + y^2$	$YY_i = YY_i + YY_j$
	$XY_{i+1} = XY_i + x * y$	$XY_i = XY_i + XY_j$
Perimeter	$P_{i+1} = P_i + 1$ or 1.4	$P_i = P_i + P_j$
Bounding rectangle	$X_{\max_{i+1}} = \max(X_{\max_i}, x)$	$X_{\max_{i+1}} = \max(X_{\max_i}, X_{\max_j})$
	$Y_{\max_{i+1}} = \max(Y_{\max_i}, y)$	$Y_{\max_{i+1}} = \max(Y_{\max_i}, Y_{\max_j})$
	$X_{\min_{i+1}} = \min(X_{\min_i}, x)$	$X_{\min_{i+1}} = \min(X_{\min_i}, X_{\min_j})$
	$Y_{\min_{i+1}} = \min(Y_{\min_i}, y)$	$Y_{\min_{i+1}} = \min(Y_{\min_i}, Y_{\min_j})$

Note that these terms should not be evaluated until the region is completely connected, since the normalized features for region components do not combine by simple addition as for the features in Table 2. The normalized matrix can be diagonalized by forming an eigenvalue problem in terms of the rotation matrix in the image plane.

$$\begin{bmatrix} \cos(\theta) & \sin(\theta) \\ -\sin(\theta) & \cos(\theta) \end{bmatrix} \begin{bmatrix} xx & xy \\ xy & yy \end{bmatrix} \begin{bmatrix} \cos(\theta) & -\sin(\theta) \\ \sin(\theta) & \cos(\theta) \end{bmatrix} = \begin{bmatrix} U & 0 \\ 0 & V \end{bmatrix}$$

The solution of this matrix equation gives the angle θ between the major axis of the region and the x axis of the image coordinate system. The eigenvalues U and V correspond to the lengths of the major and minor axes of the region, respectively. The geometry of this is shown in Figure 18. The angle and the lengths are given by

$$\tan(2\theta) = \frac{2 * xy}{xx - yy}$$

$$U = \frac{S + R}{2}$$

$$V = \frac{S - R}{2}$$

$$S = xx + yy$$

$$R = \sqrt{(xx - yy)^2 + 4 * (xy)^2}$$

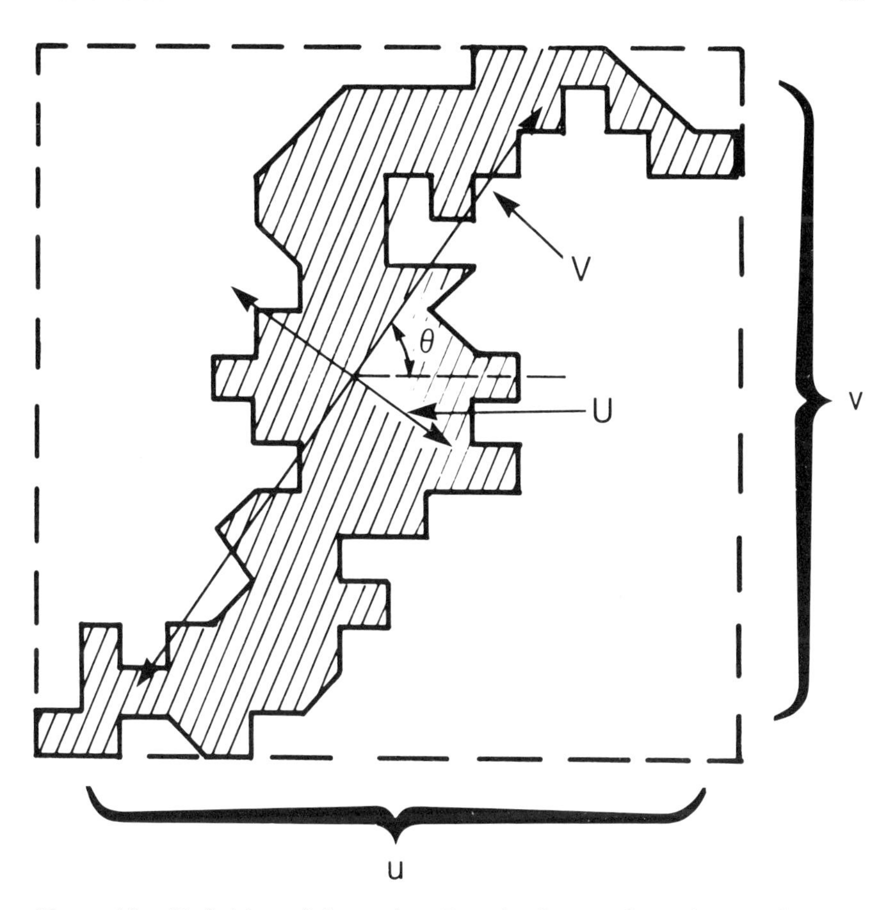

Figure 18. Definition of eigenvalues U and V for a region. The coordinates u and v of the bounding rectangle are also shown.

A related measure is the bounding rectangle, which is also shown in Figure 18. The width and height of the rectangle are given by u and v, respectively. The ratio of the lengths, u/v, gives a measure of the extremes of the region rather than the average distribution. The bounding rectangle requires considerably less computation than the accumulation of moments. However, it can give a misleading representation of the shape of the region. An example of this problem is shown in Figure 18. As shown, the aspect ratio of the region is large, but the ratio of length to width of the bounding rectangle is unity. This condition can be detected by computing the ratio of region area to bounding rectangle area, which is known as the fill ratio. If the fill ratio is nearly unity, then the bounding rectangle is a good indication of region shape. If the fill ratio is low, the area moments defined earlier are required to give an accurate indication of shape.

The boundary description initially consists of a closed sequence of chain codes which gives a detailed description of the region shape. This code description is usually too detailed to provide efficient use directly in any shape analysis procedure. Ideally, the boundary should be segmented with functional forms such as line segments and arcs as shown in Figure 19. The sequence of such boundary primitives can be analyzed to arrive at a classification of the global region shape. In most robotic applications using binary vision, it is adequate to identify key boundary points which represent important discriptive of the object such as corners, tabs, and slots. The most common set of primitives is a sequence of straight lines. There are a number of line segmentation approaches [9] which can be applied to boundary approximation. One approach is based on accumulating sums of boundary coordinate terms similar to the area moment forms introduced earlier. A method requiring less computation is based on the idea of enclosing the boundary in a box representing the maximum distance that the points along the boundary can lie from the line segment [10]. The segmentation of a portion of a boundary according to this idea is shown in Figure 20.

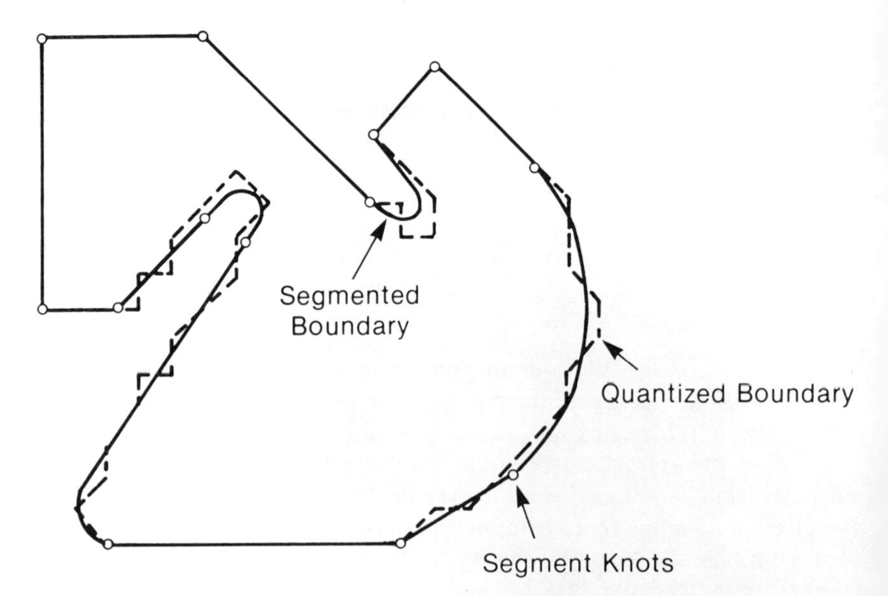

Figure 19. The approximation of a quantized boundary with curve segments. The end points of segments are called knots.

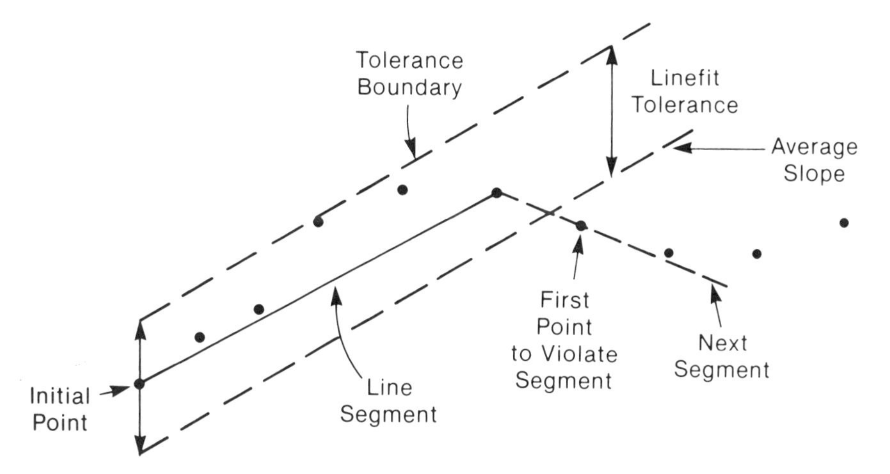

Figure 20. The Tomek method of linear segmentation. The tolerance boundaries are free to ratchet about the initial point. When the boundaries become parallel or divergent the segment must be terminated and a new segment started. The upper boundary is ratcheted upward and the lower boundary downward by the data points.

2.5 Decisions Based on Region Features

2.5.1 *The Decision Tree Method*

The recognition of object identity, location, and orientation can be based on the relatively simple region features derived in the previous section. The most straightforward approach to this is through the use of a decision tree. An example of such a tree is shown in Figure 21. The nodes of the tree represent decision points based on the region features. The tree is typically a binary tree where the left leaf corresponds to the feature value lying below a threshold and the right branch is taken if the feature value is not below the threshold. The decision process can proceed quickly since any branch can be reached in time proportional to the depth of the tree.

Another advantage of the decision tree method is that the features can be computed sequentially. Thus if a feature is expensive computationally, it does not have to be evaluated if the decision sequence does not lead to the section of the tree containing a decision involving the feature. Along the same lines, the tree can be optimized so that the recognition of the objects can be carried out with the least number of feature comparisons. The optimum sequence is one where such feature tested divides the remaining object classes as evenly as possible. This in turn can be determined from the statistical distribution

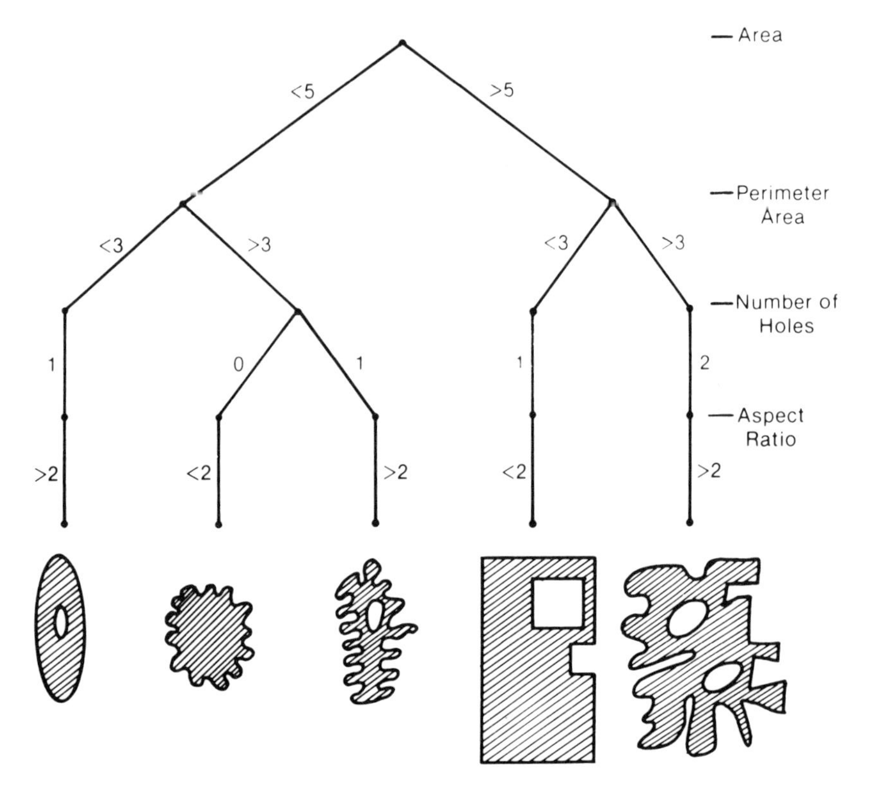

Figure 21. A decision tree based on features of regions. Shapes corresponding to feature decisions are shown at the leaves of the tree.

of classes in an n-dimensional space where each coordinate corresponds to the value of a feature. The threshold of a feature corresponds to a plane in the n-dimensional space which is orthogonal to the feature. This plane intersects the feature axis at the threshold value. The set of planes for all the features will ideally segment feature space into regions, each region containing only one class. This is illustrated in for $n = 3$ in Figure 22.

2.5.2 Graph Matching

One major drawback to the decision tree approach is that it does not provide a convenient way of representing the relationships between edges and vertices on region boundaries. A more natural representation is the graph shown in Figure 23.

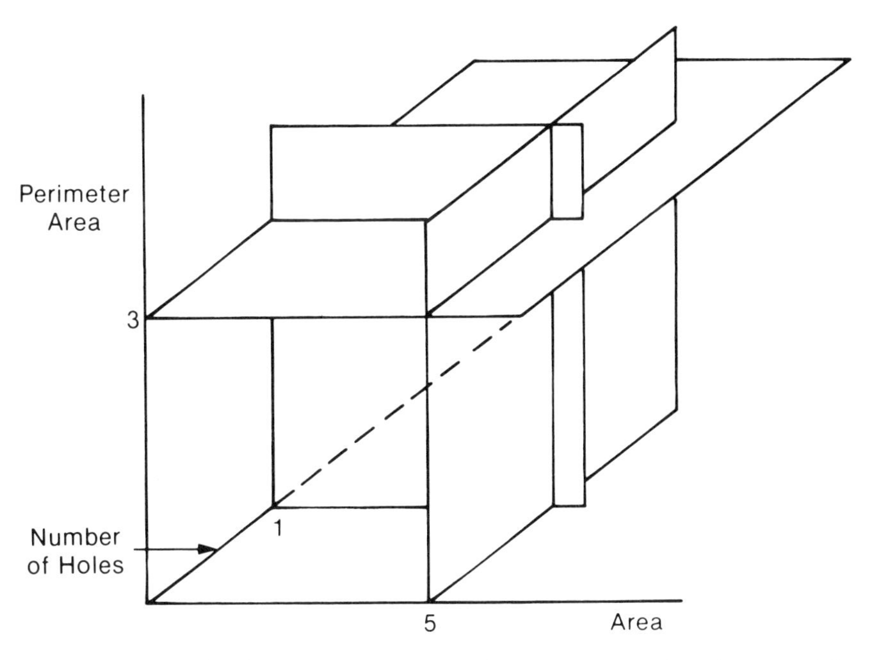

Figure 22. A decision space corresponding to three region features. The planes represent decision surfaces.

In this graph structure, the nodes correspond to the segments and vertices in an object boundary. The edges in the graph correspond to relations which hold between the graph nodes. For example, if one segment is perpendicular to another, then an edge would exist between the two region nodes indicating the perpendicular relation. This graph is also able to present properties of boundary features by using a unary form of the relation edge, which depends only on a single node. In this way the graph structure can be represented uniformly. An example of such a representation is shown in Figure 23.

It is necessary to derive a relational graph for each object which is to be classified in the domain of the application. The recognition procedure proceeds by the matching of the ideal graph for each object and the graph obtained from the image. In effect, the recognition process is carrying out a search for graph isomorphism between the graph obtained from the image and the ideal object model graphs. This search in general will require exponential time based on the N-factorial number of combinations of potential matches between the graph and edges (N is the number of nodes). This search can be reduced by defining key features in the object model which are significant in distinguishing the objects. This is analagous to the decision tree

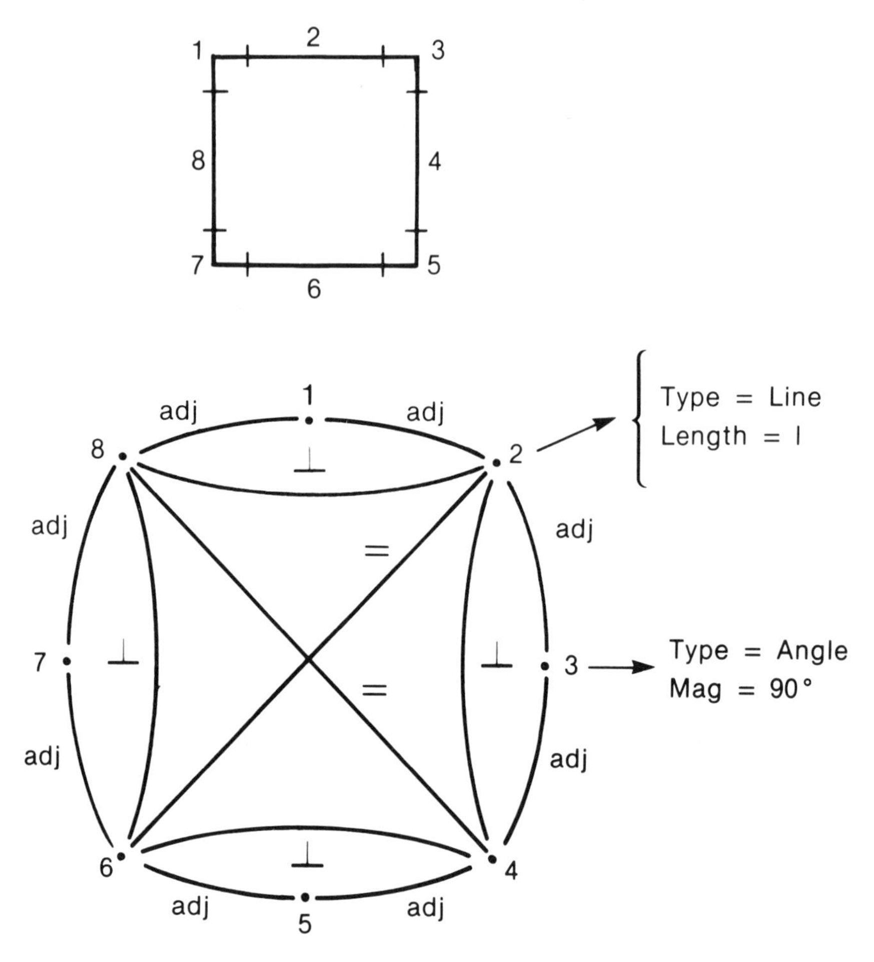

Figure 23. A relational graph for an ideal square boundary. The arcs in the graph are labeled according to the relation they represent. Only two nodes are labeled for clarity. The boundary features of the square are numbered at the top of the figure. The nodes of the graph are numbered accordingly.

method where the features which divide the object classes the most powerfully are considered first. If a key feature is present in the image graph, then it is matched to the corresponding node in all the model graphs which contain the key feature. Next, the image graph is searched for all the features and relations which lead from the key feature node in the model graph. If this match is successful, then the search proceeds from the secondary features and so on until the match is complete. The search can be made as efficient as possible by ordering all of the features in a model so that significant mismatches are detected early and irrelevant object classes are ruled out right away.

The matching process discussed so far tacitly assumes that all of the features and relations required by the object model are present in the image representing the object. This is not usually the case. First, the derivation of image features is not always reliable; for example, sensor noise or poor illumination can lead to missing or inaccurately represented featues. Even more likely is the case where one object obscures the features of another. In this case even key features may be missing, so that ordering of features is necessary to insure robust recognition [11].

A measure for determining the degree of success for remaining matches must be defined to decide between the objects. In general, the partial-match problem can be represented in terms of maximal cliques, which are sets of correspondences between the object model graph and the image graph. The largest maximal clique is the best match between the objects and the image [12].

3. GRAY-LEVEL METHODS

3.1 Introduction

The use of binary image analysis is restricted to images with high contrast so that a single threshold can serve to segment the objects in the scene. As pointed out earlier, this is usually not successful without extreme care in the control of illumination and the reliance on some unique optical property of the objects and the background. The typical industrial vision application requires the observation of objects with reflected illumination and without special preparation of their surfaces.

These conditions lead to a full range of gray levels in the image with low contrast or gradual variations in intensity which eliminates the possibility of using a single threshold. A range of techniques are available to analyze such images, and the use of such techniques is referred to as gray-level image processing. The main goal of such analysis is to locate the boundaries of the objects in the scene and segment the objects from the background. This section will concentrate on the problem of locating boundaries in the gray-level image. In this context the concept of a boundary is taken to be a change in image intensity which is spatially coherent on a local basis. For example, the intensity change along a vertical boundary in the image will be approximately uniform in a vertical direction along the boundary.

3.2 Edge Detectors

The development of edge detection procedures in gray-level images has sustained the efforts of many researchers in machine vision for the past several decades. While there are many variations in the specific details of the methods, the basic approaches are very similar in general. The idea is to form a region around each pixel in the image, which is referred to as the pixel neighborhood. Each pixel in the neighborbood is multiplied by a weight, and the resulting products are accumulated over the neighborhood. A new image is created where each pixel has the value of the accumulated neighborhood products for the corresponding pixel in the original image. This can be expressed mathematically as follows:

$$C(i,j) = \sum_{m,n} \{I[(i-m),(j-n)] \cdot M(m,n)\} \tag{1}$$

where the indices m, n range over the neighborhood surrounding each pixel. The result, $C(i,j)$, is refered to in the signal-processing literature as the convolution of I with M and is indicated by

$$\mathrm{C} = \mathrm{I} * \mathrm{M} \tag{2}$$

The array of neighborhood weights M is refered to as the convolution mask. The convolution C can be thought of as the result of filtering the image I with a two-dimensional filter with impulse response M. This interpretation permits the application of Fourier transform methods to guide the design and analysis of the convolution process. For example, if the mask M consists of elements which are all equal, the convolution C is equivalent to multiplying the Fourier transform of the image by the function $\mathrm{sinc}(f_i, f_j)$. Here, f_i and f_j refer to the spatial frequency components in the horizontal and vertical image directions, respectively. The sinc function represents effectively a low-pass filter which reduces the high spatial frequency content of the original image. This Fourier domain viewpoint leads naturally to the development of a set of masks which are considered the most effective in detecting edges in images with noise and variable spatial scale.

Consider Figure 24, which shows a typical sepectral distribution of spatial frequencies in a gray-level image. The low-frequency portion of the spectrum corresponds to the gradual variations in the image intensity due to illumination and reflectance nonuniformity in the scene. The high-frequency portion is mainly noise and sharp-edge transitions. The valuc of the spectrum at $f = 0$ is the average value of the intensity for the entire image. The portion of the spectrum which is most closely related to the spatial variation due to object boundaries is in the middle.

From this Fourier perspective, it is seen that a reasonable filter to highlight the object boundaries is one which forms a bandpass near the middle of the image spectrum. The first approximation to this is simply to define a passband interval in spatial frequency. This is a box function of width w and center f_0 which multiplies the image function spectrum. The Fourier transform of this function is the impulse function for a filter which highlights the object boundaries. The fourier transform for a box centered on the origin is

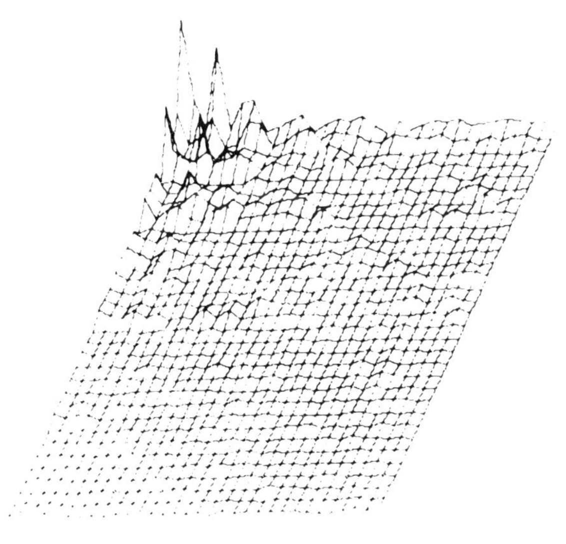

Figure 24. The two-dimensional Fourier power spectrum of an image. The surface height is proportional to energy at a given spatial frequency. The two axes are f_x and f_y corresponding to horizontal and vertical spatial frequency.

simply sinc(x). For the box centered at f_0 it is necessary to apply the shifting theorem [13] which leads to the complex filter impulse response given for one dimension:

$$M(x) = M_{re}(x) + iM_{im}(x) \tag{3}$$

The components of this complex response are given by

$$M_{re}(x) = \text{sinc}(w \cdot x)\cos(f_0 \cdot x) \tag{4}$$

$$M_{im}(x) = \text{sinc}(w \cdot x)\sin(f_0 \cdot x) \tag{5}$$

This simple analysis shows that both an even and odd spatial impulse response function is needed to implement a bandpass function. These impulse functions correspond to step and peak boundary edge intensity variations. Each of these boundary types can occur in images of natural or industrial scenes. The step edge is related to reflectance contrast or shadowing; the peak edge is usually a highlight due to specular reflection along the object contours.

The mask forms developed so far are based on a viewpoint that the image and filter functions are continuous and of infinite spatial extent. In fact, both the image and the mask are discretely sampled and of finite size. The sampling problem can be minimized by chosing a sample interval which corresponds to significantly higher spatial frequencies than those involved in the detection of boundaries. The problem of finite size is not so easily dispensed with. The arbitrary truncation of the mask function by the boundaries of the mask neighborhood will introduce error and spurious response in the convolution result.

The large spatial extent of the mask function is due largely to the restriction of the bandpass function in the spatial frequency domain. This is directly analagous to the classic space–momentum observation limit for wave functions, the so-called Heisenberg uncertainty principle. The product of the filter bandwidth and the spatial extent of the impulse response can be minimized by an appropriate choice of the band-limiting function. The solution to this minimization problem is the Gaussian function

$$\exp\left[\frac{-(f-f_0)^2}{W^2}\right]$$

Note that W is the width of the Gaussian passband and f_0 is the center. The Gaussian filter also provides some additional properties which are useful:

- The shape and extent of the bandpass function is controlled by only two parameters, f_0 and W.
- The function is a close approximation to the ideal estimate for a band-limited signal, the prolate spheroidal wave function [14].
- The bandpass function can be represented by a difference of two Gaussians of different width [15].

Using the Gaussian as the bandpass function, the impulse responses for the complex mask are given below, where $w = 2/W$:

$$M_{re} = \cos(f_0 \cdot x) \exp\left[-\left(\frac{x}{w}\right)^2\right] \tag{6}$$

$$M_{im} = \sin(f_0 \cdot x) \exp\left[-\left(\frac{x}{w}\right)^2\right] \tag{7}$$

These functions are illustrated in Figure 25.

Since the Gaussian is nearly optimum for estimating band-limited noisy signal values, it is possible to arrive at similar expressions for M_{re} and M_{im} from a slightly different perspective. It is well known that the optimum estimate for a linear operation on a signal is achieved by applying the linear operator to the estimation function and then applying that operator to the signal [16]. It can be observed that the step edge and peak edge represent

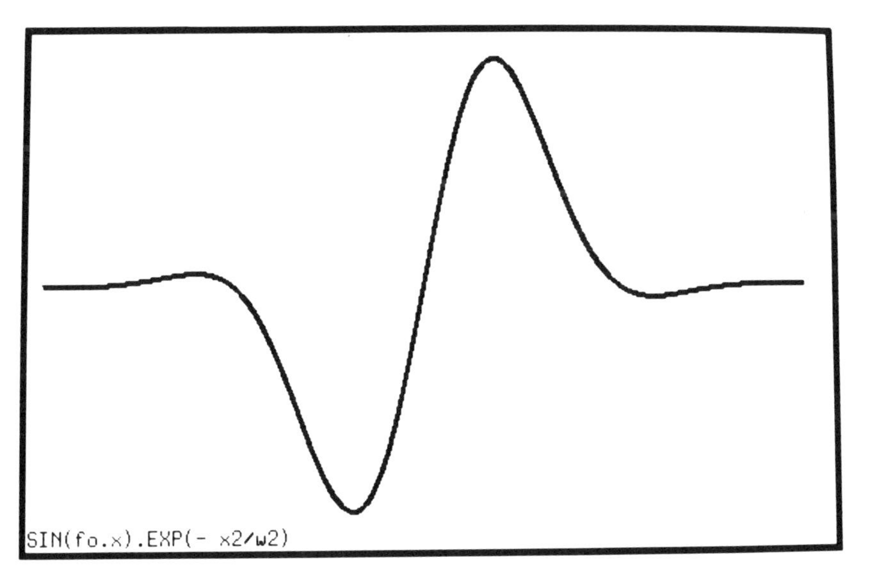

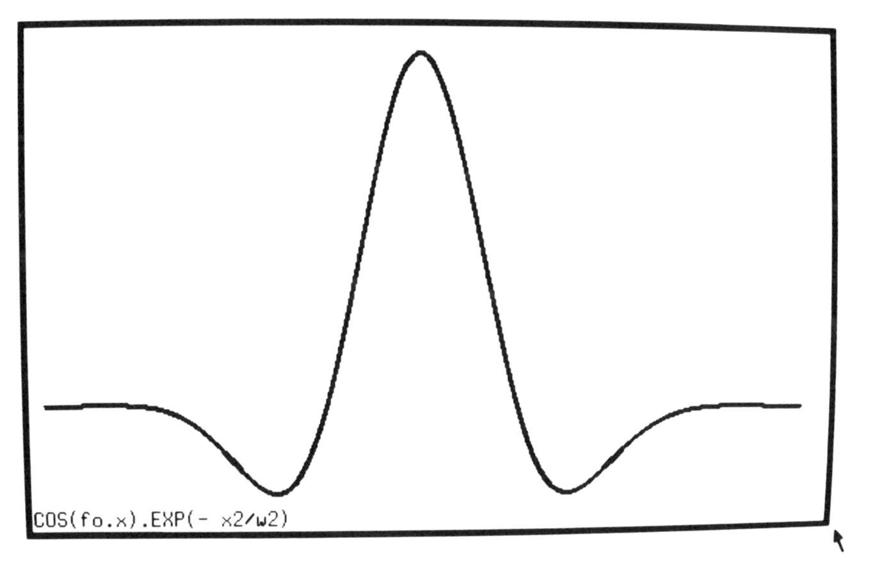

Figure 25. The spatial impulse responses required to implement a one-dimensional bandpass filter. Note that the top response is an odd function, the lower an even function.

extremes in first and second derivatives, respectively. Thus an alternative approach to define optimum filters for these features is to form impulse responses based on the first and second derivatives of the Gaussian function. These forms are

$$\frac{\partial G}{\partial x} = -(2x/w^2) \exp[-(x/w)^2] \tag{8}$$

$$\frac{\partial^2 G}{\partial x^2} = (2/w^2)\{[2(x/w)^2] - 1\} \exp[-(x/w)^2] \tag{9}$$

It is seen that there is close correspondence between the two sets of expressions in that the optimum derivative estimates (8) and (9) are similar to the series expansions for (6) and (7) carried out to second order.

By selecting the value of bandwidth W and center frequency f_0 it is possible to derive masks which are a good approximation to the continuous filter function for any mask size. The extension of these one-dimensional functions to two dimensions is achieved by expressing the Gaussian in polar form:

$$G(x, y) = \exp[-(x^2 + y^2)/w^2] \tag{10}$$

The second derivative or M_{re} filter function generalizes to two dimensions by adding the second derivatives for the orthogonal directions x and y. Here x corresponds to the direction along the row and y along the column in image coordinates. This is equivalent to the Lapacian of $G(x, y)$. The final form for M_{re} which is suitable for discrete sampling is given by

$$M_{re}(x, y) = [1 - k(x^2 + y^2)] \exp[-(x^2 + y^2]/w^2 \tag{11}$$

Note that this is a symmetrical function in x and y. Also, the constant multiplier has been eliminated, which will be accounted for by normalization later.

The first derivative or M_{im} function cannot be achieved in two dimensions with a single mask. Instead, a number of masks must be defined which correspond to directions in the image plane along which the slope is to be estimated. In most cases, only two directions need be considered, namely, the row and column directions in the image. Thus, the two-dimensional forms for M_{im} are

$$\frac{\partial G}{\partial x} = x \exp[-(x^2 + y^2)/w^2] \tag{12}$$

$$\frac{\partial G}{\partial y} = y \exp[-(x^2 + y^2)/w^2] \tag{13}$$

The discrete masks, based on the equations for the first and second derivatives, are obtained by a value for w so that the mask values are negligible outside the array boundary. The value of k is selected so that the sum of the coefficients in the mask for M_{re} is zero. This follows from the observation that the convolution of the mask with an image of constant intensity should yield a second derivative of zero. The first derivatives masks also sum to zero, as insured by the odd form of the derivatives. The discrete masks for size 3 × 3 are

3 × 3 masks:

$$\frac{\partial G}{\partial x} = \frac{1}{8}\begin{bmatrix} -1 & 0 & 1 \\ -2 & 0 & 2 \\ -1 & 0 & 1 \end{bmatrix}$$

$$\frac{\partial G}{\partial y} = \frac{1}{8}\begin{bmatrix} 1 & 2 & 1 \\ 0 & 0 & 0 \\ -1 & -2 & -1 \end{bmatrix}$$

$$\frac{\partial^2 G}{\partial r^2} = \frac{1}{12}\begin{bmatrix} 2 & -1 & 2 \\ -1 & -4 & -1 \\ 2 & -1 & 2 \end{bmatrix}$$

In these masks, the values have been rounded to the nearest integer. The scale factor in front of the mask is multiplied by the convolution result. This scale factor is calculated to give a unity value to the resulting first and second derivatives if they are unity in the image. It is interesting that the forms for $\partial G/\partial x$ and $\partial G/\partial y$ are identical to the widely used Sobel edge detector masks [17].

Filters can be easily derived for larger masks. The results for 5 × 5 are

5 × 5 masks:

$$\frac{\partial G}{\partial x} = \frac{1}{152}\begin{bmatrix} -1 & -2 & 0 & 2 & 1 \\ -4 & -10 & 0 & 10 & 4 \\ -8 & -16 & 0 & 16 & 8 \\ -4 & -10 & 0 & 10 & 4 \\ -1 & -2 & 0 & 2 & 1 \end{bmatrix}$$

$$\frac{\partial G}{\partial y} = \frac{1}{152}\begin{bmatrix} 1 & 4 & 8 & 4 & 1 \\ 2 & 10 & 16 & 10 & 2 \\ 0 & 0 & 0 & 0 & 0 \\ -2 & -10 & -16 & -10 & -2 \\ -1 & -4 & -8 & -4 & -1 \end{bmatrix}$$

$$\frac{\partial^2 G}{\partial r2} = \frac{1}{2136}\begin{bmatrix} -18 & -33 & -32 & -33 & -18 \\ -33 & 2 & 64 & 2 & -33 \\ -32 & 64 & 200 & 64 & -32 \\ -33 & 2 & 64 & 2 & -33 \\ -18 & -33 & -32 & -33 & -18 \end{bmatrix}$$

A local maximum of the appropriate convolution value corresponds to the location of an intensity edge. More specifically, the maximum in the derivative estimate occurs in the vicinity of a step and the maximum second derivative corresponds to a peak edge. The decision process for the boundary is thus to consider a boundary point to be any location where a convolution value exceeds an appropriate threshold. Figure 26(a) shows a gray-level image. The convolution results for the 5 × 5 masks are shown in Figures 26(b)–(d).

The location of zero crossings in the convolution value provides a more accurate estimate of the edge location. For example, the location of a step edge corresponds to the zero crossing of the second derivative convolution value. If the image is of high quality, the edge points form a continuous connected curve surrounding each region. This curve can be described by a piecewise linear curve as in the binary image case. The extraction of region and boundary properties can proceed identically to the methods developed in Section 2. In most practical cases, the quality of the gray-level image is not sufficient to allow the formation of well-defined boundaries. The effects of shadows, noise, and low contrast lead to corrupt detection of the object boundaries.

(a)

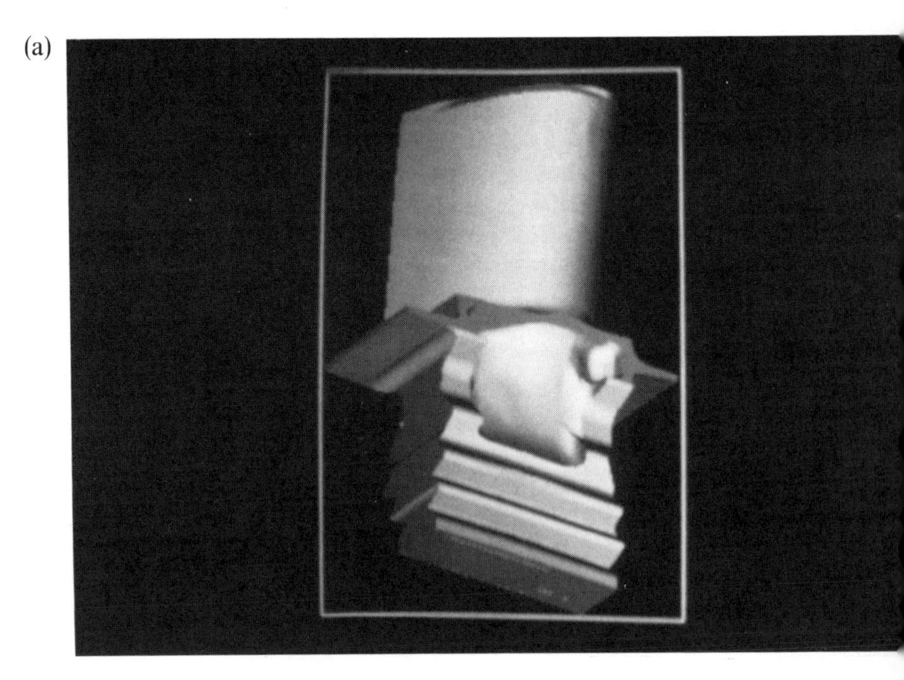

Figure 26. The result of applying various convolution masks: (a) the original image;

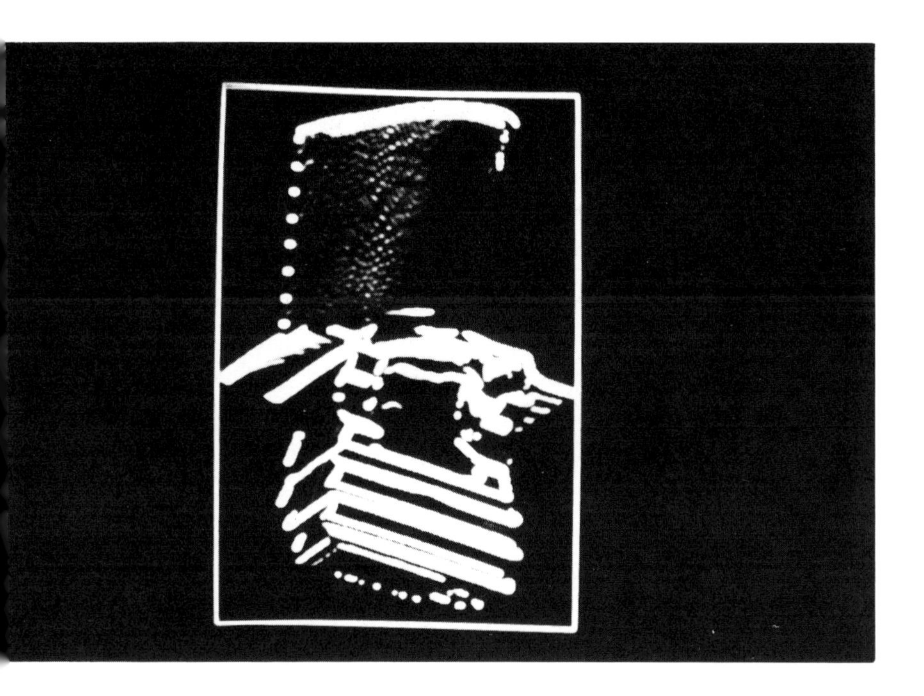

Figure 26 (continued) (b) the derivative with respect to y; (c) the derivative with respect to x;

(d)

Figure 26 (continued) (d) the second derivative with respect to polar radius.

3.3 Model-Driven Edge Detection

Model-driven edge detection problems can be overcome only through the use of prior knowledge of the object shape. In this approach, a structural model of the object is known in advance, and the image analysis is used to determine default parameters of the model. It is not necessary to have complete object boundaries. All that is required is to find a partial fit of the edge locations to a model. This allows the determination of the location of the object and the spatial relationship between important object features.

This model-driven method of image analysis is illustrated by an example taken from automatic inspection [18]. The problem is to locate the boundary and hole pattern of the trailing edge of the airfoil shown in Figure 27(a). Once the hole and boundary locations have been determined, then the diameter and distance between holes can be easily found. The wall thickness between the hole and boundary is also of interest.

Due to the rough surface finish of the trailing edge of this casting, it is difficult to obtain an image which is well described by two discrete levels. A typical gray-level image for this part using slightly oblique illumination is

(a)

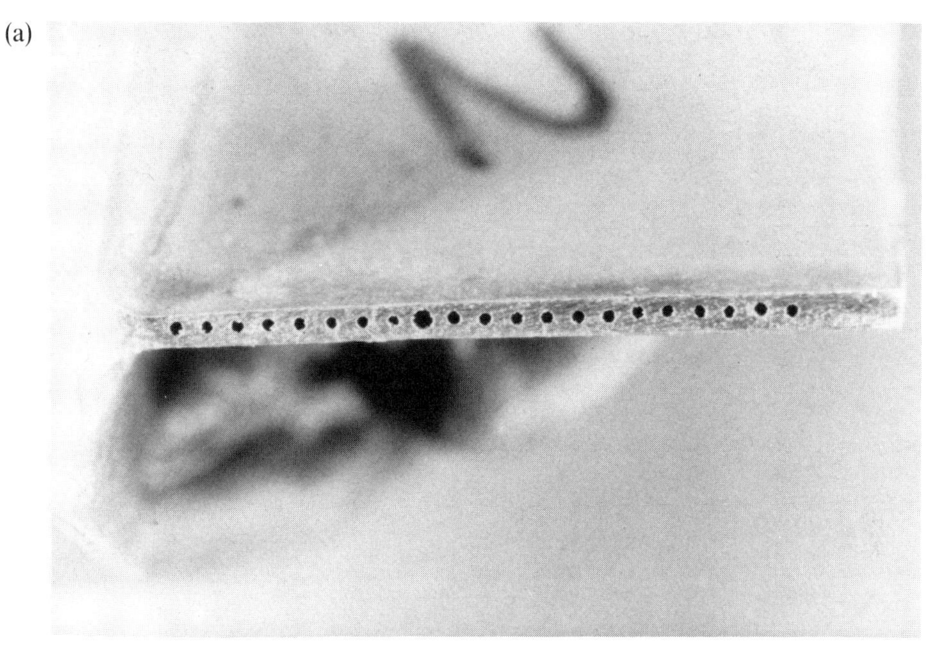

(b)

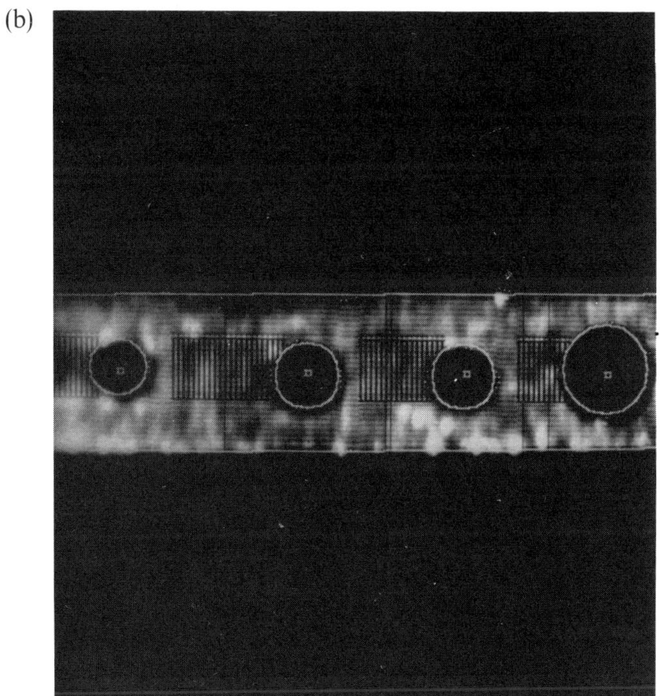

Figure 27. An example of model driven vision: (a) a metal casting with cooling holes; (b) the detection of the holes using a geometric model.

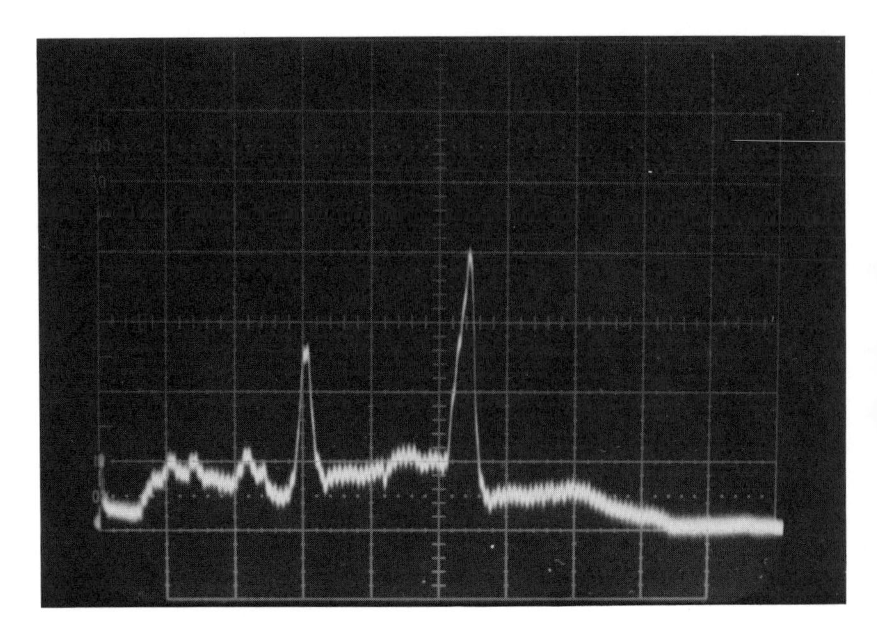

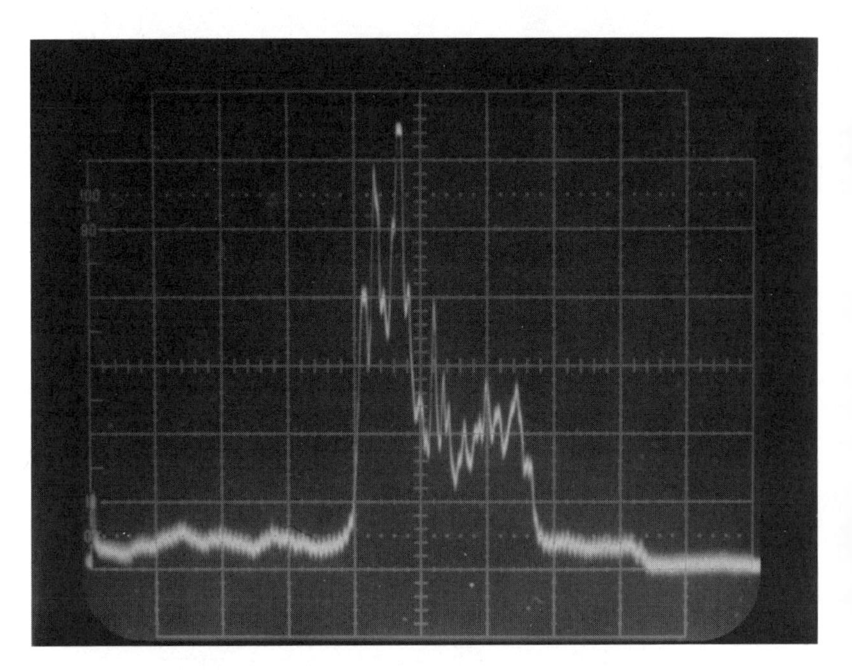

Figure 28. Two intensity cross sections taken across the image of the casting shown in Figure 27.

shown in Figure 27(b). Several cross sections taken through the image, presented in Figure 28, show the impracticality of thresholding the image to obtain a reliable segmentation of the hole boundaries. Instead, the methods of gray-level edge detection are used to find the most likely location. Since the image is noisy and of poor contrast, the edges are not perfectly detected.

In order to arrive at an accurate measurement of the important casting features, it is necessary to interpret these edges with an ideal model for the casting geometry. This model is shown in Figure 27(a) superimposed over the image of the casting. The model consists of two straight lines, which represent the boundary of the trailing edges, and a series of holes with unspecified diameter and location. The edges detected in the gray level image are interpreted in terms of this model by using the initial set of edges to form an approximate location for the features in the model. Then each edge location is compared to the location of the boundaries in the initial model interpretation. Any edges which deviates significantly from the model are deleted, and the location of the model is recomputed. This process is iterated until satisfactory agreement is achieved between the model and the edge points in the image. This approach allows the reliable detection and measurement of holes in relatively poor images. The accuracy of image derived model parameters is on the order of 0.0005 in. in a field of view of about 1 in.

3.4 Deriving Surface Information from Intensity

In the proceding section it was observed that the pattern of light intensity reflected from a metal surface is quite complex and difficult to interpret in terms of the material boundaries of the object in the image. These difficulties can be alleviated to some extent by taking into account the basic scattering properties of the surface and by providing a number of controlled illumination conditions. This allows the interpretation of the gray-level intensity in the image in terms of the local surface shape. This idea was mentioned earlier in reference to directional illumination. The concept of interpreting image intensity in terms of surface geometry was first introduced as a method for determining the three-dimensional shape of a diffusely reflecting surface by solving the differential equation of illumination [19]. These ideas can be used to simplify the interpretation of gray-level images.

The relationship between image intensity and the surface position of a diffuse surface which is observed along the direction of illumination is given by

$$I(x, y) = \frac{KRP}{\sqrt{1 + |dz/dx|^2 + |dz/dy|^2}}$$

where $z(x, y)$ is the surface position; R is the surface reflectance; P is the incident power (assumed constant with respect to x, y); and K is a constant. In effect, this relation states that the image intensity is inversely proportional to the cosine of the angle between the surface normal and the direction of illumination (observation). Thus the image intensity can be inverted to determine the surface shape by deriving a differential equation between z and (x, y). This assumes that the surface reflectance is independent of position. In practical situations this cannot be assumed, so it is necessary to acquire images with more than one condition of illumination and observation. This concept will be illustrated in the case of metal surfaces using the theory of rough surfaces [20.].

The basic model for this case is a random distribution of surface orientations described by the surface height variance and the surface correlation distance. The basic equation describing the scattered illumination is

$$\langle pp^* \rangle = \alpha \left[\frac{(1 + G)^2}{(1 + E)^4} \right] \exp\{-\beta[2(1 - G)/(\mathrm{I} + E)^2 - 1]\}$$

where $\langle pp^* \rangle$ = scattering reflection coefficient;

α = a constant depending on surface properties and the distance of the observer from the surface;

G = the cosine of the angle between the incident and reflected illumination;

I = the cosine of the angle between the surface normal and the illumination direction;

E = the cosine of the angle between the surface normal and the observation direction;

$\beta = [t/4\sigma]^2$;

t = correlation distance of the surface profile;

σ = the variance of surface profile.

This expression is greatly simplified for the condition of illumination along the direction of observation. In this case, $I = E$ and $G = 1$. With this simplification the expression for scattered power becomes,

$$\langle \mathrm{pp}^* \rangle = \frac{\alpha}{4\mathrm{I}^4} \exp[-\beta(\mathrm{I}^2 - 1)]$$

This expression is dependent only on the angle between the direction of illumination and the surface normal; the cosine of this angle is I. Note that the variation is exponential, which is consistent with the rapid variation in reflectance with angle exhibited by metal surfaces. The sharpness of this variation is reduced with decreasing β, which is also consistent with the observation that rough surfaces are not as specular as smooth surfaces. As the surface roughness approaches zero, the expression above converges to an impulse in reflectance about the specular direction. This corresponds to the condition where the illumination direction is along the surface normal. Thus the model is in good qualitative agreement the reflectance properties of metals. The expression has also been verified on a quantitative basis [21].

Another simplifying condition is achieved when the illumination is applied at essentially at right angles to the direction of observation. Let the I for this case be defined as I', and in terms of the I for the normal illumination case,

$$G' = 0$$

$$E' = I$$

$$I' = \sqrt{1 - I^2}$$

This results in a slightly more complex form for the scattered power,

$$\langle pp^* \rangle = \frac{\alpha}{(I' + 1)^4} \exp\left\{ -\beta \left[\frac{2}{(I' + I)^2} - 1 \right] \right\}$$

Again the behavior is exponential in the angles between the surface normal and the illumination I' and the observation I. The same dependence on β holds as in the normal illumination case. Both expressions involve the surface reflectance, which is imbedded in the α term. If we take the ratio of the scattered power for the grazing illumination to that for normal illumination, we have

$$r = \left[\frac{4I}{I' + I} \right]^4 \exp\left\{ -\beta \left[\frac{2}{(I' + I)^2} - \frac{1}{I^2} \right] \right\}$$

This expression is dependent only on the direction of the surface normal and the surface roughness. The effect of surface reflectance has been eliminated. An example of the use of this result is shown in Figure 29. A photomicrograph of some scratches in a rough surface is shown in 29(a). The resulting ratio is shown in 29(b). In this figure the quantity $\log(r)$ is plotted to reduce the dynamic range of the result. Note that the scratches have been well represented and that most of the fluctuations due to the surface roughness have been eliminated. In fact, the resulting image could be successfully thresholded to segment the scratches. This result is shown in Figure 29(c). This result could not be achieved in the original image since the intensity fluctuations due to the surface roughness are of equal contrast to those caused by the surface slope variation in the scratches. The ratio image, however, has retained effects mainly related to the global surface normal. In effect, the ratio r provides the geometry of the surface and could be used to solve for the surface coordinates as described by Horn [19]. In practice, the use of the ratio image is to enhance the contrast of surface features so that conventional segmentation methods can be used to analyze

(a)

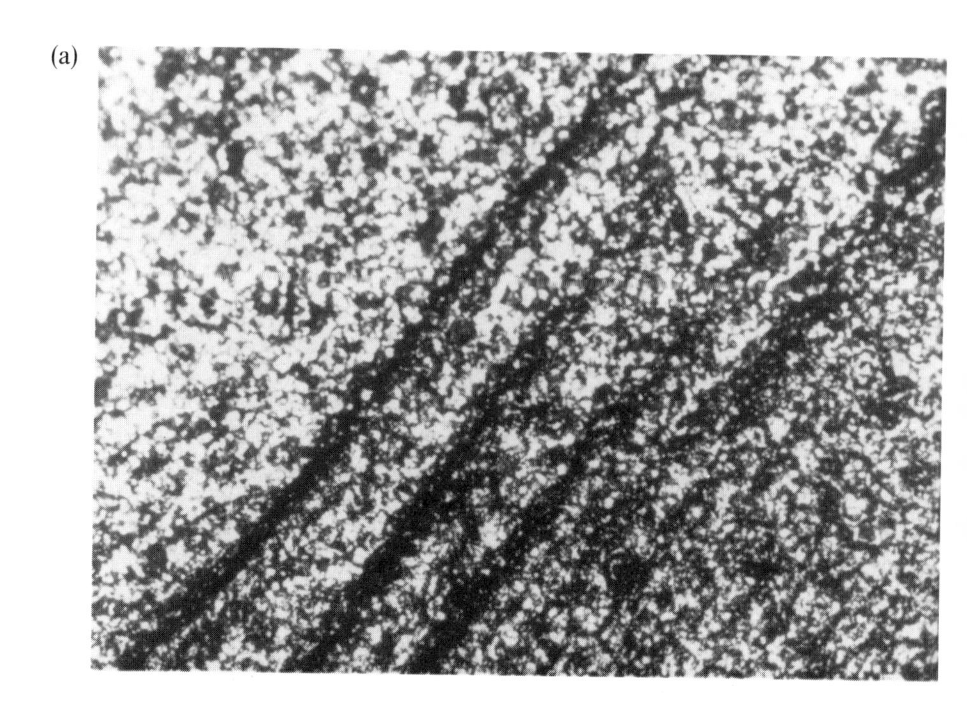

(b)

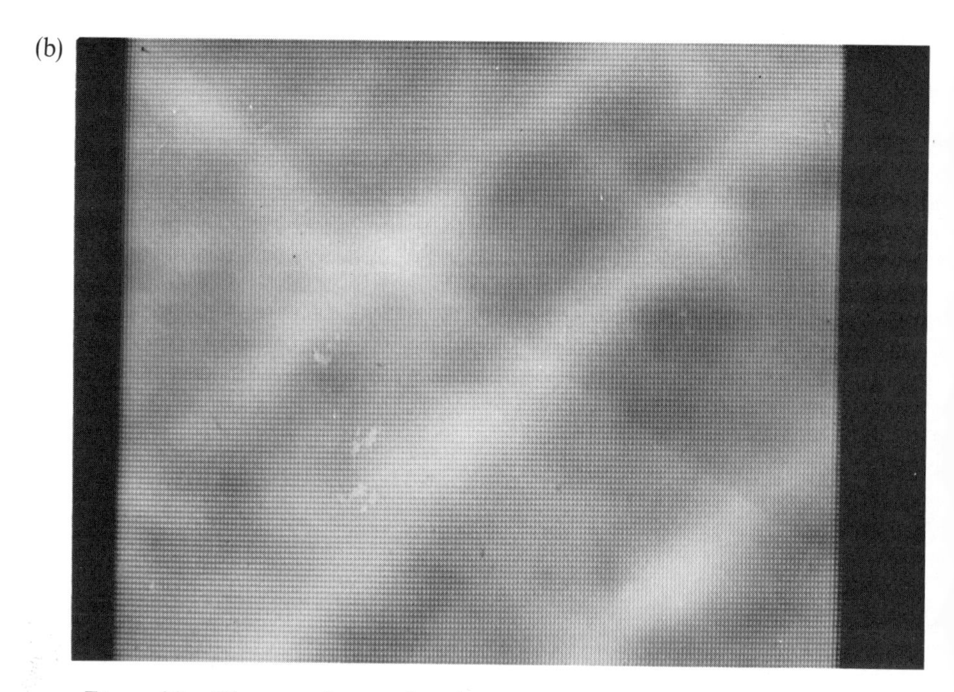

Figure 29. The use of normal and grazing illumination: (a) a rough surface with scratches; (b) the value of log(r) shown as intensity;

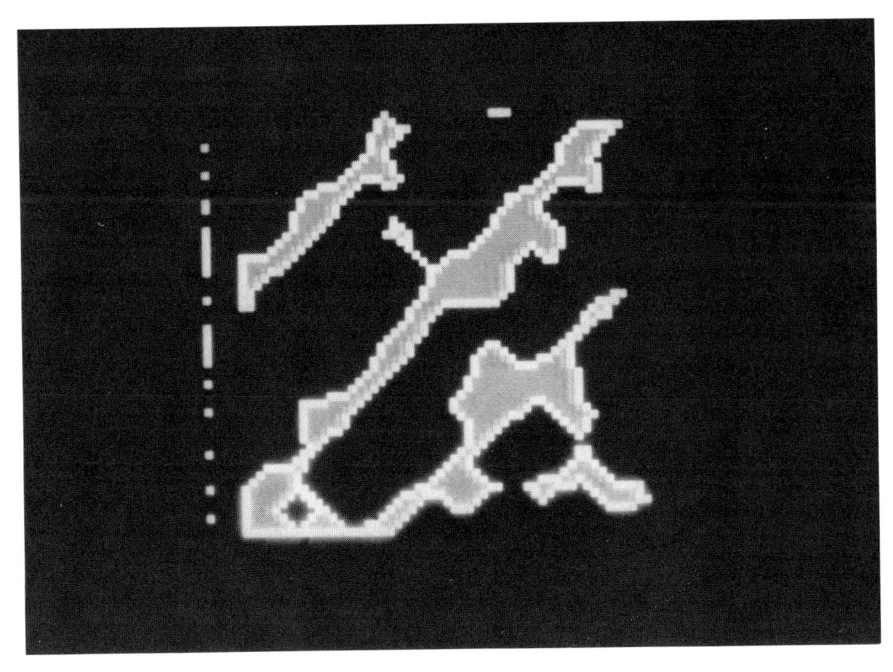

Figure 29 (continued) (c) the image in (b) threshold to show the location of the scratches.

images derived from metalic surfaces. Without such measures the images of rough metal surfaces present many irrelevant intensity fluctuations which are easily confused with object boundaries and genuine surface defects.

4. THREE-DIMENSIONAL VISION

4.1 Introduction

The direct acquisition of three-dimensional surface coordinates of objects eliminates many of the problems of surface reflectance and ambiguity of object location and orientation due to the distortions of projective geometry. The location of object surfaces in space can be accomplished by a number of techniques such as triangulation and time of flight. The use of direct three-dimensional information is rapidly becoming an important source of data for the control of robot motion and thus will be treated in some detail in this section.

The basic methods for the acquisition of three-dimensional images have been available for at least a decade, but little practical use of these techniques was made until the last few years. The most extensive use is in the area of seam-welding robot guidance, but other applications in inspection and three-dimensional modeling are also reported in recent papers. Also, the situation is in a state of rapid flux, since some commercial three-dimensional sensors are now available which will accelerate application activities.

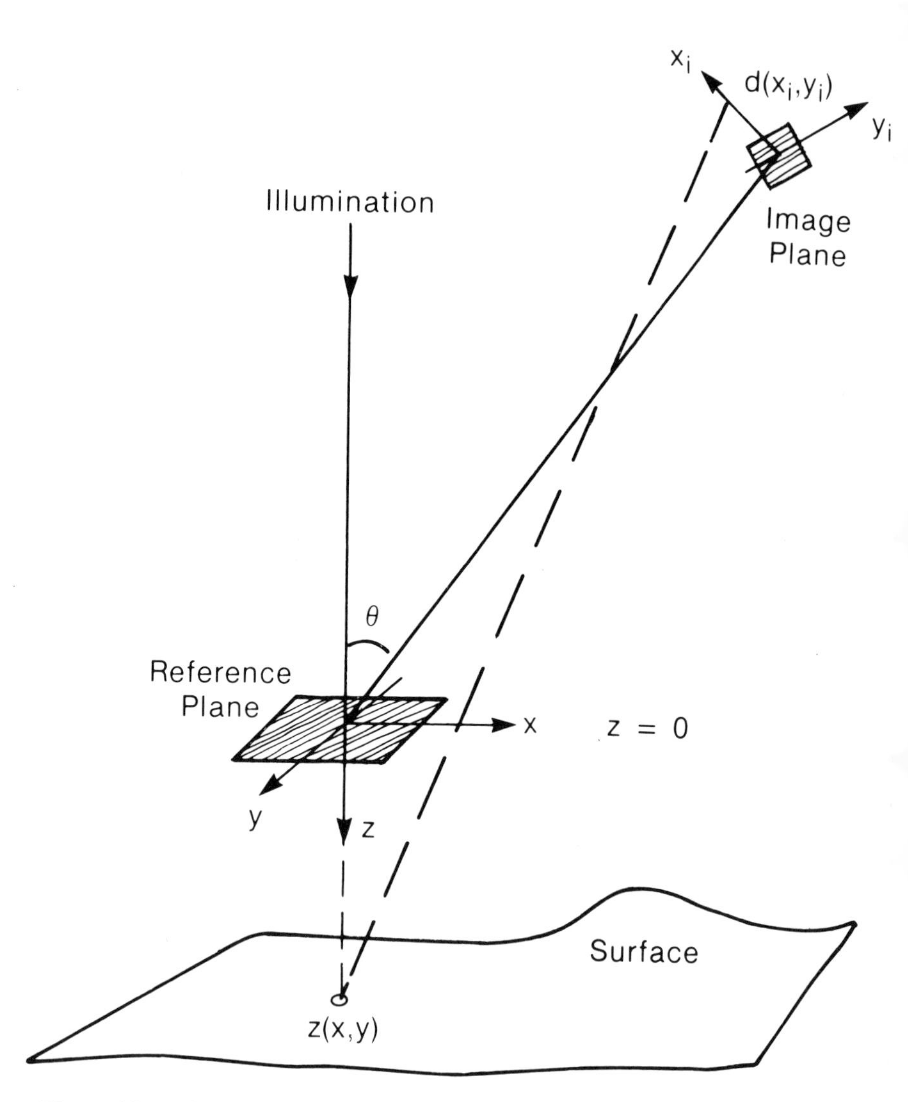

Figure 30. The basic geometry of triangulation. An illumination ray is observed from an angle. The vertical position, $z(x, y)$ is seen as a lateral shift $d(x, y)$at the sensor. This shift is zero at the reference plane.

4.2 The Triangulation Principle

The most direct method for obtaining three-dimensional surface coordinates is illustrated in Figure 30. A single ray of light is incident on the surface. The reflected illumination from this ray is imaged by the lens onto the image plane. The optical axis of the image system is inclined with respect to the ray by the angle θ, known as the parallax angle. Now if the surface is translated along the axis of the incident ray, the image of the ray is shifted in the image plane by an amount given approximately by

$$d = z \sin(\theta)$$

where z is the amount of surface translation. Since θ is known, the absolute position of the surface can be determined directly from d.

This triangulation principle can be used to obtain the three-dimensional coordinates of the surface of an object by sweeping the ray over the surface and comparing the image coordinates of the ray with those of a reference surface. It is necessary to remap the coordinates of the surface points because the parallax angle mixes the effects of translation in the surface and the position of the surface along the ray direction. The reference surface which is the most straightforward is a plane which is orthogonal to the incident ray direction as shown in Figure 30.

Here it is assumed that the ray is scanned over the surface with the direction maintained along z. This is an approximation, since the ray is scanned about a central axis which causes a change in direction of the ray with transverse position. For the moment we will assume that the scan axis is far enough from the surface so that this direction change can be ignored. It is necessary to remap the transverse coordinates of each point depending on the location of the surface in z since this variation in range causes a lateral shift in the image plane.

In general, is not possible to image all points on a surface from a single orientation. There are two conditions which prevent the measurement of a surface point, occulusion and shadowing. As shown in Figure 31, the illumination source has limited angular extent with respect to the subtended angle of the camera image plane. The angular domain which is within the field of view of the camera but not within the illumination wedge lies in shadow. This size of the shadow domain increases with parallax angle. On the other hand, the sensitivity of the shift in the image plane to surface distance also increases with parallax angle. A typical compromise selection for θ is 25° [22].

4.3 A Practical Implementation of Triangulation

The basic approach presented in the preceding section has been reduced to practice in a number of implementations; a typical example is now discussed. The illumination source is a HeNe laser, which is directed by a scanning

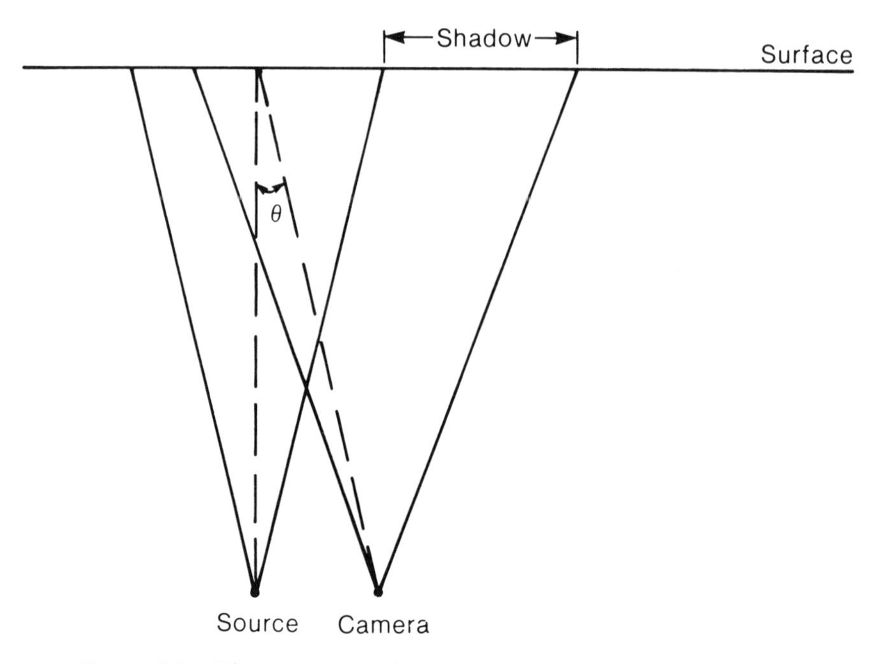

Figure 31. The geometry of shadow formation in range images.

mirror to scan in one plane. The location of the laser spot image is determined from two vantage points as shown in Figure 32. The location of the laser spot image in one sensor relative to the other is the basis for triangulation calculations.

The location of the laser spot in the image is determined from analysis of the signal from a linear photodiode array. The linear array consists of typically 1024 or 2048 diodes on 0.001-in. centers. The signal level is proportional to the number of photons collected at each diode during the time since the last reading. The time between readings is known as the integration time. There is a trade-off between the sensitivity of the array to low light levels and the rate at which the array is sampled. For a typical case, the arrays are read in a time on the order of a millisecond. This provides enough sensitivity to obtain a reasonable signal-to-noise ratio with a 5-mW laser source.

The diameter of the laser spot is on the order of 1 mm. If the field of view of the detector arrays is 10 in., then each diode covers about 0.01 in. Thus, the image of the laser spot covers 5 diodes. This situation is shown in Figure 33. As the figure indicates, the laser intensity pattern is approximately a Gaussian function. Since the spot covers a number of diodes, it is necessary to use interpolation to obtain an accurate location of the laser image. To

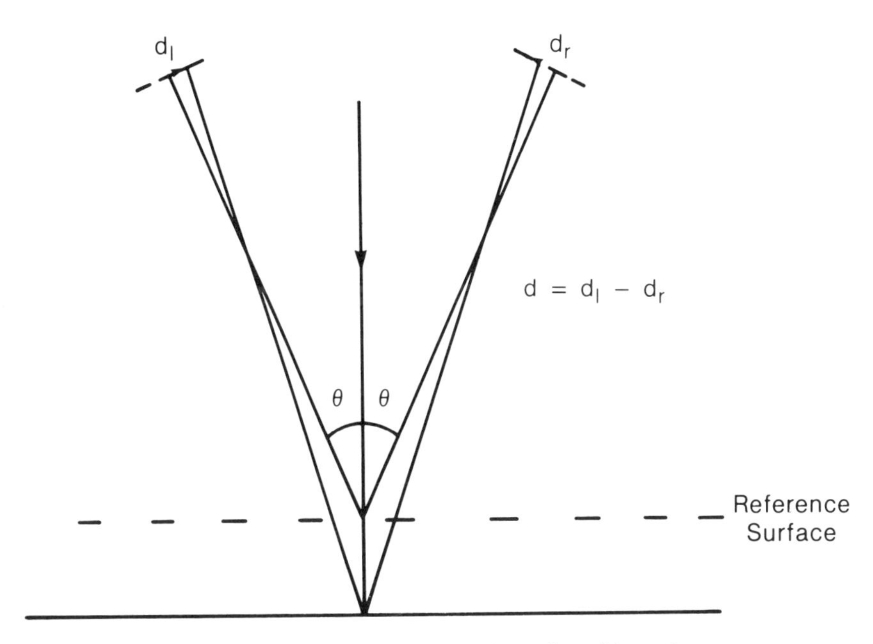

Figure 32. Occulusion an shadowing can be reduced by using two cameras. A laser ray is incident at the intersection of the two fields of view.

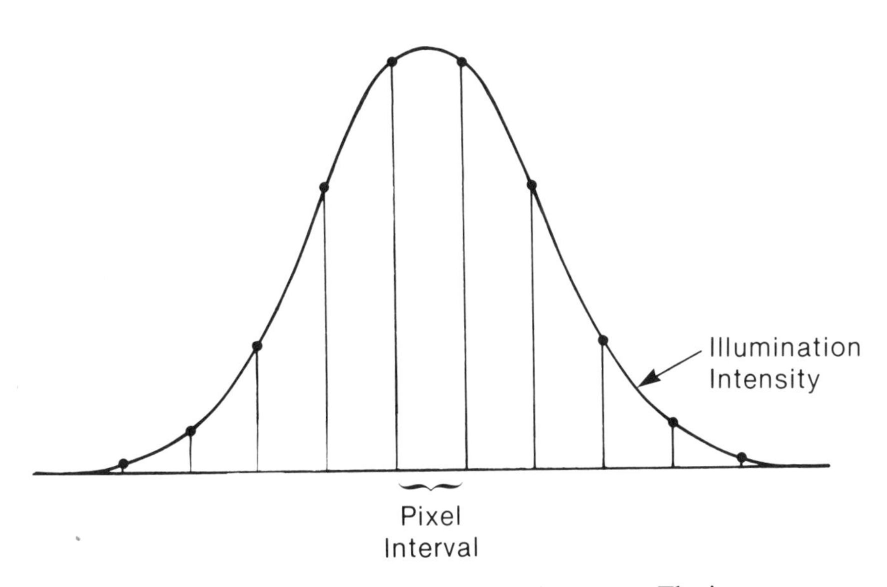

Figure 33. The intensity of the image of the laser spot. The image spans a number of detector elements. The image response is approximately a Gaussian function.

get an estimate of the required accuracy, let us assume that the parallax angle is 25° and it is necessary to determine the range to 0.01 in. The shift in the image location for a 0.01-in. shift in range is approximately 0.04 in., or about one-half a diode interval. Therefore, the location of the center of a Gaussian shape must be estimated to a accuracy of one-tenth its width.

The interpolation can be accomplished by an approach similar to the edge detection methods described in Section 3. The approximate location of the spot can be determined by thresholding the diode signal. The approximate center is given by the average location of the diodes above threshold. This location is then the center of the search for a zero crossing of a convolution of the signal with the odd impulse function shown in Figure 23. This zero crossing is located at the center of the Gaussian shape, and the zero crossing can be determined to subpixel accuracy using Newton's method. If the signal-to-noise ratio is adequate, it is possible to refine the center location to at least one-tenth pixel.

This performance cannot be achieved if the surface has rapid variations in reflectance. In particular, if the reflectance varies at a rate which is significant compared to the beam width, then the estimation of the Gaussian center is disturbed. This situation is shown in Figure 34. Here the surface is assumed to reflect very little for a small distance and then return to a normal level. The fluctuation causes an apparent shift in the center of the Gaussian distribution, which is not related to the parallax shift of range variation. This distortion is not distinguishable from a range shift and therefore appears as a range error.

To see the magnitude of this effect, consider a condition which produces a false shift in the distribution of one-half the beam width. Using the conditions established earlier, this shift corresponds to an error in range of 0.05 in. Since this is due to unpredictable variations in surface reflectance, it is impossible to eliminate this noise without reducing the overall accuracy of the system to the worst-case value. In some cases, a priori knowledge about the surface can be used to remove the range noise. For example, if the surface is known to consist of planar regions, then any range values which deviate significantly from a plane can be restored to the correct value by interpolation from surrounding range values. This assumes that the noise is characterized by local fluctuations rather than a global systematic curvature or other distortion. Since the range noise due to reflectance variations is related to regions of reflectance change, it is likely that most surfaces will obey the local continuity assumption. Global errors are due to nonlinear terms in the range calculations and optical distortions. These effects can be removed by calibration with respect to reference surfaces. The use of multiple channels of illumination, described later, allows some separation of the effect of reflectance from range variations. The next section considers methods to increase the rate of range readings over that achievable from a single-ray illumination.

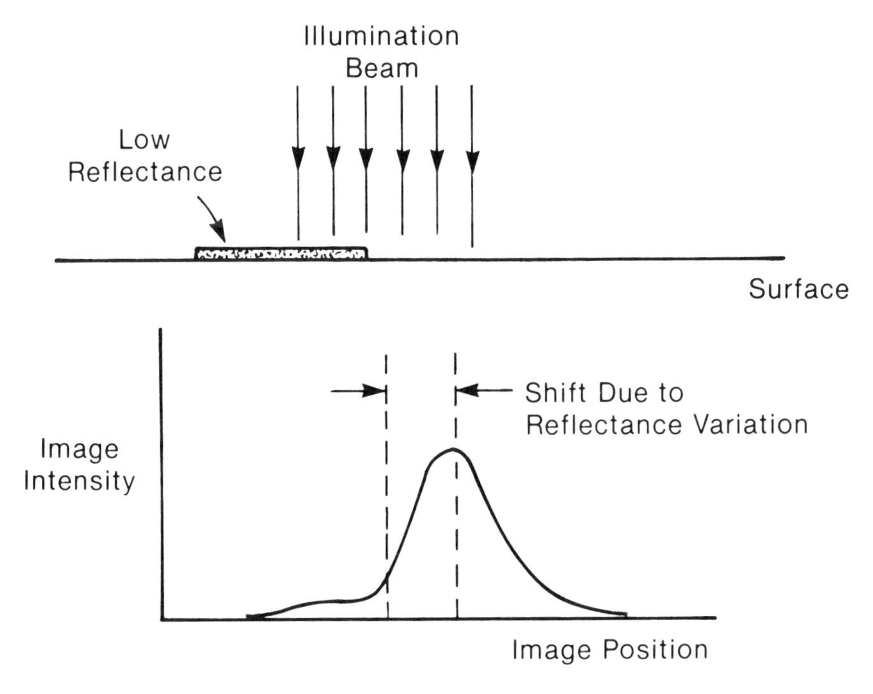

Figure 34. The effect of reflectance variation on the accuracy of laser spot location. The spot appears to be shifted due to low reflectance over a portion of the beam.

4.4 Structured Illumination

In order to increase the throughput of the range data collection process, it is necessary to collect data points in parallel. One method for achieving parallelism is shown in Figure 35. Here the illumination is in the form of line or stripe, which extends across the entire height of the image. The direction of the line is normal to the plane of the angle of parallax. The line will thus appear to shift in the image plane according to the range of the surface illuminated by the stripe. This is also illustrated in Figure 35.

The advantage of this illumination pattern over that of a single point is that a range reading can be made on each line of the image instead of only one reading per frame. On the other hand, no advantage is gained over the case of the linear array since one range reading per line is made for that implementation with single-ray illumination. The main advantage is that the integration time of the two-dimensional array is much longer than that of a one-dimensional array, and thus the sensitivity is much better. For example, compare a 1024-diode linear array with a 256 × 256 two-dimensional CCD

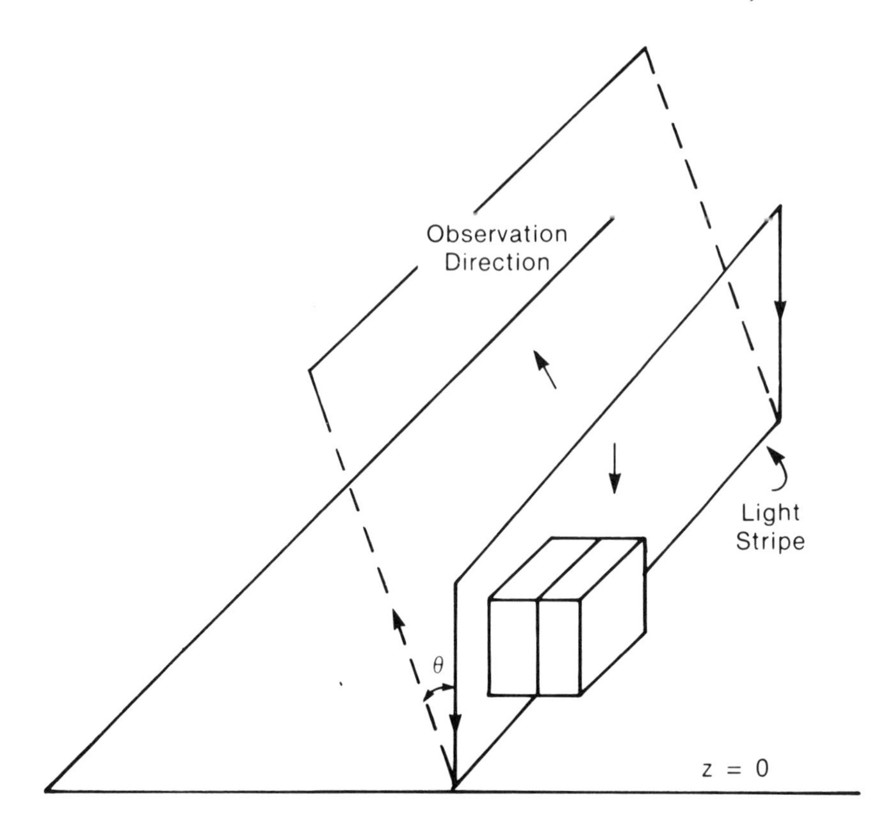

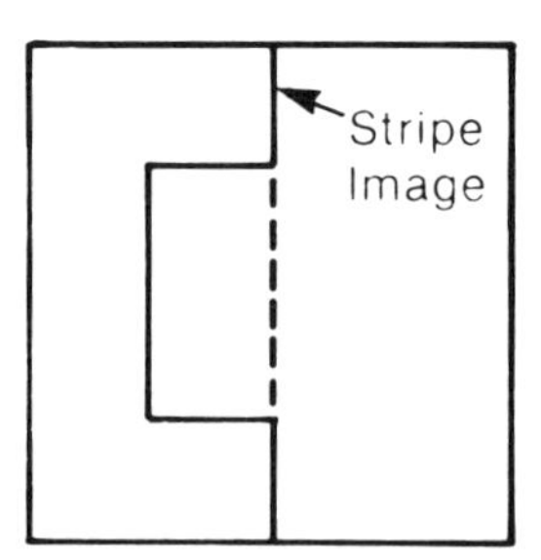

Figure 35. The use of a light stripe to apply triangulation a many points in parallel. The resulting image is shown at the bottom of the figure.

imager. If we assume that the readout rate is 1 million pixels/sec for each sensor, the integration time for the two-dimensional array is 64 times longer for the two-dimensional array.

There are some disadvantages to the two-dimensional array. First, the available resolution is much less. The largest two-dimensional array available is 388 × 480, while it is possible to find a number of commercial examples of 2048-element linear arrays. Another problem with the two-dimensional stripe case is indirect illumination. This effect is caused by concave regions in the object surface which act as corner refectors which return light to the sensor which is not related to the parallax-shifted image. This situation is shown in Figure 36. Secondary illumination can happen with a single-spot

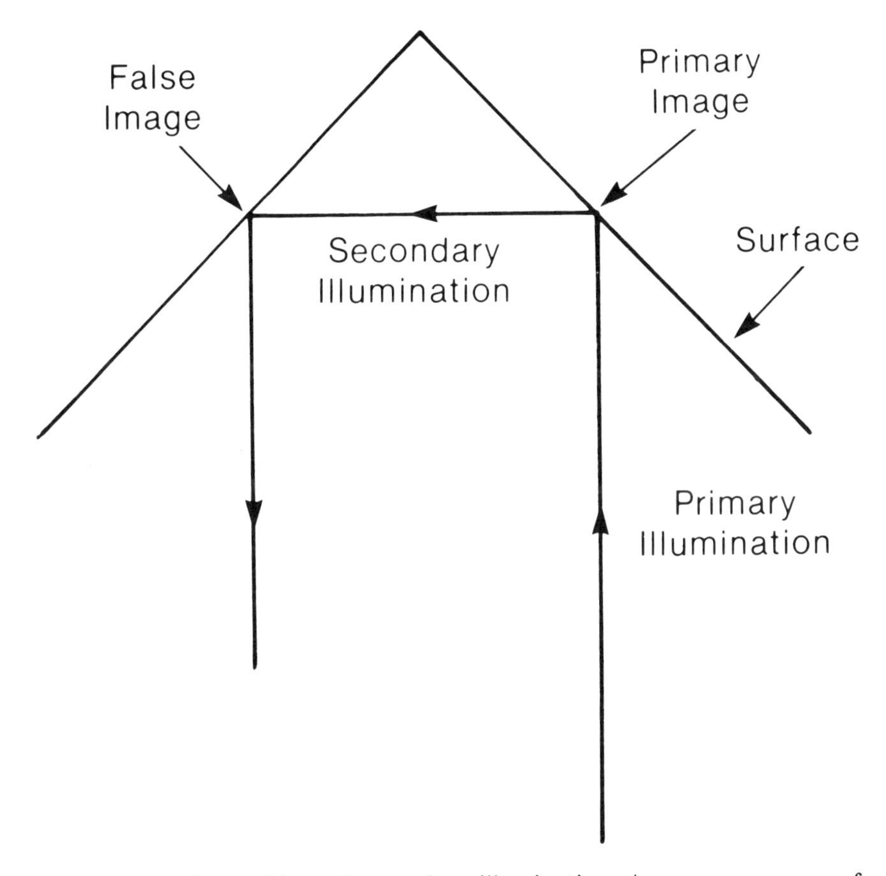

Figure 36. The problem of secondary illumination. A common source of secondary illumination is the corner reflector. This causes false position detection in the formation of the range image.

illumination as well, but the probability of indirect reflection is reduced because of the limited spatial extent of the ray.

This problem is quite severe for the case of specular metal surfaces. In this case, the secondary reflection can be more intense than the primary image. It is possible to reduce this effect by tilting the camera axis out the plane of the illumination; assuming that the specular reflections are returned within the illumination plane, the secondary effect can be much reduced. For diffuse (Lambertian) scattering surfaces, this will not provide a significant reduction in the problem since the secondary illumination is distributed uniformly in angle.

Even more parallelism can be obtained by illuminating with multiple stripes. This immediately leads to a problem of ambiguity, since it is impossible to distinguish the image of one stripe from another. If the range variation causes a shift of one stripe interval, then the range reading becomes ambiguous and rolls back to the value corresponding to an unshifted stripe. The situation is shown in Figure 37, where a multiple-stripe image is shown illuminating an annular ring. Each stripe shifts as it passes across the ring. Figure 38 shows the image position of two adjacent stripes as a function of

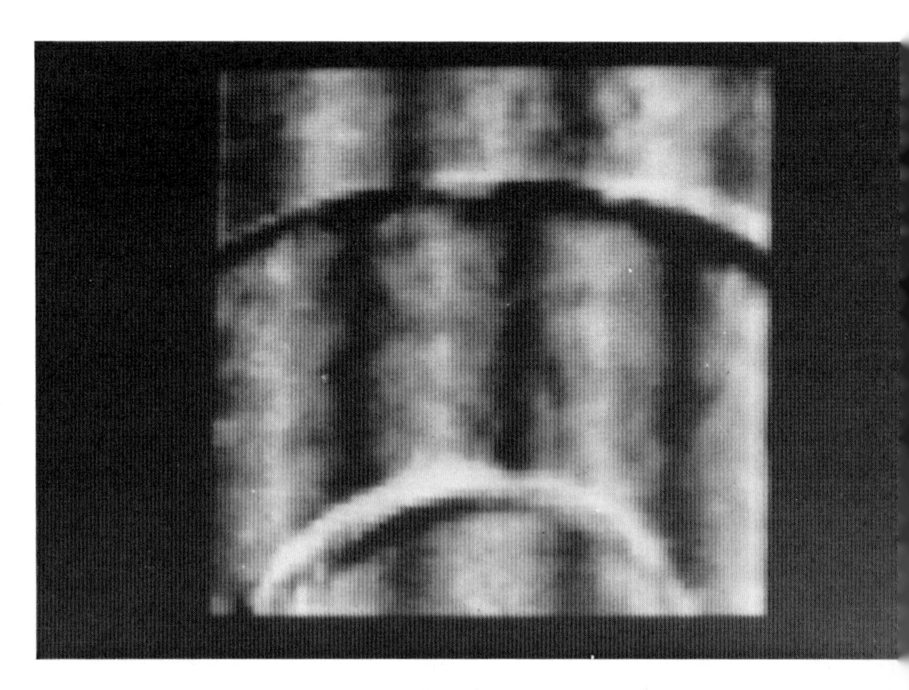

Figure 37. The use of multiple stripes. Here an annular ring is illuminated with a series of stripes. The parallax angle causes a shift as the stripes pass over the ring.

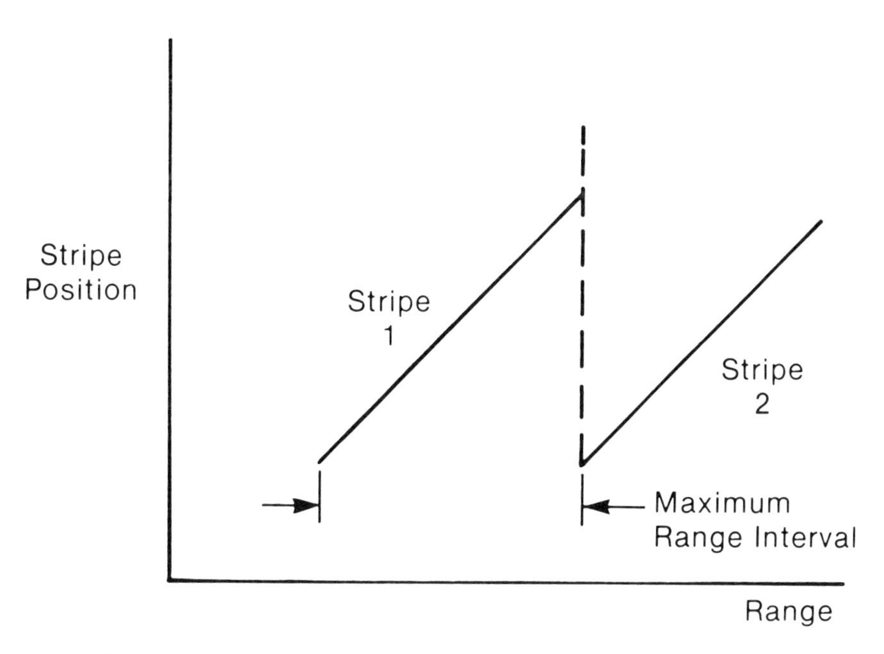

Figure 38. An illustration of phase ambiguity in multiple-stripe illumination. This plot shows the image position of two stripes as a function of surface range. Note that the identity of the stripes is confused as the range variation causes a shift equal to the stripe spacing.

range. This figure clearly shows the ambiguity as the range shift exceeds the spacing of the stripes.

The phase ambiguity of a uniform stripe pattern can be overcome by projecting a number of stripe patterns with differing periods. The most efficient set of patterns are related by a binary sequence in spacing as shown in Figure 39. The sequence represents a gray code which has property that no more than one bit in the sequence can change at a transition. This is essential so that large errors in range do not result from small errors in transition location. The gray code has been used for many years in the implementation of encoders for shaft motion for the same reason.

The main point of the multiple sequence of stripe patterns is that each point of the field of view has a unique code. Thus each point can be indentified regardless of the amount of parallax shift. The trade-off for this is that now a number of images must be processed in order to achieve one range image. For example if the field of view is quantized to a resolution of 512 × 512, then nine binary patterns must be processed in order to uniquely label each resolution element across the image. It is estimated that using this

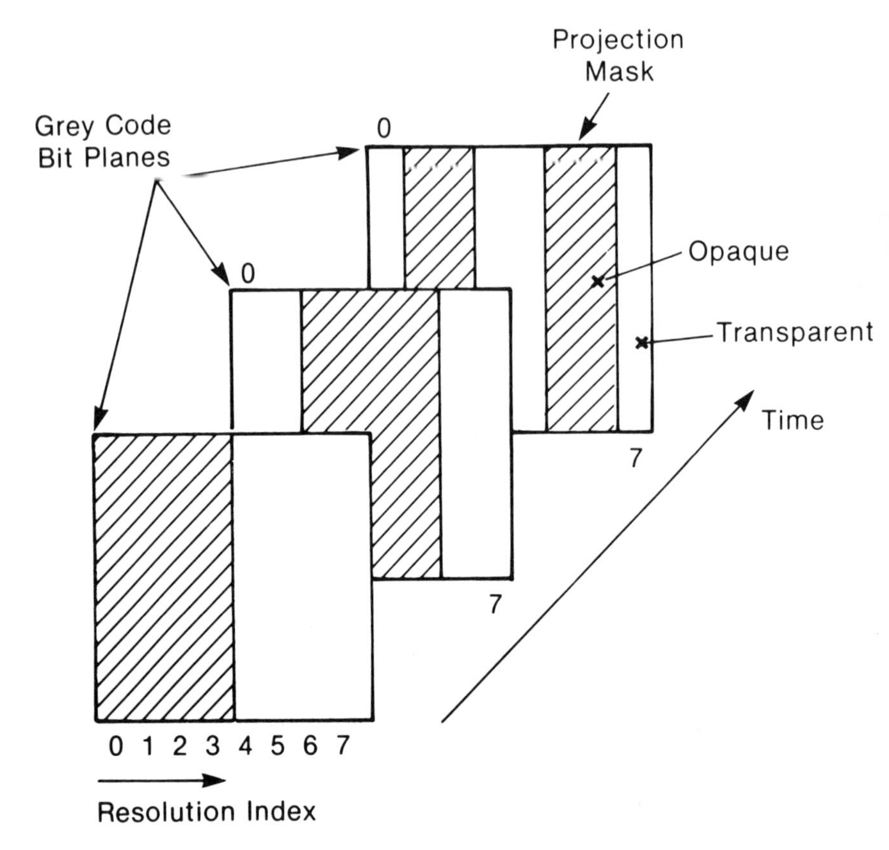

Figure 39. The use of binary sequences to remove the phase ambiguity in stripe illumination. Each code is applied in a time sequence. Each point has a unique code sequence.

method, it is possible to acquire about three range views per second with a standard 30-frame-per-second sensor acquistion time. A sensor based on time multiplexed gray-code labeling is currently under development by the National Bureau of Standards [23].

It should be pointed out that it is not necessary to provide a code with unique identity for all possible shifted resolution elements. Most optical implementations have limited depth of field so that the amount of range variation which can be successfully imaged is limited by this effect. For example, for a $f/1$ 50-mm lens operating at a object distance of 1 m, the depth of field is less than 2.5 cm. Naturally, by reducing the lens aperture, it is possible to extend the depth of field, but at the expense of image sensitivity. Since the excursion of range is limited, it only necessary to provide enough code bits to uniquely identify the stripe shifts over the depth of field.

4.5 Multiple-Wavelength Illumination

It is possible to provide multiple channels of information about the object surface through the use of different wavelengths of illumination. Ths use of color coding can extend the spatial bandwidth of the light pattern. This extended bandwidth can be used in a number of ways. One use is the formation of the spatial codes mentioned in the preceding section. Instead of using sequential projection of binary patterns, the patterns can be coded in individual colors and separately decoded using wavelength filters and multiple-image sensors.

The bandwidth can be put to more effective use as a means of determining the reflectance of the surface. The implementation of a range sensor which is designed on this principle is now described. The purpose of this sensor is the measurement of the surface profile of metallic surfaces [24].

The sensor is able to resolve 0.001-in. profile variations with a total depth of field of 0.05 in. The operations described below have been implemented in special-purpose hardware, and up to 60,000 range readings per second have been achieved. The sensor is based on the projection of two interleaved stripe patterns illuminated by an incandescent source. One pattern is projected with wavelengths in the infrared range (800–1200 nm), the other in the visible band (300–800 nm). The motivation for this selection is that each band contains approximately equal power from the illumination source.

Figure 40 shows the basic optical arrangment, including the sensor optics. In this implementation, the image sensors are linear diode arrays with 512 elements. The source power is split into the two wavelength bands using a dicroic mirror which reflects visible light but passes infrared. The two wavelength beams are recombined using a patterned mirror. The two channels can be used in a number of ways to reduce the effects of reflectance on the computation of range. In this case, the detection of the parallax shift is equivalent to the determination of the phase of the essentially sinusoidal patterns. Thus the basic problem is the accurate measurement of the location of prominent features on the illumination pattern.

A straightforward approach is to form the difference of the signals divided by the sum. That is, if the visible channel signal is labeled V_a and the infrared signal V_b then a normalized signal can be derived, V_n which is given by

$$V_n = \frac{V_a - V_b}{V_a + V_b}$$

If the reflectance of the surface for visible and infrared is the same, then this expression becomes independent of reflectivity. To show this, suppose that

$$V_a = Rv$$

$$V_b = R(1 - v)$$

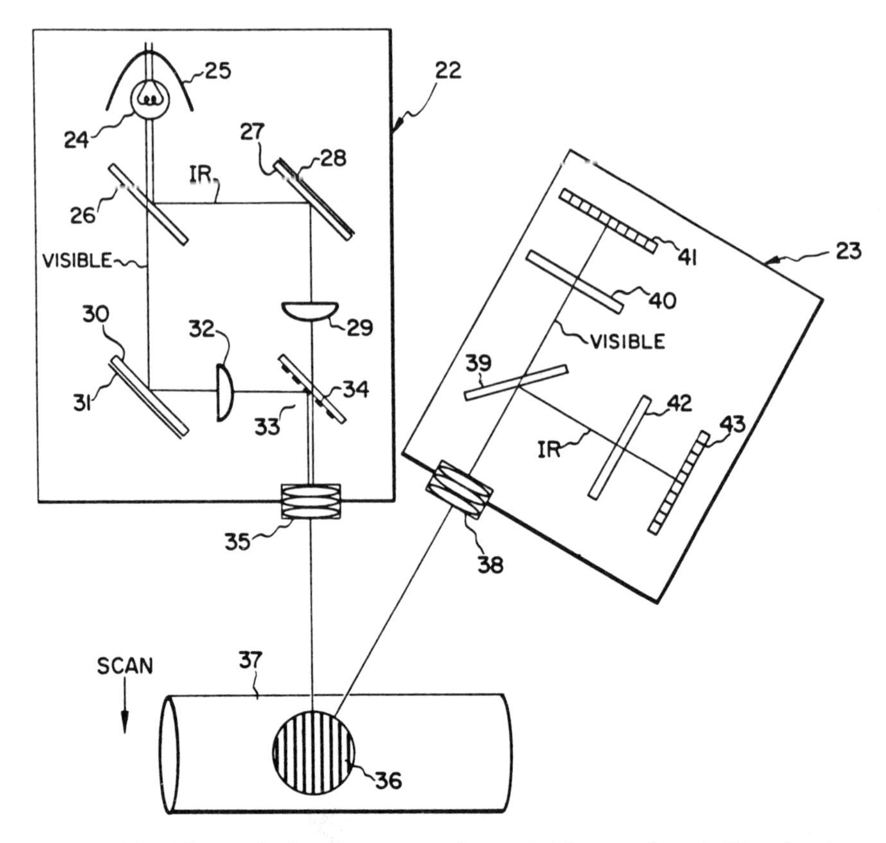

Figure 40. The optical arrangement for multiple-wavelength illumination. The source is split into two paths by wavelength filter mirrors. The paths recombine at a patterned mirror to form registered illumination stripes. The sensor splits the image back into two separate channels for signal normalization.

where v is the spatial modulation function and R is the surface reflectivity. Using these quantities in V_n, it follows that

$$V_n = 1 + 2v$$

which is not a function of R.

It is possible to extract four unique features from the normalized waveform, the two zero crossings and the two extrema. These locations can be determined accurately using edge detection methods. The zero crossings are found by making an optimum estimate of V_n using a least-mean-square filter and then taking the absolute value of the result. In this implementation, the five-point estimator is formed by the convolution of the mask

$$[-3 \quad 12 \quad 17 \quad 12 \quad -3]$$

With V_n. The absolute value of this result is illustrated by the dotted line in Figure 41. The exact location of the minima of this signal correspond to the zero crossings of V_n. It is possible to interpolate to find the minimum quite precisely by the use of the Newton formula:

$$\delta x = \frac{\partial V/\partial^2 x}{\partial^2 V/\partial x^2}$$

where δ is the interpolation distance to the exact location of the zero crossing.

The differential quantities in this formula can again be computed by using optimal convolution masks based on least-mean-square estimation. In the current implementation of this sensor, the masks are

$$\frac{\partial V}{\partial x} = [-2 \quad -1 \quad 0 \quad 1 \quad 2]$$

$$\frac{\partial^2 V}{\partial x^2} = [-2 \quad 1 \quad 2 \quad 1 \quad -2]$$

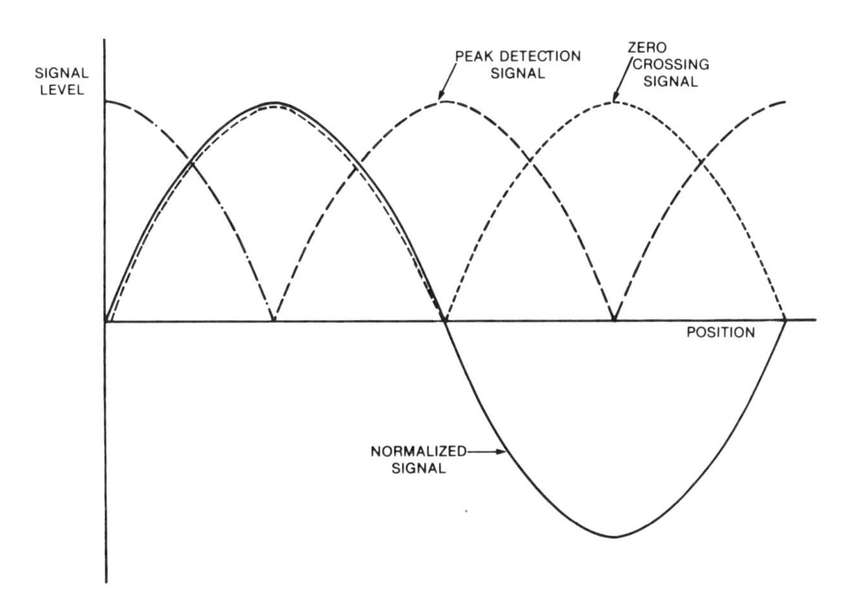

Figure 41. The detection of zero crossings and peaks in the normalized signal. The solid line is the normalized signal. The dotted line is the absolute value of an intensity estimate. The dashed line is the peak detection signal.

These five-point masks provide a reasonable trade-off between accuracy and smoothing of noise in the estimate. Using the Newton interpolator it is possible to estimate the location of the zero crossing to better that $\frac{1}{32}$ pixel for noise free signals and to better than $\frac{1}{2}$ pixel for extremely noisy data. In this case, extreme noise is such that the sinusoidal character of the signal is not apparent.

The detection of the extrema can be carried out in a similar manner by forming a peak detection linear filter and then taking the absolute value of the result. In the current implementation the peak is defined by the absolute value of two quantities:

the slope—

$$[-2 \quad -1 \quad 0 \quad 1 \quad 2]$$

and the sum of slopes—

$$[-2 \quad -1 \quad -2 \quad 0 \quad 2 \quad 1 \quad 2]$$

These are essentially measures of the symmetry of the peak which pass through zero at the point of symmetry. Thus the sum of the absolute values of these filters form a similar second-order interpolation neighborhood to the zero crossing signal. The resulting waveform is shown as a dashed line in Figure 4.

The form of V_n used above is limited to surface reflectances which are neutral with respect to wavelength. If the surface does not reflect visible and infrared energy equally, then the normalization is not valid. For example, assume that the reflectance in the visible channel is zero and unity in the infrared channel; then the expression for V_n above becomes constant and equal to unity.

An alternative form of normalization which is not sensitive to color is given by

$$V_n = \frac{1}{V_a}\frac{\partial V_a}{\partial x} - \frac{1}{V_b}\frac{\partial V_b}{\partial x}$$

This expression allows unequal reflectances in the two channels and is independent of reflectivity except for the term

$$\frac{1}{R_a}\frac{\partial R_a}{\partial x} - \frac{1}{R_b}\frac{\partial R_b}{\partial x}$$

This term is usually small except near regions of large and unequal spatial variations in R. The normalization with respect to R helps to reduce the value of this term. In practice, the effectiveness of this normalization method is limited by the precision of the spatial alignment of the two image sensors. If they are not perfectly aligned, then the cancelation required by V_n is not achieved. The effect of this misalignment is the most severe in the case of metalic surfaces with glint-forming surface imperfections. It is possible to create spikes of intensity which are several orders of magnitude above the background over an interval of one or two pixels. This high rate of change is not perfectly canceled by the normalization and leads to noise in the range data. The three-dimensional profile of a metal-casting surface is shown in Figure 42. A step of 0.001 in. was introduced by moving the sensor during the scan. The noise level due to reflectance variations is seen to be on the order of 0.0005 in.

The future development of three-dimensional sensors based on triangulation will be concerned with improved normalization and light-coding methods. This is a rich area for further research, and progress is likely to be rapid over the next several years. It is likely that sensors which can digitize three-dimensional space at essentially video rates will soon be available. The ability to analyze and interpret three-dimensional data must progress rapidly in order to match the rate of sensor development.

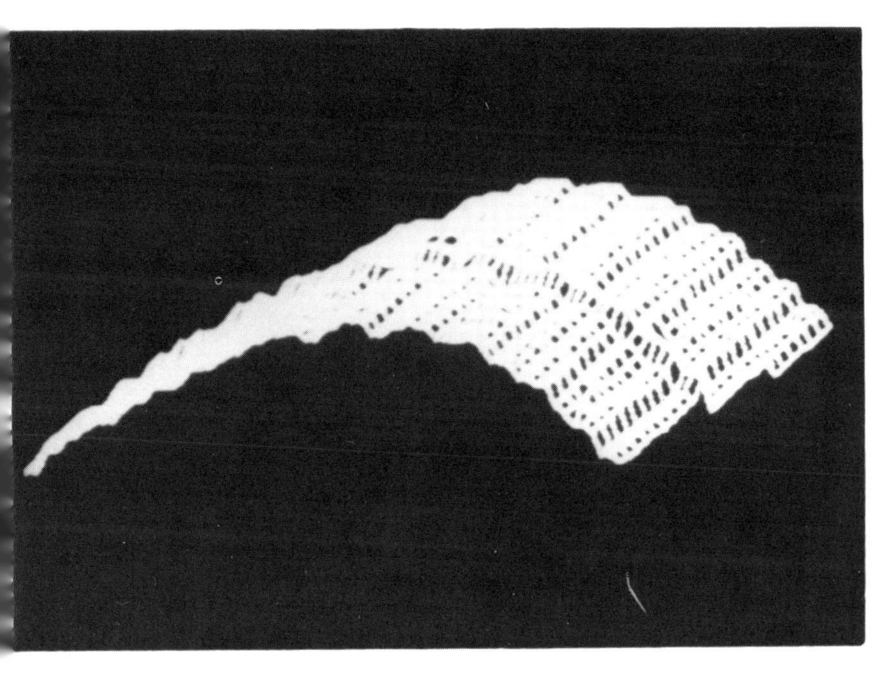

Figure 42. The surface profile of a metal surface. A 0.001-in step is introduced to show the scale. The noise is due to reflectance variations and sensor element misalignment.

REFERENCES

[1] Mason, M., Compliance and force control for computer controlled manipulators. *IEEE Trans. on Systems Man and Cybernetics, SMC*-11, 1981.

[2] Nevins, J. L., and Whitney, D. E., Robot assembly research and its future applications. In *Computer Vision and Sensor-Based Robots*. New York: Plenum Press, 1979, p. 275.

[3] Agin, G. J., and Duda, R. O., SRI vision research for advanced automation. *Proc. 2nd U.S.A.-Japan Conf.*, 113.

[4] Corby, N., Private communication.

[5] Horn, B. K. P., Obtaining shape from shading information. In *The Psychology of Computer Vision*. New York: McGraw-Hill, 1975, p. 115.

[6] Otsu, N., A threshold selection method for gray level histograms. *IEEE Trans. on Systems Man and Cybernetics, SMC*-9, 62, 1979.

[7] Zucker, S. W., Region growing: Childhood and adolescence. *Computer Graphics and Image Processing*, 5, 382, 1976.

[8] Hanson, A., and Riseman, E., Segmentation of natural scenes. In *Computer Vision Systems*. New York: Academic Press, 1978, p. 129.

[9] Pavlidis, T., *Structural Pattern Recognition*. New York: Spinger-Verlag, 1977.

[10] Tomek, I., Piecewise linear approximation. *IEEE Trans. on Computers, C*-23, 445, 1974.

[11] Bolles, R., Recognizing and locating partially visible objects. *Int. Journal of Robotics Research, 1* (3): 57, 1982.

[12] Ambler, A. et al., A versatile computer-controlled assembly system. *Proc International Join Conference on Artificial Intelligence*, 1973, p. 298.

[13] Papoulous, A., *The Fourier Integral and Its Applications*. New York: McGraw-Hill, 1962.

[14] Shanmugam, K. et al., an optimal frequency domain filter for edge detection in digital pictures, *IEEE Trans. on Pattern Analysis and Machine Intelligence PAMI*-21, 37, 1979.

[15] Marr, D., and Hildreth, E., Theory of edge detection. *Proc. Royal Society London, B 207*, 187, 1980.

[16] Papoulous, A., *The Fourier Integral and Its Applications*. New York: McGraw-Hill, 1962.

[17] Ballard, D., and Brown, C. M., *Computer Vision*. Englewood Cliffs, NJ: Prentice-Hall, 1982.

[18] Porter, G. B. et al., Automatic visual inspection of blind moles in metal surfaces. *Proc. IEEE Computer Society Conference on Pattern Recognition and Image Processing*, 1979, p. 83.

[19] Horn, B. K. P., Obtaining shape from shading information. In *The Psychology of Computer Vision*. New York: McGraw-Hill, 1975, p. 115.

[20] Beckmann, P., and Spizzichino, A., *The Scattering of Electromagnetic Waves from Rough Surfaces*. New York: Pergamon Press, 1968.

[21] Mundy, J., and Porter, G., Visual inspection of metal surfaces. *Proc 5th International Conference on Pattern Recognition*, 1980, p. 232.

[22] Porter, G., and Mundy, J., A non-contact profile sensing system for visual inspection. *Conference Record IEEE Computer Society Workshop on Industrial Applications of Machine Vision*, 1982, p. 119.

[23] Unpublished. Presented at *American Association of Artificial Intelligence Workshop on Three Dimensional Vision*.

[24] Porter, G., and Mundy, J., A non-contact profile sensing system for visual inspection. *Conference Record IEEE Computer Society Workshop on Industrial Applications of Machine Vision*, 1982, p. 119.

Chapter 6

SHAPE FROM TOUCH

Ruzena Bajcsy

ABSTRACT

The aim of this chapter is to present a coherent approach to the problem of determining shape from touch. This problem can be divided into two subtasks: One is the gripper as a data acquisition device and the other is the sensory processing. We have structured the content of the chapter into the paradigm according to the complexity of the gripper. They are: one-finger, two-finger, and multiple-finger hand scenarios. There is an additional difference in these scenarios: The one finger is rigid and can move only in a Cartesian coordinate system. On the two-finger gripper, again the fingers are rigid but the gripper is on a wrist which can move in a polar coordinate system. Finally, on the multifinger hand, the fingers are not rigid and of course the wrist can move rotationally. In all cases, we assume that the fingers are equipped with pressure sensors. After we review the presently available different designs for fingers and hands and pressure and/or force sensors we propose several algorithms for shape descriptions.

Advances in Automation and Robotics, Volume 1, page 209–258.

ISBN: 0-89232-399-X

INTRODUCTION AND MOTIVATION

The purpose of this chapter is to explore three-dimensional form perception and description via dextrous hands and tactile sensation. In the past most of the work concerned with form and shape perception and recognition in the psychological literature as well as in machine perception literature considered only visual sensory information. This type of work was covered in chapter 5 of this volume. Nevertheless, there is plenty of evidence that people and other primates use hands and tactile sensation as well [16] for form perception. For an excellent recent review see Ref. [31]. Hence there is no reason that intelligent robots should not have similar capabilities. Since our primary goal is to study the three-dimensional (3D) data acquisition systems using hands and touch and then how this data can be used for recognition purposes, we shall not be concerned with various control issues and the dynamics of the hand. We also are going to skip the problem of reach space [18,30], though it could be sometimes of importance, especially when the object to be recognized must be picked up, turned around, and so on. For our studies we shall assume that the examined object is rigid and is stationary and could be grasped if so desired. We also assume that the finger or fingers used in this study are equipped with tactile and/or force sensors and can change their position under computer control. For analysis purposes one can structure the problem shape from touch into three scenarios: one-finger scenario, two-finger scenario, and finally a hand scenario. This division is analogous to that of Binford [5].

The one-finger scenario basically follows surfaces. It records the position of the touching finger and detects surface texture. From the position information one can compute the local surface normals and hence compute the shape. The results of experimenting in this mode were reported in Ref. [1] and are summarized in Section 3. While with one finger we cannot lift an object, in the two-finger scenario we can pick up an object and detect the thickness (another dimension) of the shape of the object. Of course, all the other measurements of one finger can be executed with two fingers as well. Most of the current grippers belong to this category, but they are usually not equipped with tactile sensors. A few in research laboratories are beginning to have tactile arrays, and we shall report some preliminary experiments with some of them in Section 4.

A hand scenario is truly the future. We shall report what exists in Section 5. To summarize the organization of our paper: We start with the review of the tactile and force sensors, their construction and physical properties, and their processing (Section 2). Then we shall describe the one-finger scenario followed by the two-finger scenario. In each mode we shall discuss the appropriate hardware, software, and pattern recognition algorithms. Finally, we shall describe the hand and touch for shape recognition. Here we shall start with a short presentation of the anatomy of the human hand. We feel that it is a good start since the human hand is so far the best model there is. Then we shall discuss the design considerations for an artificial hand from a kinematic point of view only. This of course is important since it determines the flexibility, generality, and applicability of mechanical hands in comparison with the human hand. Some examples of existing hands and algorithms for pattern recognition will be shown.

2. SENSORS

Manipulator's most researched sensors are of three types (from Kinoshita et al. [28]): force, inertial and tactile sensors. The primary interest of this chapter is in tactile sensation. The force sensing is going to be briefly and only partially reviewed because it may serve as a detector of contact between the hand and the examined object. The inertial sensing will be ignored here completely. Existing force-sensing devices can be classified into one of three categories [51] according to their placement relative to the manipulator with which they are used: Force sensors are mounted on the joints or at the end of the manipulators (at the hand and/or fingers) or on the support platform of the manipulator. For best accuracy of sensing the load, the most desirable placement of the sensor is at the very end of the manipulator. However, that requires a miniature, lightweight implementation of the sensor and the associated electronics. And this still leaves unresolved the problem of where to locate the servo motors. Again, if the motors are closer to the sensors that provide the feedback, then the control is simpler than otherwise. At the same time, however, the motors add weight to each link, which complicates the control issues. As usual, the actual implementation is a compromise, and typically the sensory devices are distributed throughout the manipulator. Sensing the torques that are produced at each of the manipulator's joints is perhaps the oldest of the three techniques and has been very successful in master–slave systems. Joint-torque sensing has an advantage of detecting forces and moments not only applied at the hand but also those applied at other points on the manipulators. The drawback is the uncertainty in predicting the friction, and joint damping is sufficient to preclude the possibility of accurately detecting small hand forces. Over the years the wrist has evolved as the most desirable place to put force sensing [15,49,61]. As an example, we shall describe the "Scheinman wrist." Fashioned like a Maltese cross, the wrist performs motions that are measured by a series of semiconductor strain gauges mounted on the cross-webbings. There is one gauge on each side of each of the four webbings, hence 16 gauges. A similar system at MIT with different geometry was used for some delicate peg-in-hole assembly tasks [24,50,52]. Shimano [51] developed a grip matrix G which allows us to compute the net forces between the fingers and the grasped object as well as to determine the net object velocities from sensed contact velocities. This measurement actually is not necessary for our form recognition, but it is important when one considers that the studied object must be grasped for further examination purposes.

Interest in touch sensing for a flexible robot dates back to the mid-1960s [55]. Even then researchers recognized the need for contact feedback, especially in remote manipulation such as in radiation-contaminated areas or other dangerous and/or remote places. The problem, however, was and still is the lack of adequate tactile sensors. Binford [5] presented a coherent review of the state-of-the-art of sensors and stated specifications for desirable visual, tactile, and force sensors that are needed for future robots.

To our knowledge, Kinoshita et al. in 1973 [28] were the first to report pattern classification of a grasped object by an artificial hand. The artificial hand had an array of 20 pressure-sensitive elements (5 rows × 4 columns), and the shape of the surface was identified by one single grasp. Similar tactile array arrangements placed on tongs of the hand were reported by Hill and Sword in 1973 [22]. Later we found reports from Bejczy [4] on a tactile array of 32 elements, and from Clot and Falipou [10] and Briot [7] on an artificial skin of variable resolution (10 × 10 and 24 × 24 elements). These reports differ in type of physical principle on which the sensors are based. The most common experimental sensors are based on the following:

a. Strain gauges
b. Microswitches
c. Piezoelectric changes
d. Magnetostriction
e. Change of resistance with compression
f. Piezoresistance

The other considerations for a desirable sensor are, citing Binford [6], as follows:

a. Spatial resolution <2 mm
b. Deformable surface—sensors in surface
c. Time response—reaction time <0.1 sec
d. Mechanical requirements for ruggedness and reliability
e. Number of wires—small
f. Uniformity among the individual sensors
g. Small or no hystereses
h. Pressure sensitivity <10 g
i. Localized force sensing
j. Dynamic range 1000 overload capacity

Harmon recently [19–21] updated the specification for a desirable tactile sensor based on questionaires from about 55 representatives from industry, research laboratories, and academia. Below we cite his specs:

1. A touch-sensing device is assumed to be an array of 10×10 force-sensing elements on a 1-sq-in. flexible surface, much like a human fingertip. Finer resolution may be desirable but not essential for many tasks.
2. Each element should have a response time of 1–10 ms, preferably 1 ms.
3. Threshold sensitivity for the element ought to be 1 g, the upper limit of the force range being 1000 g.
4. The elements need not be linear, but they must have low hysteresis.
5. This skinlike sensing material has to be robust, standing up well to harsh industrial environments.

These are of course rather stiff requirements, and the prediction is that perhaps 5–10 years of research may lead to such a sensor. Briot [7] has developed a touch sensor based on changing electrical resistance of contact between an electrode and a conductive silicone rubber. This sensor is very similar to the one developed by Purbrick [41] and later advanced by Hillis [23], both from MIT. The advantage of the MIT sensor is finer spatial resolution (256 sensors fit on the tip of a finger) and fewer wires. For a matrix of $N \times M$ the MIT sensor requires only $N + M$ wires as opposed to the French artificial skin, which requires as many wires as there are sensors.

A true solution to the wire problem has very recently been put forward by Raibert [44]. His approach is based on the VLSI technology. This technology allows coupling the sensor with the computer, which is necessary for processing the signal, and thus reducing the information into fewer bits than the raw signal, hence reducing the number of wires.

From this and other reviews [19–21] it is clear that (a) touch is a useful and desirable modality for a robot, (b) there are available some tactile sensory devices for experimentation, and (c) work needs to be done in improving the currently available tactile sensory devices as well as investigating their applications.

3. ONE-FINGER SCENARIO

This is the simplest setup for obtaining sensory information for the task of shape from touch. It has only three components: a rigid finger equipped with an array of tactile sensors; a Cartesian, 3-D positioning device; and a computer controller and processing unit. The block diagram for the one-finger sensing system is shown in Figure 1. It shows the major logical units of the

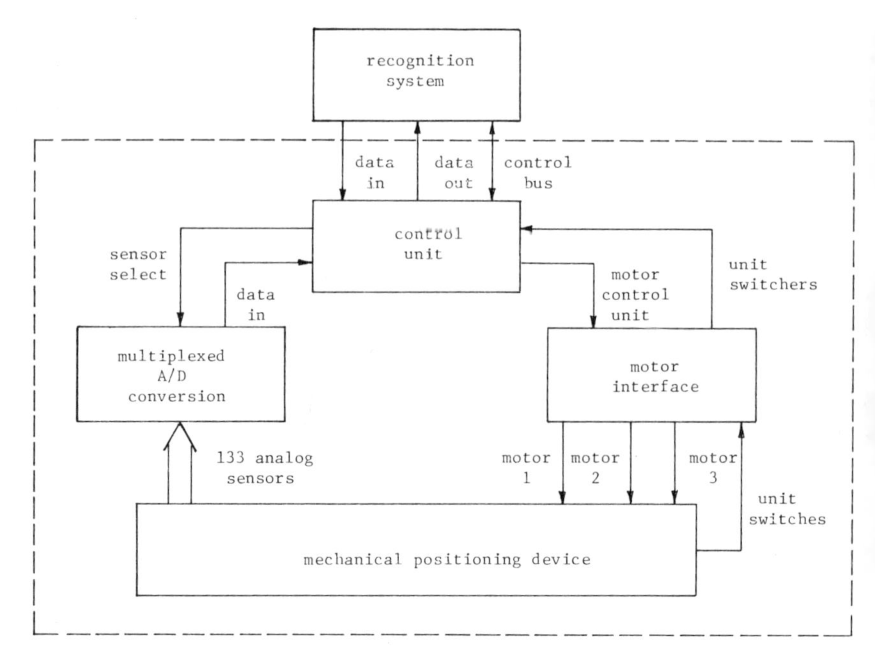

Figure 1. Block diagram of software design of the tactile system.

system. The picture of the actual entire device is displayed in Figure 2(a), and the detailed tactile sensor is in Figure 2(b). The implementation of the one-finger sensing system architecture includes hardware at three levels. From lowest to highest, these are as follows:

Level

Mechanical:	Mechanical device performing two primitive functions: (1) Positioning/movement of the sensor (2) Tactile sensing
Interface:	Interface of mechanical device to digital control unit
Electronics:	(1) Provide digital interface to the motors which perform device positioning/movement (2) Provide digital interface to the analog tactile sensor
Digital Control:	Digital control system for tactile sensing device

At the lowest level we provide a mechanical system, which allows the tactile sensor to be positioned anywhere in a constrained three-dimensional region (about 0.5 m per axis) with reasonable precision and accuracy (within 1.0 mm). The mechanical system's three degrees of freedom consists of orthogonal rectangular axes along which the arm containing the tactile sensor is moved by stepping motors.

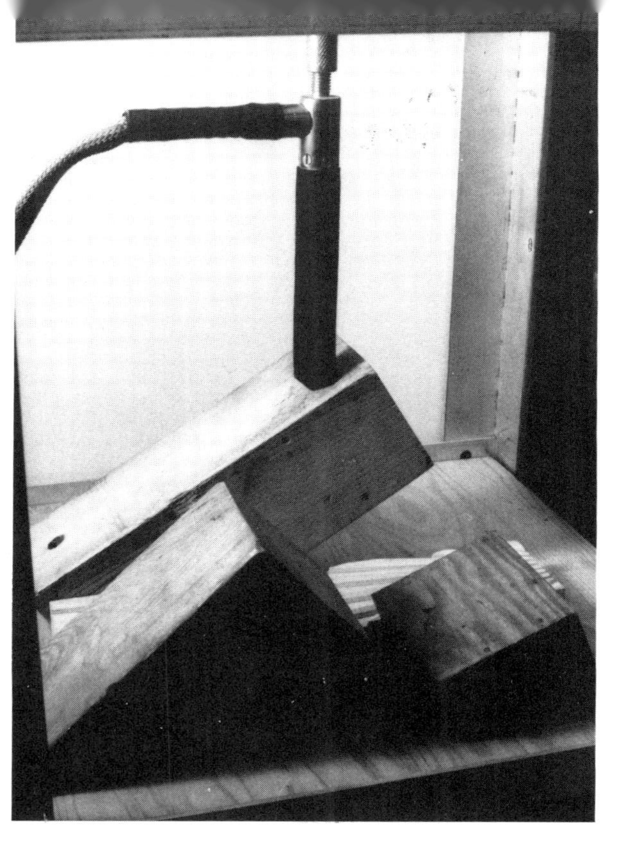

Figure 2(a). Positioning device.

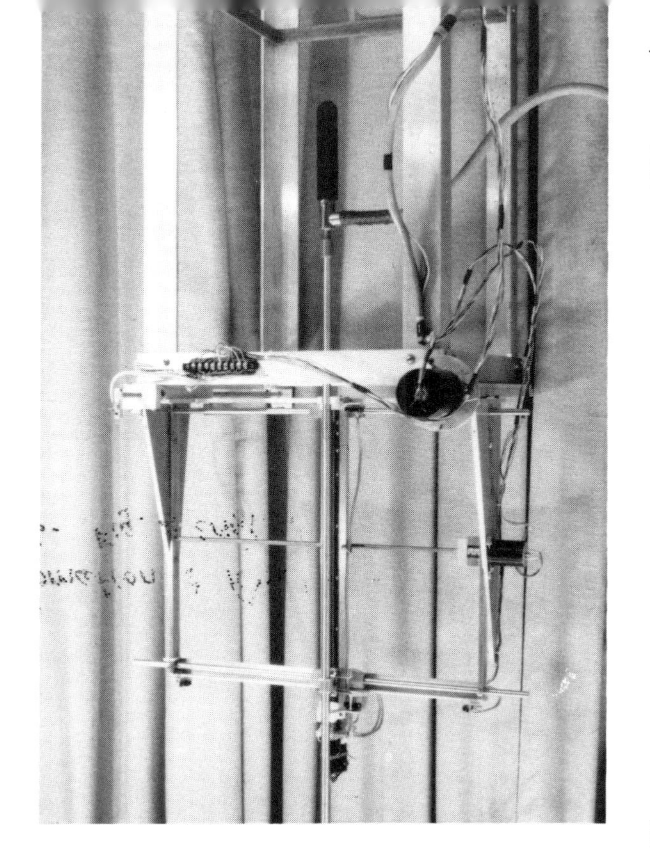

Figure 2(b). Close-up of the finger sensor.

At the conceptual level directly above the mechanical system and sensors are two units that function as the interface between the mechanical system and the control unit. The first device, the motor interface unit (MIU), provides all the electronics necessary for control of the motors which position the tactile "finger." It also provides interrupt driven end-of-travel detection on each of the three axes of motion. The second device, thee tactile interface unit (TIU), supplies the analog-to-digital conversion for the interface to the 133 pressure sensors on the "finger."

Finally, at the architectural apex of the tactile sensing system is the control unit. It provides the digital control of all that is below it in the system hierarchy. It is comprised of two Z80 microprocessors. One, the motor control processor (MCP), is responsible for driving and positioning the stepper motors, and the other, the tactile sensing processor (TSP), is dedicated to tactile data acquisition and compression. The MCP and TSP communicate

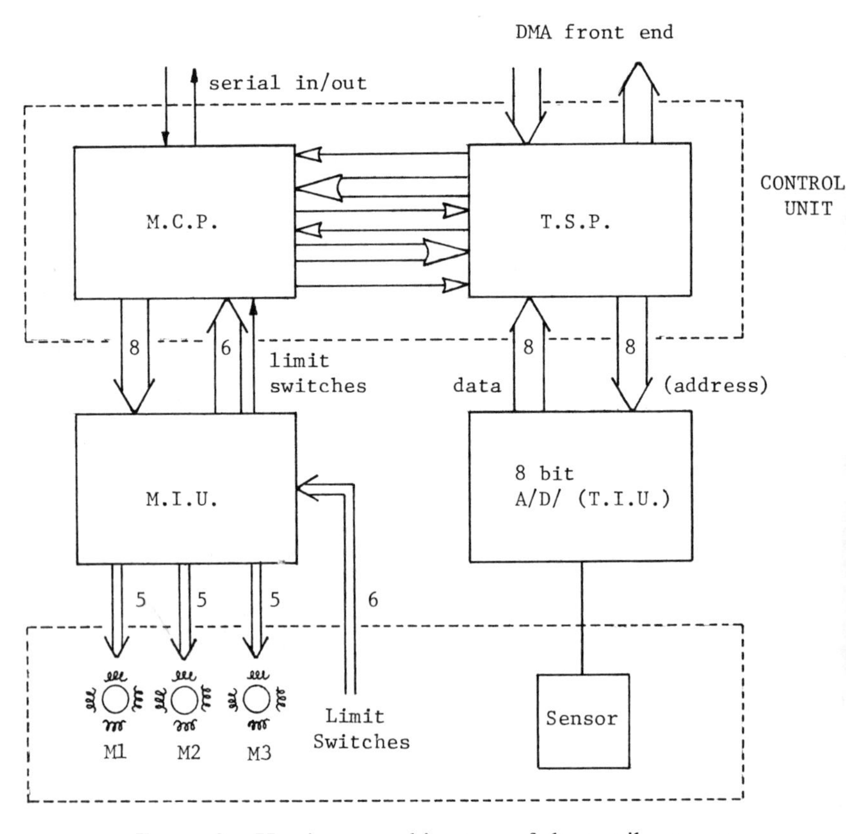

Figure 3. Hardware architecture of the tactile sensor.

with each other via a 14-bit-wide parallel data path. The PDP 11/60 issues high-level commands and receives positional information through a serial connection to the MCP. Finally, tactile data are passed to the PDP 11/60 through a DMA link from the TSP (see Figure 3).

3.1 The Software

The purpose of the developed software is to provide easy interaction with the system and deliver the desired sensory data. We have two sensors: the positional and pressure sensors. They both can be read in two modes: static and dynamic. Thus, texture will be obtained by analyzing the pressure sensors in motion while touching. The shape information will be available in both modes. In the static mode we get impressions or tactile images of the touched object, while in dynamic mode we can use the finger path for scanning of the object and thereby obtaining the shape description.

In our design we felt that it was important to delegate as many as possible of the low-level tasks to the local processors and so diminish the computational load on the host PDP 11/60. The tactile branch, in keeping with this principle, would have a set of commands which could be invoked by the host to perform various input–output and timing intensive operations, or functions involving real-time feedback. Following are some of the commands that were considered:

1. Reset the machine.
2. Move to absolute coordinates (x, y, z), stop on collision with an object. This can be used as a "find something in this direction."
3. Scan cross-section—trace the contour of an object in an arbitrary plane in 3D space. Returns to the host a list of step vectors describing the finger's path.
4. Local texture—trace around a small area on the surface of an object and produce a description of the texture.

These are rather complex instructions that are or will be composed from simpler and lower-level commands listed below:

H	Home—return to inner, upper left corner, and reset the current position to (0, 0, 0).
nX	Move n steps in the X direction (n may be positive or negative and defaults to $+1$ if omitted).
nY	Move n steps in Y direction.
nZ	Move n steps in Z direction.
@x, y, z	Move to absolute position (x, y, z).
n(	Begin nest.
)	End nest.
=	Return current position as x, y, z coordinates, ASCII-coded decimal values separated by commas.

Q	Quit the program—return to power-on monitor.
1S	Take a snapshot of the sensor, store data in memory, increment frame count.
−1S	Take as many snapshots as possible until the completion of the current motor step.
0S	Clear the frame memory.
G	Send the contents of the frame memory to the host, beginning with the frame count. All data is in ASCII-coded hexadecimal. Then clear the frame memory.
space	Null operation.

These commands are obviously very simple and are at the level of an assembly language. However they can be very powerful when grouped together. For example, the sequence

@100,100,100 50(3(20 × 20z S −20z) 20y −60x)G

takes 150 snapshots, in a 50 × 3 grid, beginning at (100, 100, 100), then sends all the collected data to the host computer.

3.2 Tactile Sensing Processor

The TSP program consists of a single loop in which each of the sensors is interrogated for its 8-bit pressure value. It has basically two functions: to report which face of the sensor is touching and provide readings from each sensor individually.

For the first function, each value of a sensor is thrown into one of three categories with respect to a low and a high threshold. The category indicates whether the sensor is not touching anything, is in contact with an object, or is pressing too hard. The sensors are then grouped by finger face and a face status is computed for each face using the following rules:

If any sensor is over range, the face is over range;
If all sensors are below range, the face is below range;
Otherwise, the face is within range.

If there were any face status changes since the last pass, the motor control processor is informed.

3.3 Motor Control Processor

The motor control processor's basic job is to control and coordinate the three stepping motors which position the finger. When it is necessary for the host computer to know the path the finger follows during the execution of a command, the MCP provides it.

Steps are taken in a synchronous fashion. That is, if the step rate is set to 125 steps/sec (the default case), the processor is interrupted every 8 ms to determine which motors are to be stepped and in which direction.

So after each interval, the MCP may pulse any combination of the three motors, and each can be in one of the two directions. This leads to 26 possible directions in which a single step can move (ignoring the case where no step is taken at all). We represent this direction as a 6-bit "step vector," organized as follows:

bit 5	4	3	2	1	0
Z direction	Z step	Y direction	Y step	X direction	X step

Since this fits in an 8-bit byte, it is very convenient now for the MCP to give a path to the host computer. It simply sends a 1-byte step vector over the serial line for each step taken. The host collects the sequence of step vectors in a buffer, and the exact path can be reconstructed very quickly at any time.

When moving from one position to another in 3D space, it is desirable to do so in a straight line. This requires varying the speeds of the individual motors so that they all arrive at their destinations simultaneously. We have developed an algorithm which uses the synchronous stepping scheme. The overall line of motion is described and stored in terms of three direction components. There are also two accumulating counters, one for the mid direction and one for the min direction. (The mid direction is the dimension which has the second largest number of steps to take. Min direction is defined similarly.) Both are preset to zero.

After each 8-ms interval, a step vector is created, and the motors are stepped accordingly. The max direction is always stepped. For each of the other two directions, the accumulating counter is incremented by the corresponding direction component value, and the result is taken modulo the max direction component. If an overflow occurs, a step is taken. Two corollary actions occur as a result of this algorithm. First, if the MCP is providing path information, the step vector is sent to the host. Second, a termination test is made.

This command also terminates if the tactile sensing processor indicates that the finger has come in contact with an object. Primarily, this is to protect the finger from damage.

3.4 Experiments and Results

The one-finger sensor is about 5 in. long, with an octagonal cross section about $\frac{3}{4}$ in. in diameter. Each of the eight rectangular faces is connected to a tapered piece, which is in turn connected to a common tip piece. There are a total of 133 sensitive sites—16 on each main face, 1 on each alternate taper, and 1 on the tip. This device was obtained as part of the U.S.–French collaboration effort between our laboratory and the LAAS, Toulouse Laboratory (Prof. G. Giralt, director). Because of the vague resemblance and its origin, we will refer to this sensor as the French finger. (See also Figure 2.)

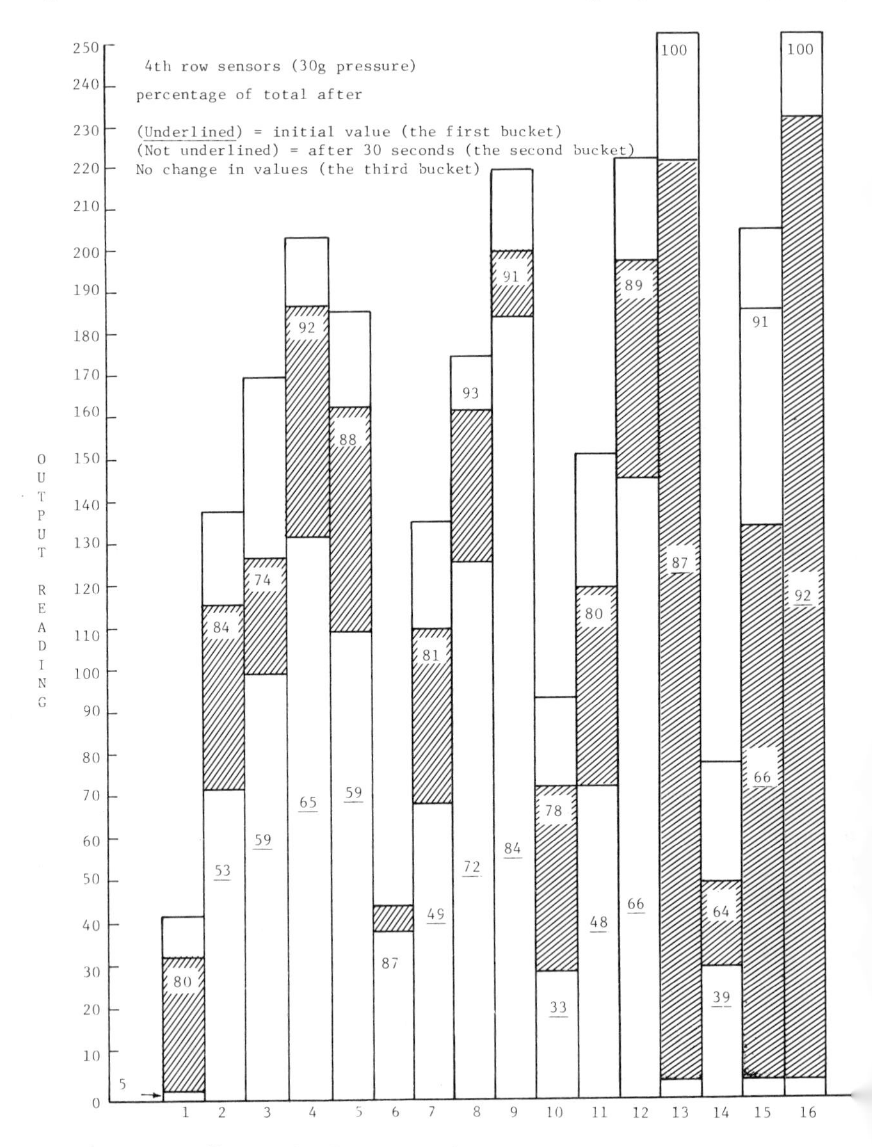

Figure 4. Homogeneity among sensors.

There have been two types of experiments performed: one aiming primarily at understanding the physical properties of the sensor, and the other using the sensor for surface description. The following studies were done:

1. Understanding of uniformity among individual sensors
2. Time response
3. Pressure sensitivity, repeatability of measurements and hysteresis

Furthermore, we have found the sensor to be capable of classifying the surface based on hardness, detecting the tactile texture, and measuring the surface normals by following the object.

3.4.1 Uniformity and Time Response of the Sensor

For the uniformity studies we pressed each sensor with a 30 g weight and took measurements immediately after touching and then after 30 sec and finally after no change in the reading occurred. These results are shown in Figure 4 for 16 sensors. The numbers on the graph denote percentage values relative to the maximum reading for each sensor under 30 g pressure changed in time. Another way of displaying the nonuniformity results is in Figure 5,

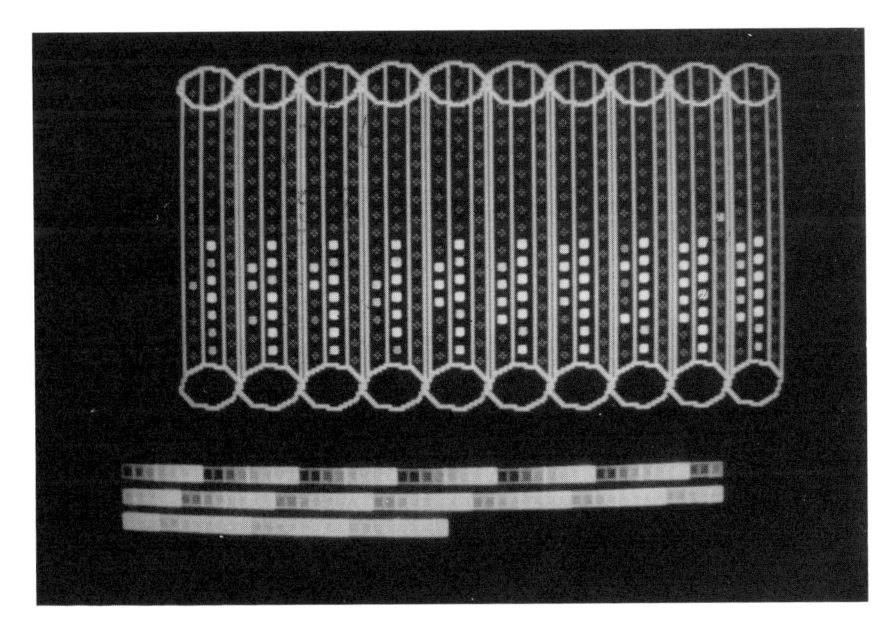

Figure 5. Graphical display of 10 frames of the French finger approaching a tested material.

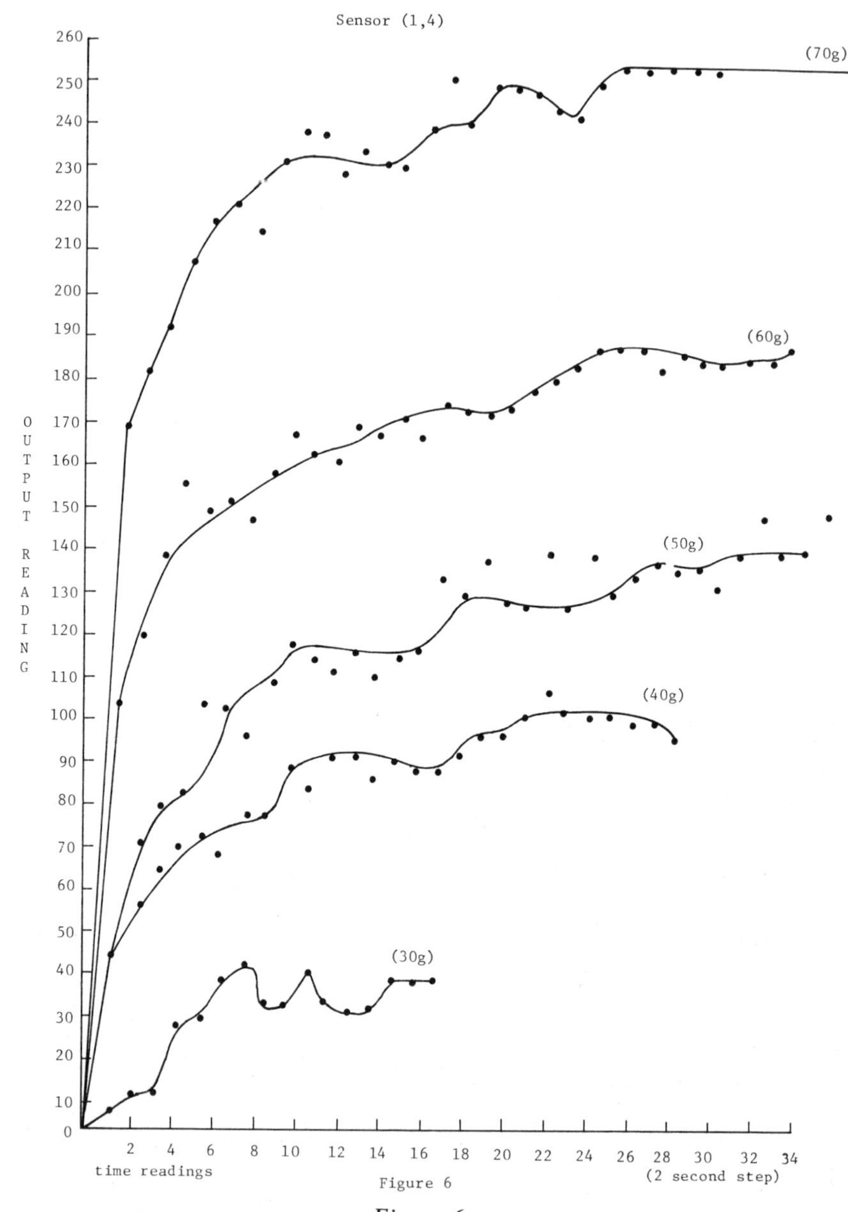
Sensor (1,4)
(70g)
(60g)
(50g)
(40g)
(30g)
OUTPUT READING
260
250
240
230
220
210
200
190
180
170
160
150
140
130
120
110
100
90
80
70
60
50
40
30
20
10
0
2 4 6 8 10 12 14 16 18 20 22 24 26 28 30 32 34
time readings
(2 second step)
Figure 6

Figure 6.

where we show 10 frames of what the sensor "feels" in approaching the felt material fixed to the board. The tip of the finger is not displayed. Note the different brightnesses on different sensors, illustrating the nonuniformity of the device.

The time response was also studied for different weights. An example of the time response of one typical sensor is shown in Figure 6.

We can also note that the sensitivity of sensors varies, but some show as less than 10 g.

3.4.2 *Experiments Testing Repeatability, Hysteresis, and Discriminability of Hardness of the Material*

The test materials used for this purpose were wood, felt, and foam. All these materials were fastened on a wooden board and positioned perpendicular to one of the finger's faces. The experiments consisted of moving the finger in steps 0.25 mm toward the surface, taking readings, and then moving away from it into the original position. The results of pressure versus the distance of the sensor to the surface are shown in Figure 7. In order to test the repeatability of the measurements, we have repeated the same experiments several times and found that the average difference is less than 10%—hence

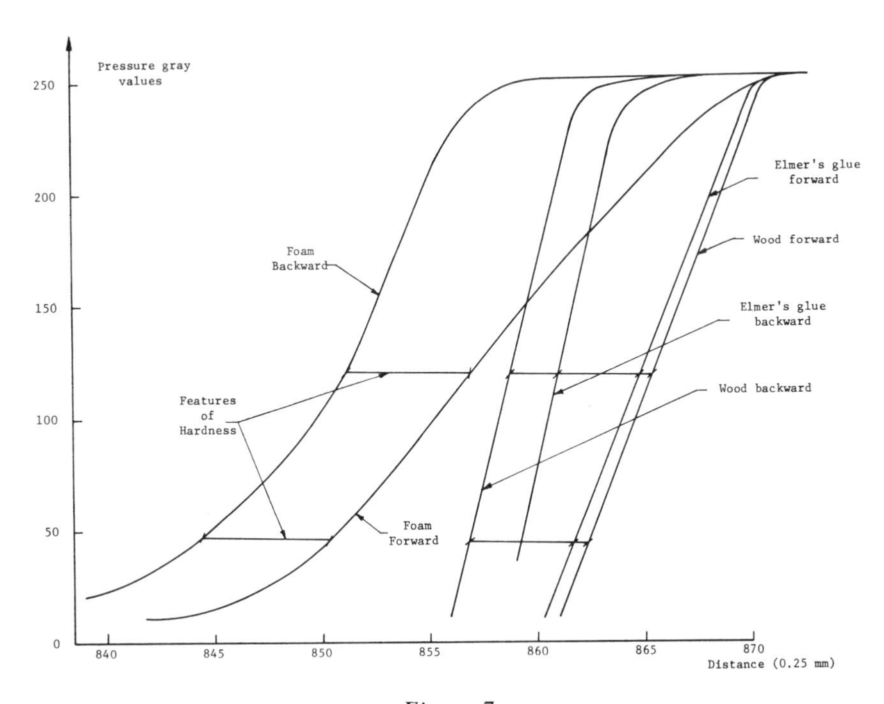

Figure 7.

Table 1.

material	*slope-forward*	*slope-back*	*width-@255*	*width-@125*	*width-@50*
wood	1.69	2.7	10	3.5	5.5
felt	1.08	1.2	10	6.5	7
foam	1.18	2.3	16	7.5	5.2

the measure of repeatability. The hysteresis curve is obtained by plotting the same values as before but when the finger is moving away from the surface. The measure of the hysteresis could be the area under the curve, the maximum average, and minimal width.

The hardness of the material can be characterized by the slope of the linear portion of the curve as measured in forward mode (moving toward the surface) and backward mode (moving away from the surface). The bigger the slope, the harder the material is. These features are tabulated in Table 1.

We did the same experiment as above with liquid material—Elmer's Glue, clay and foam, and, for comparison, hard material—wood—all in a plastic bag. The same features as in Table 1 are shown for these experiments in Table 2.

While the data from the liquid are not that different from the base board, the clay material shows an opposite hysteresis. That is to say, the hard material when pressed by the finger generates residual pressure on the foam of the finger, which causes reading of pressure even when the finger is being slowly removed from the hard surface. This behavior is consistent even when the materials are foam or felt. The opposite effect of the hystereses in the case of the clay we can explain as follows: When the finger presses the clay the clay moves under the pressure and stays that way, so that when the finger moves away there is less material and hence less pressure on the finger. This of course is not the case with the glue, which under pressure moves away, and the finger detects the pressure of the board. Once the finger starts to move away the glue returns to its original position, but it does not provide

Table 2.

material	*slope-forward*	*slope-back*	*width-@255*	*width-@125*	*width-@50*
board	2.66	6.1	1	4	2.5
wood	2.66	5	8	6.5	5.5
glue, light	2.66	5	5	4	3.5
glue, heavy	2.66	3	3.5	3.5	2
foam	1.14	2.3	13	6.5	6.5
clay*	1.6	2.5	12	10.75	13

* For clay the hysteresis is reverse that for the other materials.

sufficient pressure to counter the hysteresis from the finger's foam rubber. That is why we observe almost identical hystereses from Elmer's Glue in the light bag and from the base board. There is some difference of smaller hystereses of the same Elmer's Glue in a heavy elastic bag, which provides more force to compensate the hysteresis of the finger itself.

3.4.3 Texture Detection

The texture experiments involve the dynamic state of the finger; that is, the finger must be touching the surface and moving. So far the motion has been only translational and we have studied several surfaces: felt, cloth, wire grid (coarse texture), glue, and clay in a bag.

Again the experiments were conducted in some uniform fashion. That is, the finger was barely touching the examined surface on one of its faces. The examined surface was attached to a wooden board. The data were recorded at every reading, that is, every 5.6 ms.

The following experiments were performed:

The very coarse surface—a wire vertical grid with one horizontal bar. The results from four sensors are shown in Figure 8, and they indicate that the sensor detects the pattern without any doubt.

The medium coarse surface—a furniture cloth with medium rough texture. The corresponding results are in Figure 9. Here the pattern is not that regular and, if spectrally analyzed, has more higher frequencies than the previous one.

The fine-textured surface—felt. The tactile image is not shown because the sensor does not have sufficient resolution to detect the fine structure of the felt; hence this texture is classified as smooth. The clear distinction between this texture and the others is in the magnitude of the local variation in the surface which constitutes the fineness of the surface texture.

We have also experimented with foam, glue, and clay in a bag in the dynamic mode in order to learn about the surface texture of these elastic surfaces. Predictably, as long as the microsurface is sufficiently coarse to be detectable with the limited spatial resolution and pressure sensitivity, one obtains respectable results.

3.4.4 Measurements of Surface Normals and the "Follow" Program

For simplicity's sake we deal with polyhedral objects. The goal of this program is to follow an object while in constant contact and determine the surface normals of each face. If the plane equation is described as

$$(A(x - XO) + B(y - YO) + C(z - ZO) = 0,$$

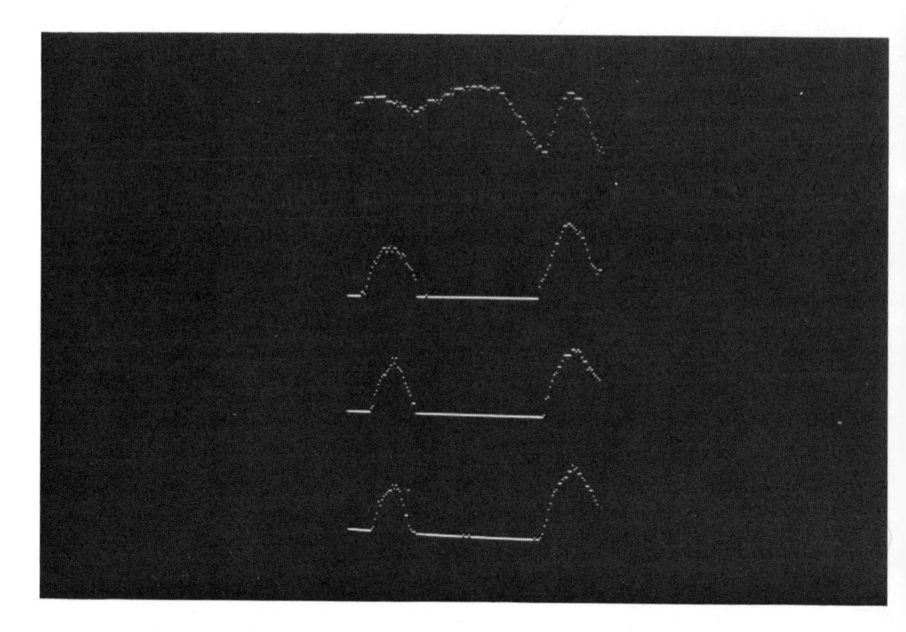

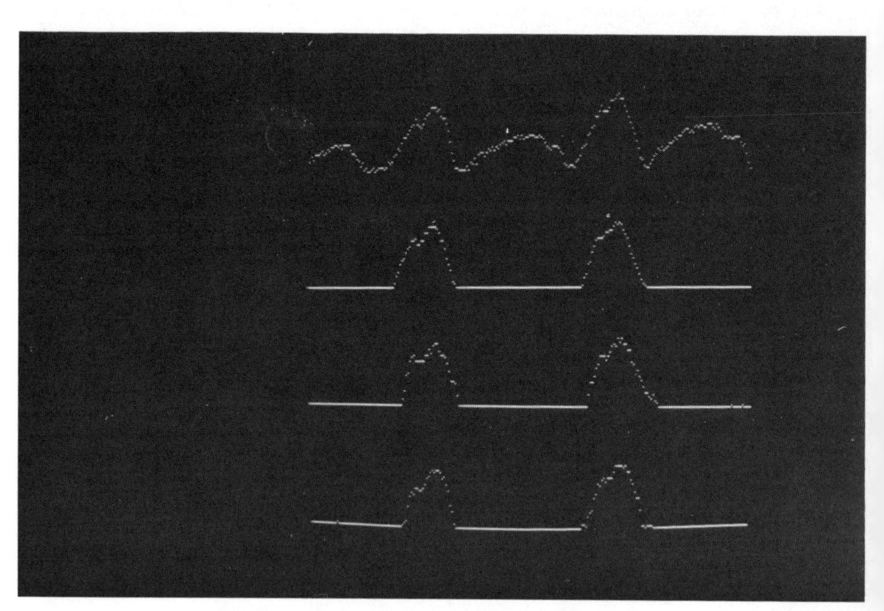

Figure 8. Display of what the four sensors of the French finger feel if dragged along a set of vertical wire bars.

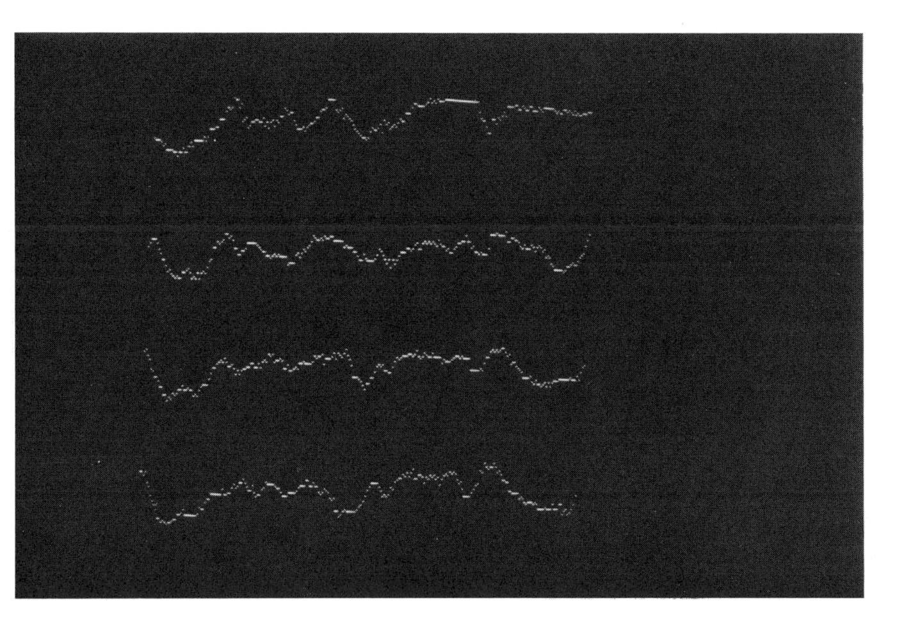

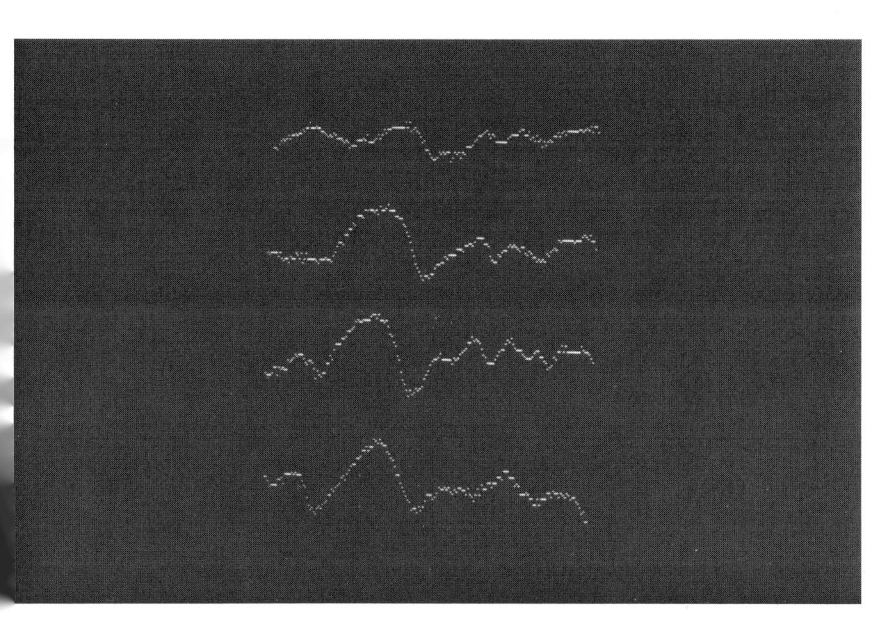

Figure 9. Display of four sensors of the French finger if dragged along a flat, coarse cloth surface.

and the variables XO, YO, and ZO are the coordinates in 3D space of a point on the plane. A vector parallel to the surface normal of this plane is given by $N = Ai + Bj + Ck$, where i, j, k are the respective unit vectors in the x, y, and z directions.

There are several methods to get the surface normals. For various reasons, we have chosen the following approach:

a. We have a prestored model of the finger (its geometry).
b. We look for a maximum pressure value among the active sensors, assuming that that sensor is in the most perpendicular touching position with respect to the examined face.
c. The actual position of the finger will be read from the positioning device.
d. The contact of the finger with the object is maintained by a stepwise search, that is, from the previous measurements follows incrementally first in the tangential direction and then in the normal direction (a kind of zigzag motion),
e. The first direction of the finger is user-determined.

The following conclusions can be drawn based on our measurements:

1. The sensors are not uniform, but this can be corrected by software.
2. The repeatability of measurments by both sensors is better than 10%.
3. There is hysteresis as documented in the analysis of the French finger. However, this hysteresis is not necessarily bad and it can be used as a parameter, as was shown in the case of the study of the elasticity property of the clay.
4. We observed some time dependency before the sensor achieves a stable value (30 g). This of course is the property of the foam and perhaps can be corrected by new materials.
5. In comparison, the sensitivity of the French finger was 10–30 g, which is about 50 times more sensitive than the Lord Corp. pad sensor.
6. The spatial resolution in both sensors is worse than we are used to from noncontact imaging devices. However, the detection of edges is unambiguous.

The results show that one can discriminate the hardness of material—for example, wood vs. foam or felt. We can also detect a certain elasticity by combining the pressure measurements and the positioning measurement. The viscous materials like glue and/or water can be detected by conjunction of pressure, position, and visual sensors (the visual detecting the motion of the liquid surface).

The texture measurements correspond to the surface roughness. Again, as the data show, the sensor properly images the coarse, medium coarse, and fine surface texture. Also, the physical edge is easily and unambiguously detected.

The program "follow" has shown that one can obtain surface information directly. From these data, it is not difficult to perform surface segmentation and later object recognition. Recently we have developed such a method [12], where using surface normals and the range information we can segment planar and quadratic convex and/or concave surfaces. An example from this work is shown in Figures 10(a)–(e). Figure 10(a) displays the original object; Figure 10(b) is its digitized version—that is, what the program sees. Figures 10(c)–(e) are the decomposed parts of the object except for the front plane, which has been recognized but not displayed.

The clear challenge in this area is the choice of the right primitives describing the physical properties of the objects and the development of algorithms that will extract them. The tactile sensation comes from the dynamic use of the sensor, that is, when the sensor is in touch with the surface and in motion. Hence, the mobility of the finger and/or having joints on the finger are crucial.

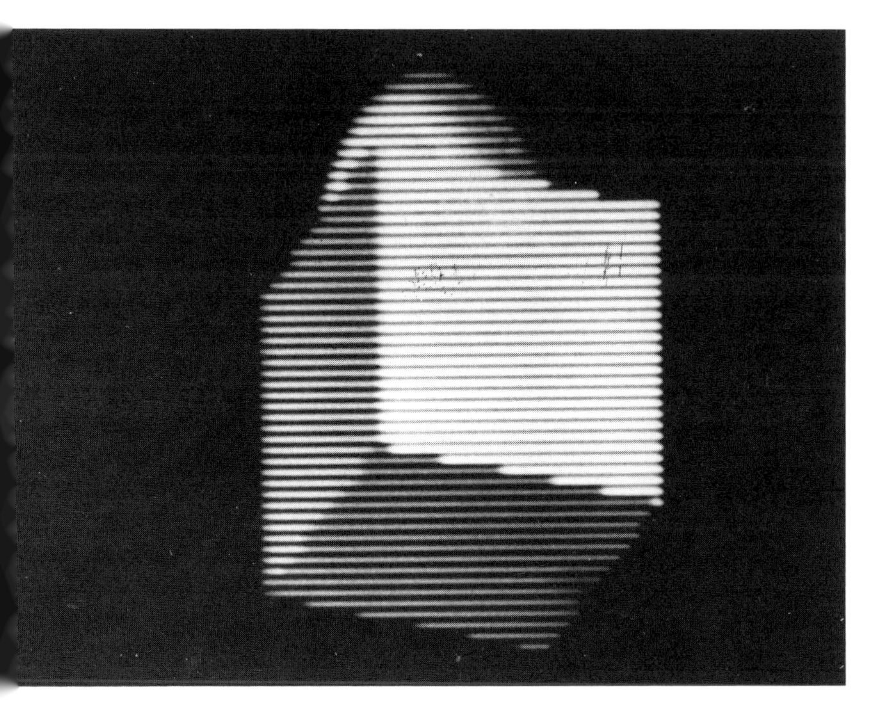

Figure 10(a). Original object

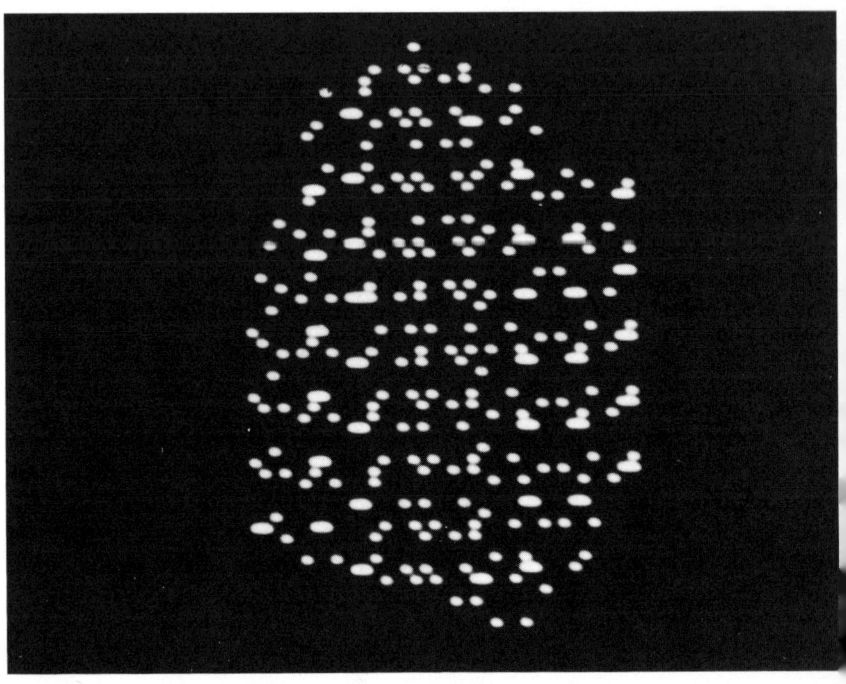

Figure 10(b). Digitized version.

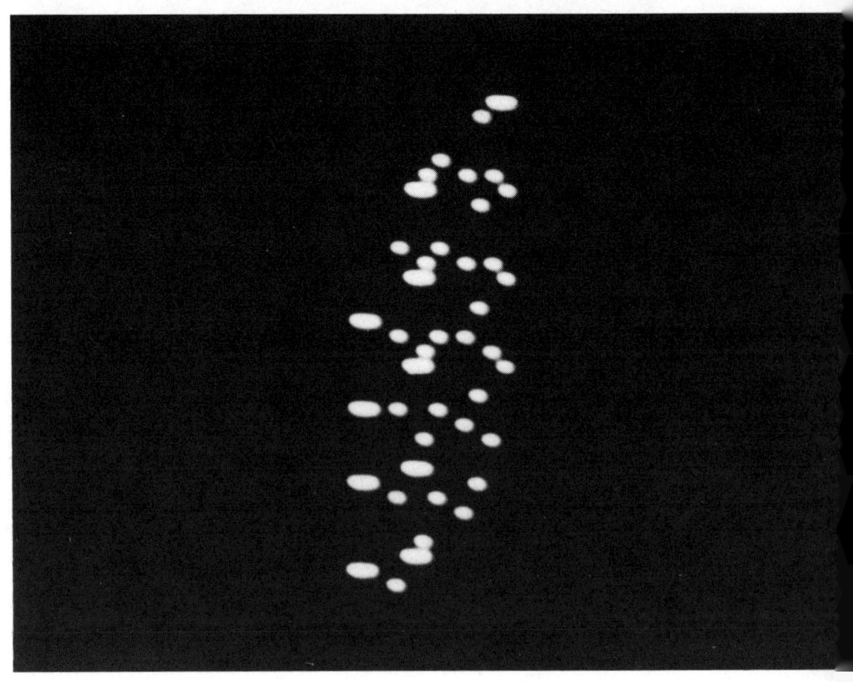

Figure 10(c). Side plane.

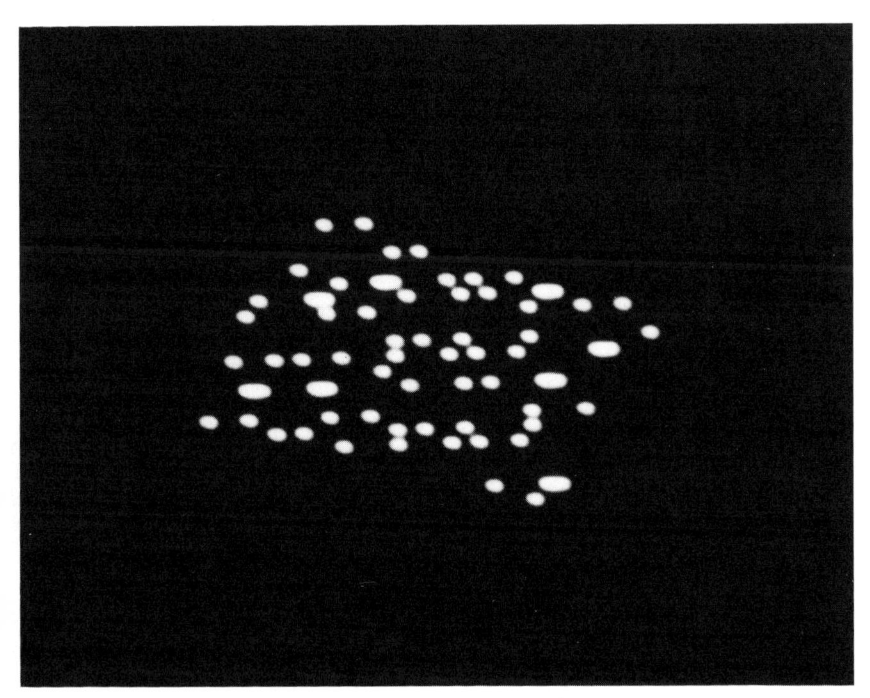

Figure 10(d). Bottom plane.

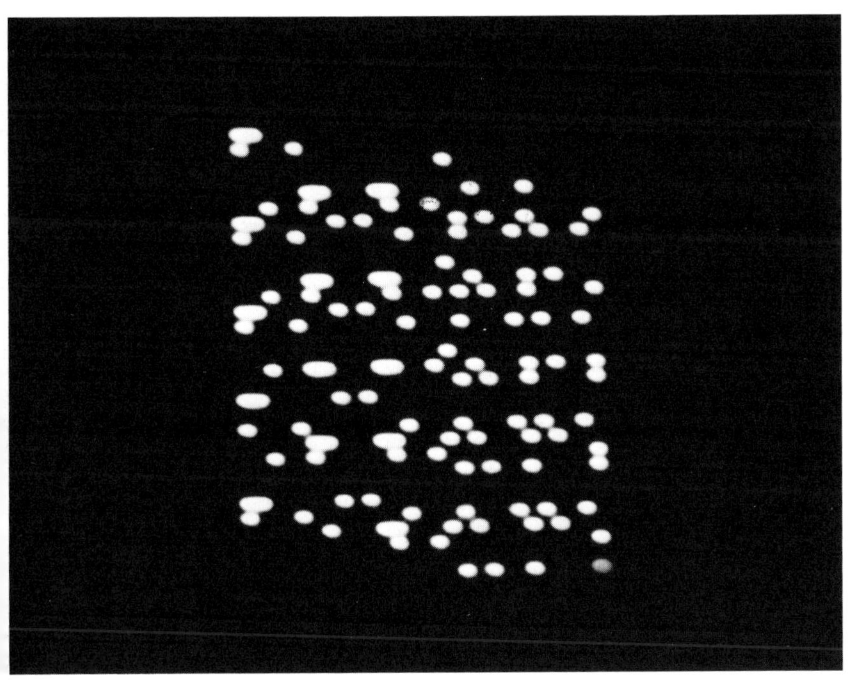

Figure 10(e). Spherical top.

4. TWO-FINGER SCENARIO

In this section we shall consider a paradigm of moderate complexity. First, we shall assume that the two rigid fingers are attached to a wrist which can rotate in 3D space and hence measure the angular position as opposed to the Cartesian positioning device in the one-finger scenario. Second, each finger has an array of tactile sensors, called pads. Third, we can record the thickness of the object from the expansion of the two fingers. The block diagram of this system is in Figure 11. As far as we know such a hand was first described by Ernst in 1962 [14]. The added complexity is in processing the second finger's tactile data and the distance measure. While we do not have yet in our laboratory a complete setup for this scenario (we do not have a manipulator with revolute joints), we have experimented with a pad sensor. So we shall report on those experiments and outline some algorithms for the use of the two fingers, each with a pad sensor.

4.1 The Pad Sensor

The pad sensor (courtesy of Lord Corp.), is a flat rubber square about $2\frac{1}{2}$ in. on a side (see Figure 12). An 8×8 grid of conical protrusions identifies the 64 pressure-sensitive sites. The pad is mounted on a square metal piece, about $3\frac{1}{2}$ in. on a side, which is in turn connected to another similar piece by four metal posts. These posts have strain gauges on them which measure the force parallel to the object's surface.

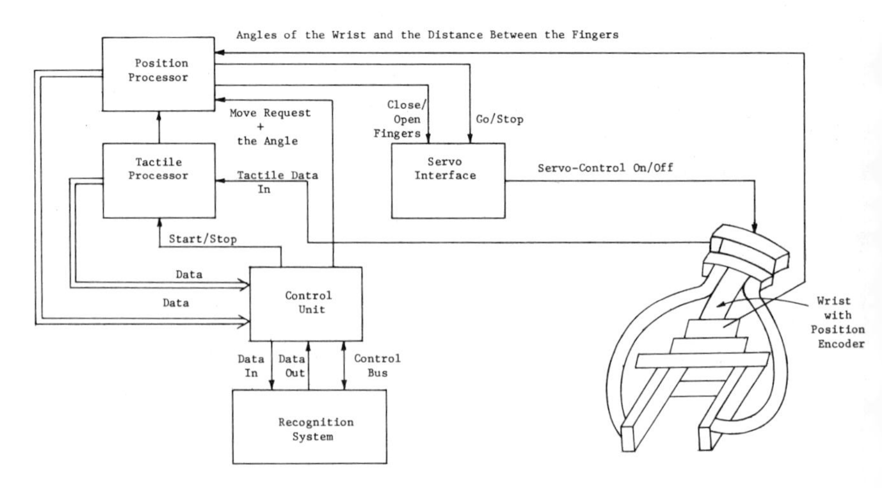

Figure 11. Two-finger scenario.

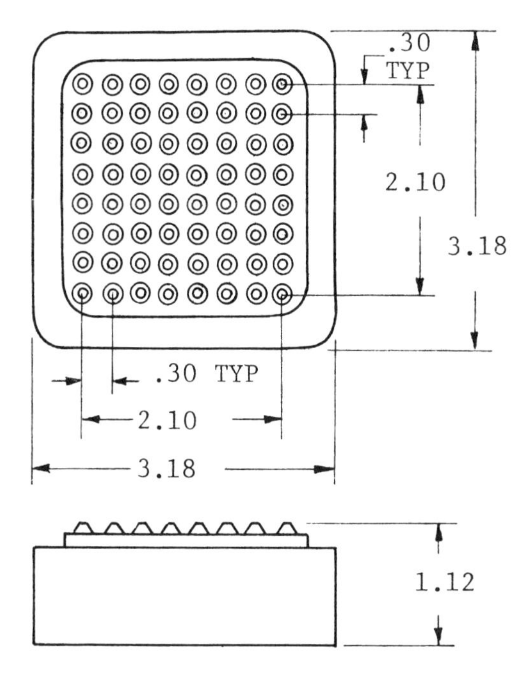

Figure 12. Lord Corp. pad sensor.

4.1.1 The Uniformity and Sensitivity of the Sensor

Each of the 64 pressure-sensitive sites puts out a slightly different range of voltage levels, hence the need for individual calibration. Each pressure-sensitive site requires roughly 1.3 lb of pressure to completely depress it. Multiplying that by 64 sites, we find that we need over 80 lb of pressure, which we did not have available. The solution was to depress each site individually, find the minima and maxima, and map all input data into a uniform range of 0–255. Because of the low sensitivity of this device we could not test the sensor for discriminability of hardness of the material. The only sensible task was straightforward detecting of tactile edges—hence the static tactile image analysis.

4.1.2 Static Tactile Image Analysis

Two different experiments were conducted, one detecting a complete boundary when the object is smaller than the array of the sensor, the other detecting a set of grooves in a surface.

The common thread in all the following experiments is that the data were taken in a snapshot fashion.

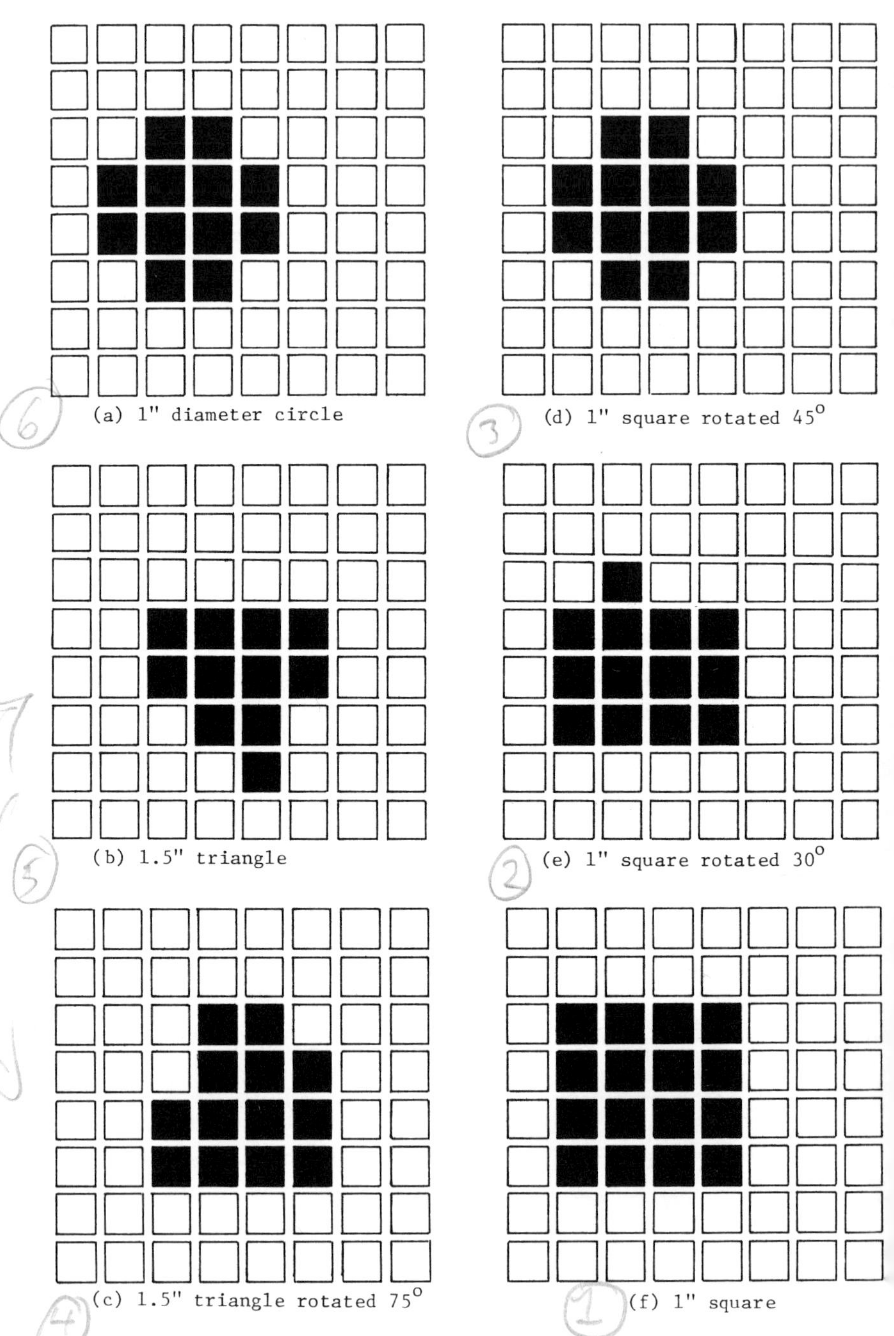

Figure 13. What the pad sensor feels.

In Figure 13(f) we used a 1-in. square, set off center but oriented orthogonally with the sensor grid axes. There is no question as to the identity of that object. A simple threshold operation would clearly distinguish it from the background.

Figures 13(e) and 13(d) show the same square rotated counterclockwise 30° and 45°, respectively. Figure 13(c) shows an equilateral triangle, point downward, and 13(b) depicts the same triangle rotated clockwise about 75°. Notice how some pixels are much lighter than others in the image with non-orthogonal edges. This phenomenon arises when the object covers less than half the area of a site. Since the site is conical in shape, the edge must be pressing on the wall of the cone. It cannot depress the cone as far as it could if it were pressing on the apex.

Finally, Figure 13(a) shows a 1-in.-diameter circle. Notice that it appears to be identical to the square in Figure 13(d). This is a question of resolution.

The next experiment dealt with an object larger than the sensor. For this we used a flat surface about 12 in. long and 3 in. wide. A set of eight grooves were cut into this surface in order to form a pattern of diverging lines. [See Figure 14(a).]

We took 50 images, stepping about 5 mm between each. The reconstruction, shown in Figure 14(b), was accomplished by superimposing the images in the appropriate positions relative to each other. Again we see the problem of the spatial resolution. We tried to improve the spatial resolution by taking three snapshots, 4 mm apart widthwise, for each of the 50 steps lengthwise. The reconstruction, Figure 14(c), shows that while angled edges are clearer at lower frequencies than in Figure 14(b), at higher frequencies this method was ineffective.

Finally, we measured the angle between the face of the pad and the touched plane. This is important for knowing whether the robot hand has a "good grip." For this experiment we used the 1-in. square as our test object. We took four snapshots. In the first image we laid the pad sensor flat on the square, as usual giving us a zero-degree standard. For the three subsequent images, we lowered the left end of the table by 1.0, 1.25, and 1.5 in., respectively, producing angles of 3.3°, 4.1°, and 4.9°

The results are shown in Table 3. For each image we arrived at a single number describing the slant. The number was calculated simply by averaging all the pressure differences between horizontally adjacent sites. However, we had to correct these numbers by a factor, which accounts for the fact that the first image exhibits a small slant value as opposed to the expected zero. The finally corrected ratios differ from the expected values by less than 2%!

4.2 Shape Analysis Using Two Padded Tanglike Grippers

The best way to state the problem is to refer to Figure 11. This is our assumed configuration: a flexible wrist with two padded tonglike grippers.

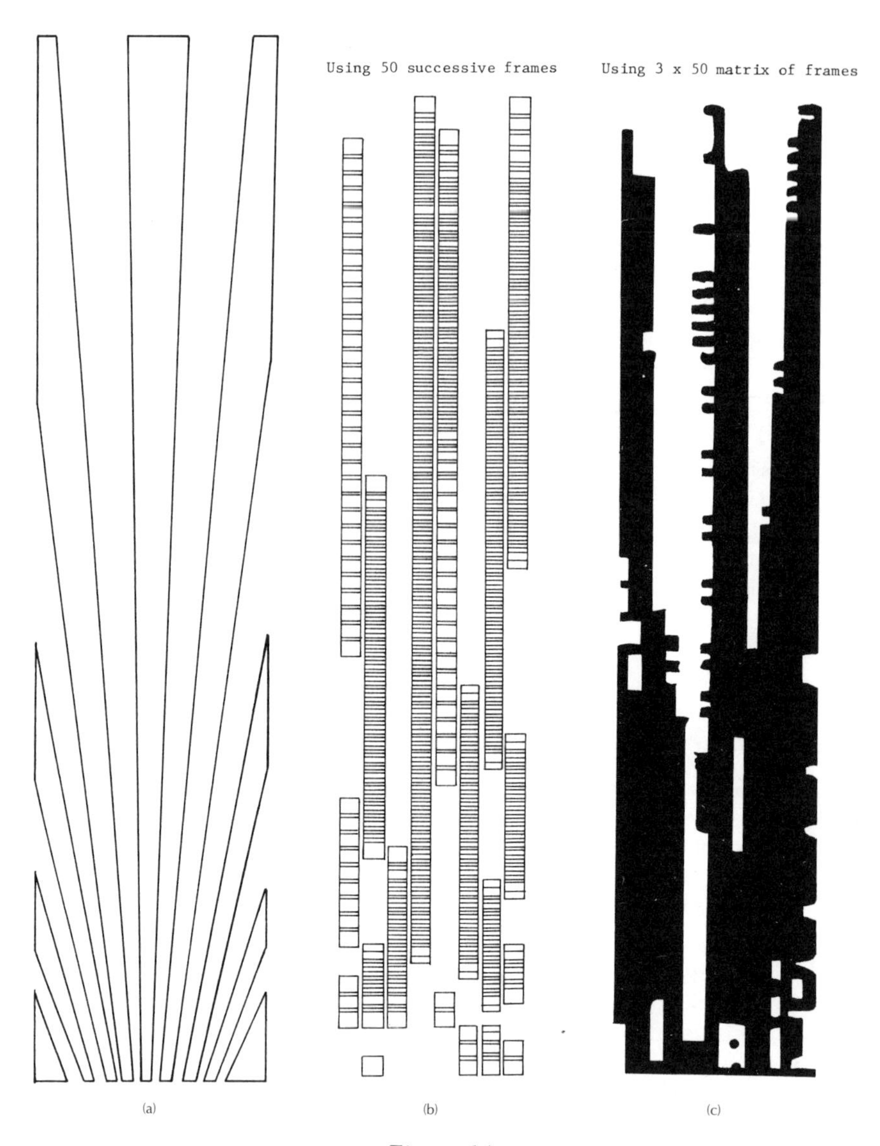

Figure 14.

There are two problems: One is to establish the angle of the gripper while in contact with the examined object, and the other is to examine the contact points and use them for shape and/or surface description.

First we have to start with discussion of how to establish a constraint of an object via points of contact between the gripper and the object, since this

Table 3. Measurement of Small Angles

Table Slant	*Data*			*Horizontal Difference*		*Average Difference*	*Ratio**
0″	45	64	80	19	16	12.625	
	42	48	64	6	16		
	48	60	75	12	15		
	34	56	51	22	−5		
1″	15	64	160	49	96	78	1.00
	28	80	192	52	112		
	16	75	195	59	120		
	17	85	153	68	68		
1.25″		48	160		112	114	1.23
		48	192		144		
		60	180		120		
		56	136		80		
1.5″		48	160		112	142	1.53
		48	240		192		
		60	225		165		
		70	170		99		

* Ratio is calculated as the vertical average divided by the vertical average at 1″ slant, multiplied by one minus the ratio of the 1″ slant to the 0″ slant. The closer this value is to the table slant, the better the results. As the reader can see, the results are exceedingly good.

is our case of the gripper in action. Here we follow the methodology introduced by Salisbury in Ref. [48]. Each contact between the grasped object and the gripper reduces the body's freedom of motion and allows forces to be applied basically in three ways:

1. Point contact (point on line and/or on plane)
2. Line contact (line on line and/or plane)
3. Plane contact (plane on plane)

Each contact type can be classified into one of five categories according to the relative freedom of motion between the bodies. A one-degree-of-freedom contact, for example, implies that only one parameter is needed to specify the relative motion of the bodies if contact is maintained; two-degrees-of-freedom contact needs two parameters; and so on. The relative freedom of motion is of course dependent on whether we consider the frictional forces or not. Below is a table summarizing this fact:

Contact	Degree of Freedom: Without Friction	Degree of Freedom: With Friction
Point contact	5	3
Line contact	4	1
Plane contact	3	0

For the analysis of stable contacts, it is customary to use a model of generalized forces, hence the following definitions. In particular we shall introduce three concepts: a screw, a wrench, and a twist. They all are based on line geometry. A *screw* is defined by a straight line in space known as its *axis* and an associated *pitch* p. A screw may be described with a 6-vector of screw coordinates $\mathbf{s} = (S_1, S_2, S_3, S_4, S_5, S_6)$ with the following interpretations. The plucker line coordinates of the axis are

$$L = S_1$$
$$M = S_2$$
$$N = S_3$$
$$P = S_4 - pS_1$$
$$Q = S_5 - pS_2$$
$$R = S_6 - pS_3$$

L, M, and N are proportional to the direction cosines of the line forming the axis while P, Q, and R are proportional to the moments of the line about the origin of the reference frame.

With any particular infinitesimal motion or velocity of a body in space there is associated a unique line, the twist axis, about which the body rotates and along which it translates. We call this motion a *twist* and identify it with a 6-vector of twist coordinates $\mathbf{t} = (T_1, T_2, T_3, T_4, T_5, T_6)$ with an interpretation analogous to the above.

Finally, with any particular set of forces and moments applied to a rigid body there are associated a unique line known as the *wrench axis*, a pitch p, and a magnitude. The set of forces and moments acting on the body is equivalent to a single force acting along this wrench axis and a moment exerted about the axis. A wrench is identified with a 6-vector of wrench coordinates $\mathbf{w} = (W_1, W_2, W_3, W_4, W_5, W_6)$, again with an interpretation analogous to the above.

The motion and force approaches are equivalent ways of looking at the same information. If s_i is the system of screws about which the body may execute twists, then the reciprocal screw system, designated s_i', is the system of screws about which wrenches may be applied to the body. To immobilize a body completely it is necessary and sufficient that the intersection of all the contact twist systems be the null set:

$$s_1\ s_2 \cdots s_n = \phi$$

Alternatively,

$$s_1'\ s_2' \cdots s_n' = 6$$

The latter equation states that an arbitrary wrench may be applied to a body thus immobilizing it if the union of all contact wrench systems comprises the full 6-space of all wrenches.

Now we can analyze the cases which could occur with the two-rigid-finger scenario. Let us assume that all the contacts are with frictions. The contact types could be as follows:

1. Each finger has one point contact.
2. One finger touches in one point and the other in line.
3. Both fingers touch the object in line.
4. One finger has a line contact and the other a plane contact.
5. Both fingers have a planar contact.

The first case is unstable since at least one wrench (force) w_1 is in a direction opposite to that of its counterpart, w_2; hence the union of the wrench system will be only a 5-system, which violates the equation. All the remaining cases are stable contacts.

Once we have established the stable contacts and we assume an initial position (x_0, y_0, z_0) of the gripper, there are two more steps left to complete the data acquisition task:

Step 1: From the pressure readings find that position of the gripper which produces local maximum, hence making sure that the pads of the gripper are making tangential contact.

Step 2: Read out the angular position of the wrist and the distance between the two fingers. These two pieces of data can be plugged into transformation, which will compute the contacted points on the object in the object centered coordinate system or in other independent reference coordinate system.

If one chooses roll, pitch, and yaw as angles that are measured and a as the distance from the midpoint between the fingers and the inner side of one finger, then the new point, denoted by PNEW, is computed as follows:

$$\text{PNEW} = \text{RPY}[\text{Roll, Pitch, Yaw}] \times E(d, a) \times \text{POLD}$$

where the multiplication is matrix multiplication; $E(d, a)$ is a matrix (4×4); d is the distance between the wrist and the fingertip, and a is the distance as described above; RPY is again a matrix (4×4) with the above angles, and the exact form is in Ref. [40]; and POLD and PNEW are vectors of the form $(x, y, z, 1)$.

4.3 Acquisition of Local Surface Information Using Two Fingers

The second part of this problem is how to utilize the contact points for more detailed surface description. Here we assume that by the constraint of the gripper (it has a planar surface but slightly elastic) and by finding locally the largest contact surface as described in step 1 above we guarantee that we are indeed measuring the surface normals. Naturally, if there is only a point contact, then we have only the surface normal available at the contact point. The more interesting problem is when there is line or planar contact. Then from the gray values of the pressure measurement one can compute the local curvature: The curvature is computed by fitting a circle into three points in several possible orientations. In Figure 15 we show two simplified cases, when the curvature needs to be computed only in two directions. Figure 15(a) is an example when the Gaussian curvature is positive (the maximal and minimal curvatures are identical), and Figure 15(b) is an example when the Gaussian curvature is zero, since the maximal curvature is the same as in Figure 15(a) but the minimal curvature is zero—hence the result is zero.

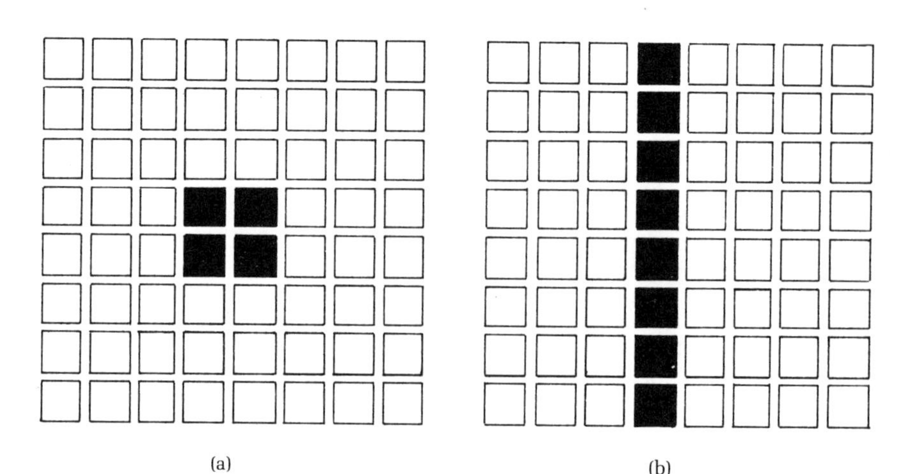

Figure 15.

The latter is the case of a cylindrical surface. In general, we shall choose the maximal and minimal curvatures as descriptors of the local surface.

4.4 Surface Description Using Two Fingers

The last part of this section discusses the acquisition of local shape information by multiple touching of the object, as shown in Figure 16. Here we are faced with a problem of how to extrapolate global shape from local measurements. Soroka [54] investigated this problem for visual data when one has local measurements in the form of slices through the object. Now assume a hypothesis verification paradigm, where given the two slices one hypothesizes a cylindrical volume. The next step is to evaluate this hypothesized object for how well it fits the data—that is, the verification phase—which prompts to take the next slice, and so forth. An example of such a result is shown in Figure 17, where the sequence from left to right represents the original figure of a *Y*, then the hypothesized object after two iterations, and finally the reconstructed *Y*. In our applications the slices would be another position of the hand—that is, another measurement. This method may resemble that one reported in Ref. [38] except that we only locally fit and/or reconstruct volume primitives and hence preserve the local properties such as size and position as well as its structure, as in the part–whole relationship.

In conclusion, in this scenario:

a. We can compute the gross approach angle from the wrist-positioning device.
b. From the tactile sensors, we can compute the local curvature of the two grasped surfaces and hence classify them into concave, convex, planar, and so on.

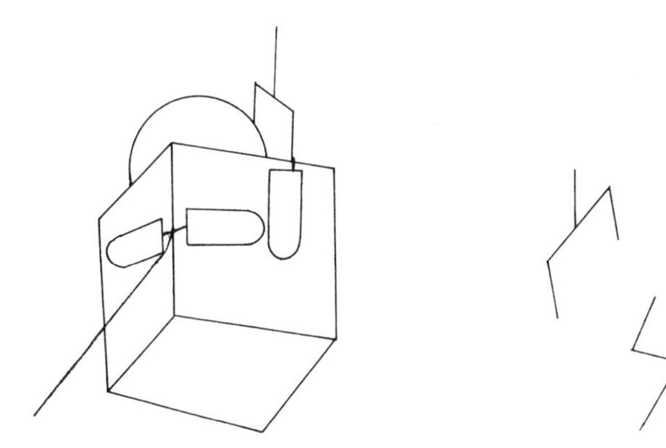

Figure 16.

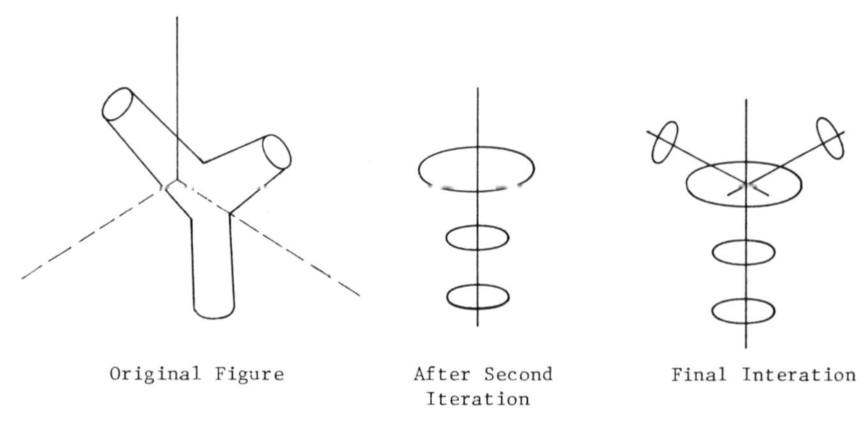

Figure 17.

c. From the multiple touching efforts which constitute steps a and b, we can extrapolate at least the gross shape descriptions.

5. THE HAND SCENARIO

In this section we shall first present a short review of the anatomy of the human hand as the best model of a flexible manipulator. Then we shall examine design considerations for a dextrous hand, and finally list some applications—specifically, how the currently available hands are used for pattern recognition tasks.

5.1 The Anatomy of the Human Hand

Following the paper of Taylor and Schwarz [57], the anatomy of the human hand can be divided into bone structures, muscles, tendons, and the skin as a multiple sensor.

5.1.1 The Bone Structures

The wrist and the root of the hand are composed of eight bones called the *carpus*. The digits are composed of metacarpal and phalangeal segments. The first, second, and third phalangeal bones comprise the fingers (see Figure 18). The carpal bones are arranged in two rows. Metacarpals II through V articulate so closely with the adjacent carpal bones of the distal row that although they are capable of some flexion and extension, independence of motion is limited. The metacarpal shafts are arched to form the palm. The metacarpophalangeal joints are similar to the interphalangeal joints in that the "virtual center of rotation" lies approximately at the center of curva-

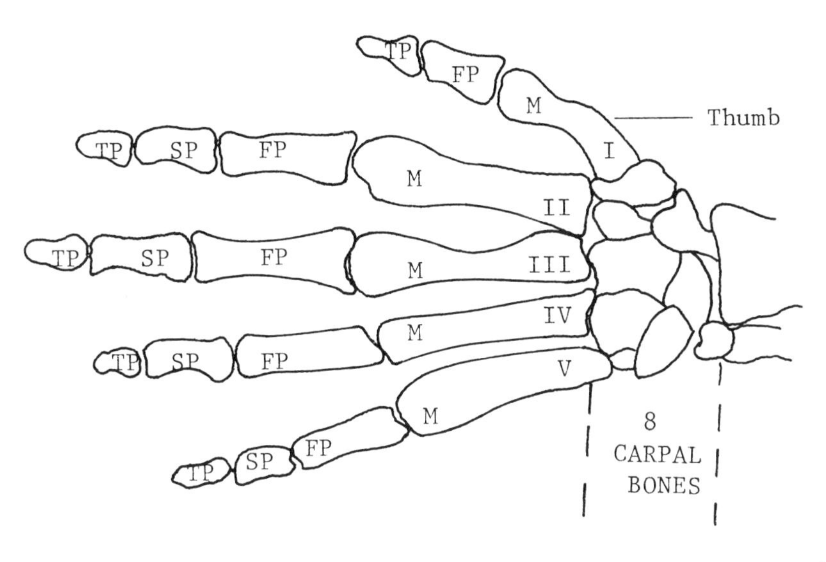

The fingers are enumerated by Roman Numerals.
M denotes Metacarpal Segments
FP denotes Phalangeal Segment
SP denotes Second Phalangeal Segment
TP denotes Third Phalangeal Segment

Figure 18.

ture of the distal end of the proximal members (see Figure 19). The lateral "aspects" of the joint surfaces are narrowed and are closely bound with ligaments so that lateral rotation is small in the metacarpophalangeal joints and lacking entirely in the phalangeal articulations. Hence the latter are typical hinge joints (revolute). The thumb differs from the other digits in that the second phalanx is missing and the thumb enjoys greater mobility in the carpometacarpal articulation.

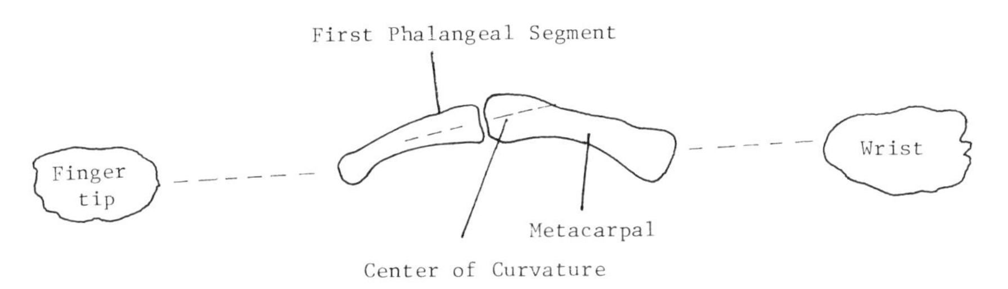

Figure 19.

5.1.2 *Muscles and Tendons*

Most of the muscles of the hand and wrist lie in the forearm and narrow into tendons, traversing the wrist to reach in sections the bony or ligamentous components of the hand. There are one flexor and one extensor for most joints. Because of the way the tendons course down the forearm, the flexors serve partly as supinators (i.e., allow the forearm to be rotated along the long axis so that the palm will be forced up); and the extensors serve partly as pronators (do the opposite of the above). Some tendons are interconnected. The tendons of the wrist and hand pass through bony and ligamentous guide systems. Some muscles are located in the hand itself. These muscles are, with the exception of the abductors (which move fingers away from other fingers) of the thumb and little finger, specialized for the adduction (bringing the fingers together) of the digits and for opposition patterns such as making a fist.

5.1.3 *The Palmar and Digital Pads and Dorsal Skin*

It is well known that the palm and digits are covered with a thick layer of fat and a relatively thick skin arranged in a series of folds allowing "give and take" for prehension. The folds are disposed in such a way as to make for security of grasp, while the underlying fat furnishes padding for greater firmness in holding. The folds are tightly secured to the skeletal elements by white fibrillose tissue. The upper side of the hand (with or without hair) has skin which yields easily under tension. In the healthy hand the degree of redundancy (the wrinkles being minute redundancies in the skin) in any given area is just such that all wrinkles are dispatched when the fist is clenched.

5.1.4 *Resting Hand Patterns*

The resting hand assumes a characteristic posture—"an ancestral position ready for grasping limbs." The resting wrist takes a mid-position in which it is bent 35°, as shown in Figure 20. This position is one of the greatest prehensive forces.

5.1.5 *Prehension (Grasping) Patterns*

The six fundamental prehensive patterns of the human hand are cylindrical (as in holding a cylinder), hook (as in carrying a suitcase), lateral (as in presenting a train ticket to a conductor), spherical (as in grasping a softball), tip (as in sorting minuscule objects with the fingertip. Of these, three main patterns predominate: palmar, tip, and lateral. For picking up and holding

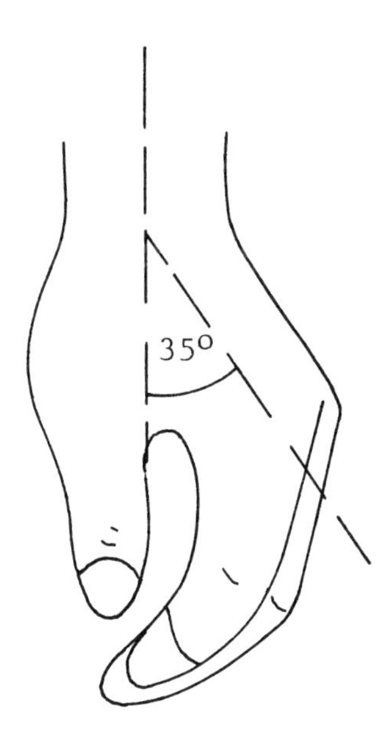

Figure 20. Resting position of a hand.

the following data are given:

Occurrence of Prehension Type

	Palmar%	*Tip%*	*Lateral%*
Pick up	50	17	33
Hold for use	88	2	10

5.1.6 Mechanical–anatomic Basis of Prehension Patterns

The fixation of carpal and metacarpal segments by contraction of flexor and extensor carpi muscles provides a firm base for independent movements and fixations of the phalangeal segments. Coordinated action between extensor and flexor groups permits fixed intermediate positions of each segment of the system. Separate flexor muscle groups attach to the second and terminal phalanges (not the first, apparently). Likewise the counterbalancing digital extensor inserts into these two most distal phalanges and on contraction rigidly extends the entire finger. It must not be forgotten that some of the intrinsic muscles of the hand (those totally located within the hand) attach to the first phalangeal segment.

The versatility of the thumb is due to variety of flexion–extension patterns which are also exhibited by the other four fingers. The thumb is unique in its mobility due to the relative mobility of its carpometacarpal joint, which allows the thumb to act in any plane necessary to oppose the digits.

5.1.7 Hand Movements

In all the types of prehension described, the hand "assumes" a fixed position. If the grasped object is hard, reactions to the hand's flexion forces are afforded by the object. Otherwise (in the case of fragile objects) the hand shape is maintained by contractions of the opposing muscle groups. In movements ranging from slow to rapid, with control of direction, intensity, and rate, there is always some degree of contraction to ensure control and so permit changes in force and velocity which allow the movement. Many familiar activities such as writing, sewing, and playing musical instruments fall into this category of "slow and rapid" movements. Ballistic movements are rapid and usually repetitive, in which active muscular contractions begin the movement, giving momentum to the member, but diminish their activity throughout the latter part of the motion. There is evidence that repetitive finger motions are more fatiguing, less accurate, and slower than motions of the forearm. Consequently, in repetitive finger activities in which there is a ballistic element, such as piano playing, wrist and elbow motions predominate, while the fingers merely position themselves to strike the keys.

5.2 Design Considerations for the Mechanical Hand

There are several aspects of the design of a mechanical hand that we shall review here, with the focus on the global task: to produce a highly versatile hand with a minimum number of moving parts, a dependable drive system, and an optimum number of degrees of freedom (D.O.F.). The hand will be an assembly of motors, sensors, and mechanisms (fingers). We start with the fingers.

Loosely speaking, fingers should have one or more bending sections (links). Each finger link should be a component of a closed linkage connected with other links via joints which can "drive" or rotate the link. Fingers should be mechanically identical and substantially attached to the hand's base. The fingers should be able to approach, contact, or pass one another during prehensile operation. There are two basic joints and two common but combination joints [45]:

- *Revolute* (R).
- *Prismatic* (P).
- *Cylindrical joint* (C) is a combination of a coaxial R and P joint.
- *Spherical joint* (S), which can be formed by a ball and a socket and are denoted RRR joints or three noncoplanar independently powered R joints.

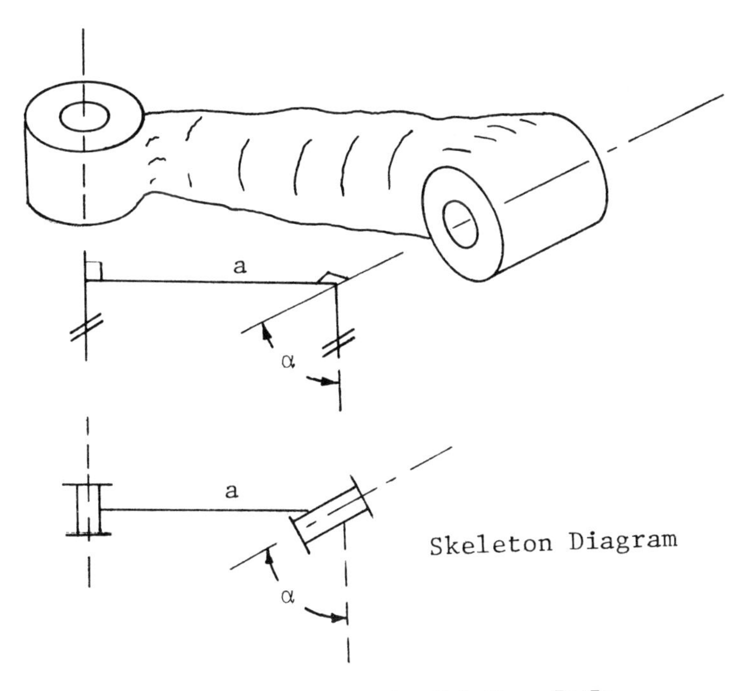

Figure 21. A connecting link (from [44]).

The connecting link may be summarized by *a*, the shortest distance between the axis, and "alpha," the twist in the link; that is, the angle between the axes in a plane perpendicular to *a* (see Figure 21). Generally, two links are connected at each joint axis. The relative positions of two such links, each on a separate normal to the axis, are given by their distance *s* along the common axis P202 and their angle "theta" measured in a plane normal to this axis (see Figure 22). Hence there are four parameters: *a*, alpha, *s*, and theta per link. Here *a*, alpha determines link structure; *s*, theta determines relative position.

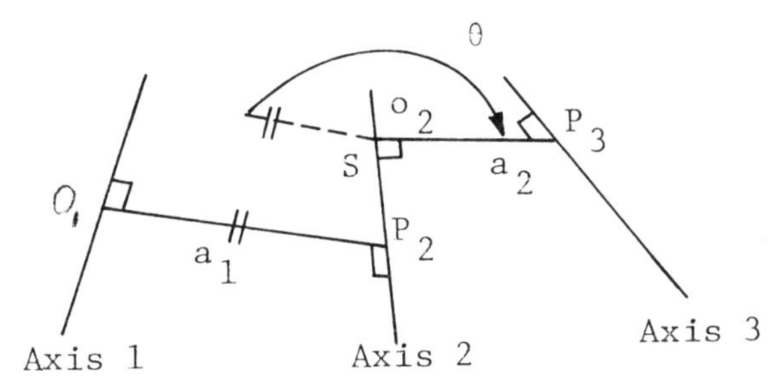

Figure 22. Two links (from [44]).

The linkages are mechanisms that have been studied in the past, for example in Ref. [39]. The mechanisms studied to bend the fingers are four-bar cross-chains, miniature compound pulleys, four-bar chains with an expanding link (for example, alternate screw threads), and tension cables (without pulleys).

Cross-bar chains are dependable, and easily built and can transmit an angular displacement w from one chain to the next, producing a rotation in a continuously compounding manner.

Miniature compound pulleys develop a high mechanical advantage and allow the finger links to bend through large angles. However, straightening a bent finger is difficult.

Expanding-link four-bar chains can be driven open and closed easily with a high mechanical advantage and reasonably low losses. This is a complex drive mechanism which requires a translating torque transmission or an expanding piston-cylinder motor in each linkage.

Tension cables flex a finger or its joints by simple direct contact, and very little space is required. To achieve high prehensive forces, high cable tension must be maintained.

The optimum degrees of freedom (D.O.F.) as it relates to the flexibility of the grasping as well as the imposed constraint on the object during grasping is as follows.

The number of D.O.F. is optimum when it is estimated that the manipulator can grasp all the basic geometrical shapes from any aspect with minimum number of external control inputs. The basic shapes under considerations are rectangular and triangular prisms, spheres and cylinders, screws, rods, and the like.

In addition to the requirement of approachability of the mechanical hand from arbitrary position during grasping, it is necessary that the hand can constrain the grasped object. Hence, the optimum number of links per finger would be the minimum number necessary to achieve this constraint.

Overconstraint would mean that a gripped object could be located in several positions within the hand between various constraint points. For example, the simplest finger is a rigid bar which bends about a single revolute joint. Additional links and revolute joints can be added. Each finger would have 1 D.O.F. of each finger link. Finger link joints can also be a combination of two revolute joints with axes 90° apart to allow "universal" bending (now 2 D.O.F. per joint). The remaining lower pair is the spherical joint (3 D.O.F. per joint). The finger now becomes so complex that great care must be given to the design. The additional third D.O.F. does not add to the finger's versability because the links are symmetrical, and revolution of the link about the central axis connecting both spherical joints does not produce any useful effect. Salisbury [48] has studied this problem more formally, and we have already presented some of his findings in the previous section. Here we wish to cite only those parts that are pertinent to his three-fingered hand.

The question is: What should be the hand design from the stable contact point of view? Salisbury took the approach of enumerating possible configurations. The assumptions are (1) it does not matter where on the link the contact occurs; (2) ignore the link dimensions; (3) the number of D.O.F. allowed by each connection between links is important. Thus a single link of a finger may contact an object with 0 through 6 D.O.F. in a total of seven ways. A finger with three links, for example, can touch an object in $7 \times 7 \times 7$ unique ways. For a three-finger hand with three links on each finger the number of possible grasps is 6,784,540 [48]. With three links per finger each finger can touch the object in one of the eight configurations shown in Figure 23.

Additional requirements can be spelled out for a well-designed hand:

1. Exert arbitrary forces or impress arbitrary small motions on the grasped object when the joints are allowed to move.
2. Constrain a grasped object by fixing (locking) all the joints.

Requirement 1 means that it is necessary for the connectivity C between the grasped object and the palm to be 6 with the joints active. Requirement 2 dictates that with the finger joints locked the new connectivity C' must be less than or equal to 0. This is a necessary condition for a grasped object to be completely constrainable at will.

In Table 4 we present 39 possible designs (from Salisbury's thesis). The configuration code refers to Figure 24. The joints are the total number of

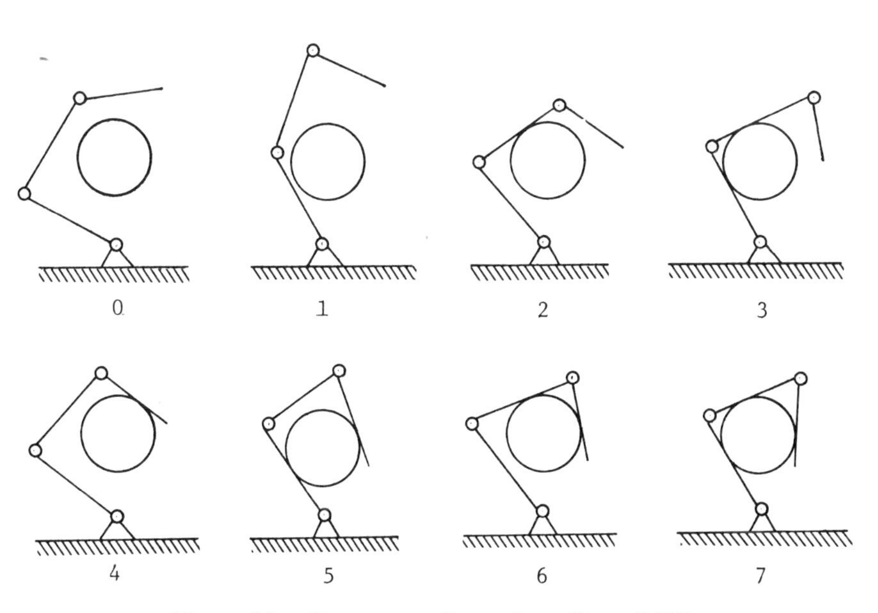

Figure 23. Contact configurations (from [48]).

Table 4. Mobilities and Connectivities for Acceptable Hand Mechanisms (from [48]).

0, 1, 2 FREEDOMS PER CONTACT:

. . . 0 . . .

3 FREEDOMS PER CONTACT:

Configuration	Joints	M/M′	C/C′
4-4-0	6	6/0	6/0
4-4-4	9	6/−3	6/−3

4 FREEDOMS PER CONTACT:

Configuration	Joints	M/M′	C/C′
2-2-2	6	6/0	6/0
4-2-2	7	7/0	6/0
4-4-2	8	8/0	6/0
4-4-4	9	9/0	6/0

5 FREEDOMS PER CONTACT:

Configuration	Joints	M/M′	C/C′
7-7-0	6	6/0	6/0
7-3-1	6	6/0	6/0
7-5-1	7	7/0	6/0
7-6-1	7	7/0	6/0
7-7-1	7	6/−1	6/−1
7-3-2	7	7/0	6/0
7-5-2	8	8/0	6/0
7-6-2	8	8/0	6/0
7-7-2	8	7/−1	6/−1
3-3-3	6	6/0	6/0
5-3-3	7	7/0	6/0
6-3-3	7	7/0	6/0
7-3-3	7	6/−1	6/−1
7-4-3	8	8/0	6/0
5-5-3	8	8/0	6/0
6-5-3	8	8/0	6/0
7-5-3	8	7/−1	6/−1
6-6-3	8	8/0	6/0
7-6-3	8	7/−1	6/−1
7-7-3	8	6/−2	6/−2
7-5-4	9	9/0	6/0
7-6-4	9	9/0	6/0

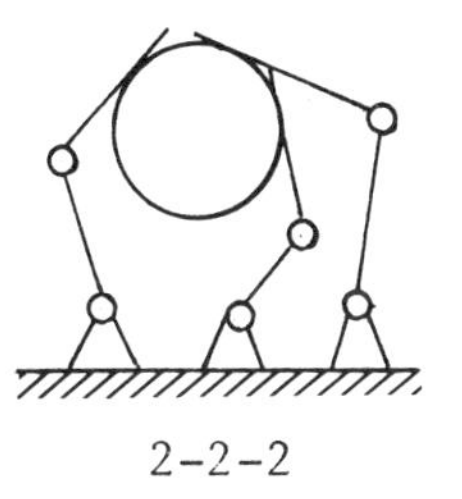

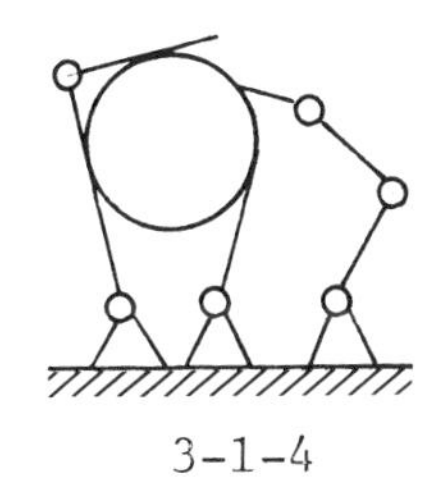

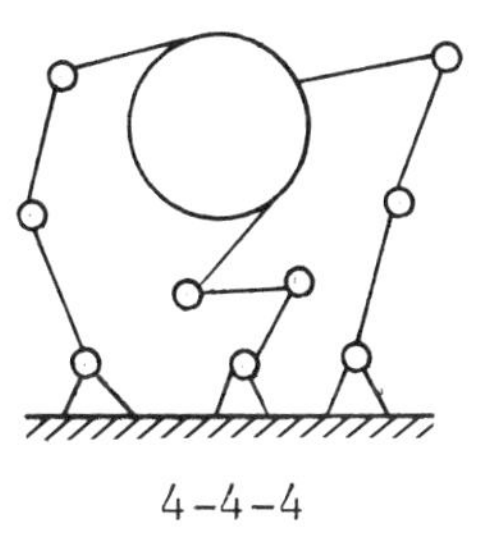

Figure 24. Examples of hand configurations.

joints, and the mobility and connectivity are complementary notions. The importance of these combinations of different configurations for our purposes is that unless we impose additional constraints, the angles of joints of the fingers are just as important information about the shape as the contact angles of the finger. Finally, we have to specify the design of the hand with respect to the anticipated load. For our purposes the load is only a secondary concern, since we shall pick up objects only for examination. Nevertheless, it cannot be ignored, especially the load that is a part of the hand, that is, the motors. Good rules for locating the motors [53] are (1) the hand should contain all its motors either in the base of the hand or in the finger units; (2) only control inputs should be able to alter the hand's prehensile patterns; (3) motors should provide rotary or linear input; (4) each motor should be reversible or double-acting and determine the position for D.O.F. of a mechanism; (5) one or more motors might be operated by a single control input when symmetrical tasks are to be performed.

5.3 Review of Different Designs of Hands with Different Applications

There have been three different applications that motivated hand development: (i) prosthetic, (ii) remote manipulation, and (iii) robotics.

Historically the first mechanical hand seemed to evolve as a prosthetic device replacing a lost hand. Indirectly we learned that in 1509 a gripping device was designed for a knight who lost his hand in battle [48]. In 1962 we heard for the first time about the Belgrade hand [59], shown in Figure 25.; subsequent reports came from the same laboratory [27,60] and a French laboratory [9,42] and from the United States [11,53,56]. Recently at the University of Utah an elaborated prosthetic hand controlled by myoelectrical impulses has been reported [26]. Almost without exception prosthetic hands have been designed to simply grasp objects, and so the dexterity of the artificial hand in itself is not that important. The control by the user–amputee comes from his or her visual source which guides the grasping process.

The second related field of applications of artificial hands has been in remote manipulations. These are the situations where one needs to work

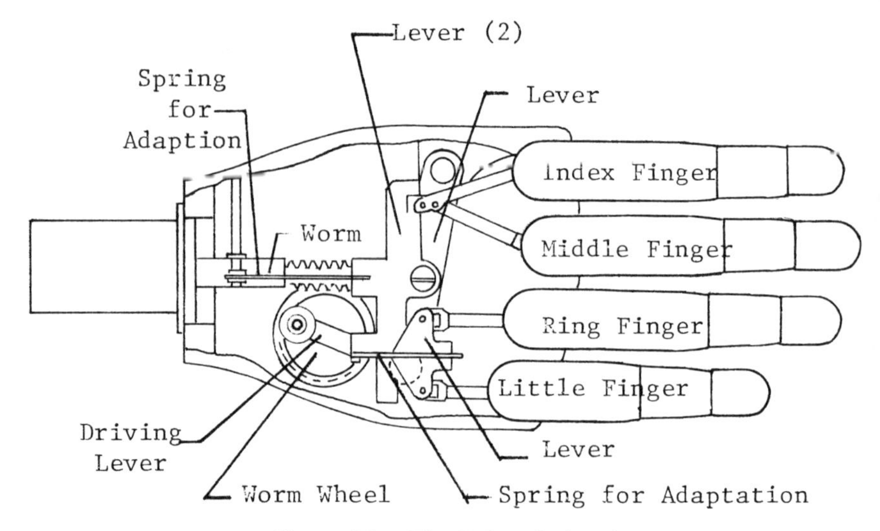

Figure 25. The Belgrade hand.

with hazardous materials such as occur in a nuclear environment or remote by distance as undersea or in outer space. Here the control is usually by man in a master–slave or teleoperator mode. Examples of these are the JPL's and others' efforts for NASA [2,53] and for robotics [22].

Okada [35–37] has reported on a versatile finger system (see Figure 26). His system is the most complex one among that generation of hands. It has three fingers with 11 D.O.F. Two fingers have four revolute joints, and the third finger has three revolute joints. The joints are driven with wire which runs inside the finger in coillike hoses. Thus it is unnecessary to set relaying points for guiding the wire at the joints of the hand, and the path of power transmission can be selected freely. It was found that fingers of circular cross section were best for 3D motions. A 17-mm-diameter free-cutting brass rod

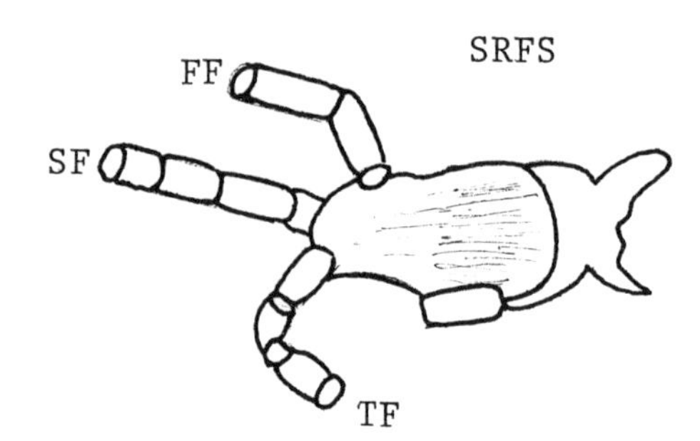

Figure 26. Okada's finger system.

was cylindrically bored. The tip of each finger was truncated at a slope of 30°. The bending and expanding of each joint was accommodated for a maximum angle of 45° bending in and out. Thus finger work space is more extensive than that of human beings. The hand is capable of rotating spheres and turning and twisting rods in the air. The control of the fingers is by a combination of position and torque control. Torque and position are both controlled by the hardware servo. A desire to design multipurpose mechanical hands in industrial robots has been pursued by Rovetta [46,47].

However, a question may be raised: Instead of a robot changing its hand for different functions, why not develop one hand which is able to successively execute the operations carried on by different interchangeable hands? This hand has N fingers; every finger, radially disposed on the palm of the hand, is performed by a linkage, and the motion of an upper rod approaches the fingertip to the center of the palm. Since every finger is "independent" of the others n of the N fingers may be used by means of a selector which actuates only n of the upper N keys. By commuting on successive sequences the actuation of the various fingers, different grasping and assembling operations may be carried out, since the fingers are differentiated for size, shape of fingertip, mechanical strength, ridigity, and pinching method. Another hand has been formed from two elements, shaped like human fingers, each of three phalanges (bones of a finger). These elements are activated by wires with springs in the joints of the phalanges. There is a plate in the middle with the function of a hand palm. The motion is obtained by means of an ac electric motor which turns a couple of pulleys. A photo-electric cell on the fingertips detects the presence of the workpiece. The hand palm can move upward, overcoming the elastic reaction of a spring; by means of a speed reducer the palm displacement is transduced to electrical equipment which introduces the amplified signal in the control system. The fingertips are interchangeable. Apparently, grasping is performed only by the fingertips of a neighboring link. A curious design has emerged from another Italian laboratory led by Nerozzi [33] (see also Figure 27) where the primary concern is how to grasp and handle delicate agricultural products like pears and peaches. Finally, the most advanced hand design published so far is that of Salisbury, shown in Figure 28. The joints and links of the manipulator having been described, the next question we must ask is how to analyze the effects of contacts between the hand and the grasped object. This is of course of primary importance for our research. That is how we gather the information about the form during its grasping.

5.4 Pattern Recognition Using Hands

The area of pattern recognition using hands is the most clearly *terra incognita*; that is, very little has been done. We have found very few references.

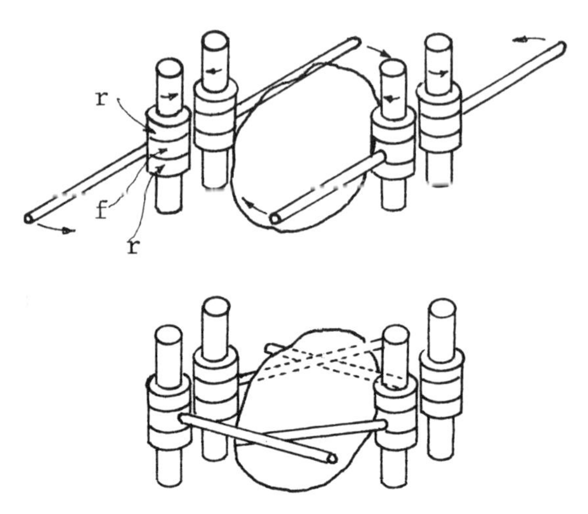

Figure 27. Operating mode of M.I.P.2 Gripper.

Figure 28. The Stanford/JPL hand.

The works reported in the literature can be divided into two categories: The first approach is the so-called bottom-up approach, where the tactile measurements are interpreted locally in terms of curved or line segments [25,34,36]. The second approach is a typical top-down approach, where there is a catalog of assumed shapes and, based on partial measurements, the shapes are inferred [26,28,29,37]. In general, the researchers in this area have concentrated on extracting edge boundaries (perhaps influenced by the work from computer vision), which is a difficult task considering the poor spatial resolution of the tactile sensors available. Hence very limited progress has been made. In 1982, Ozaki et al. [38] used a slicing approach to scan the object and then used these consecutive closed contours to recognize the object. They classify the cross section of the object into prismlike objects (with planar surfaces) and circular or elliptic cross sections. Each cross section is described as a distribution of elementary tangent vectors with unit length and angle in the local coordinate system. These vectors are called unit vectors, and their angular distribution forms the signatures for classification purposes.

The problem here is how to reconstruct the shape from sparse measurements. Various interpolation techniques must be considered, such as that of Grimson [17], who used a variation calculus for filling in data from stereo measurements and/or models of elasticity, as reported in Ref. [8,58].

6. CONCLUSIONS

In Section 1 we have outlined the motivation for this chapter. We hope that we have convinced the reader of the importance and feasibility, despite the difficulty, of using hands for recognizing shapes. We organized the chapter according to increasing complexity of various scenarios—for one finger to two fingers up to a dextrous hand or a similarly complex gripper. At the end of each section we summarized the results and formulated the state of the art. Hence here we shall make only some final remarks. From the literature it should be obvious that the most advanced area in this interdisciplinary endeavor is the mechanical design and development of the gripper and/or hand. Two-fingered grippers are commercially available devices, and there are many various mechanical hands built in laboratories. The research issues here are how to build a flexible but light weight hand, what joints to use, and how mechanically to control the joints. The less advanced area is that of building necessary sensors for a hand—primarily those for position and pressure/force. While there are commercially available position encoders, there is still a problem when accuracy and reliability are of concern. Finally, the tactile and force sensors are by and large still only made in research laboratories, and it is well known that they are inadequate in many parameters. However, progress is being rapidly made as experiments are being conducted with new pressure-sensitive materials. We believe that the least work has been done in the development of algorithms related to the hand, both in how to process the sensory data and how to computer-control a complex mechanism such as a multifingered hand. Again, some initial work in computer control has been started for grasping and manipulating, using so far simple grippers. Processing the tactile data for shape recognition has also been started, but there are still many more problems than solutions.

ACKNOWLEDGMENTS

This work could not have been accomplished without the sensor on loan provided by Mr. Jack Rebman from the Lord Corporation in Erie Pennsylvania. The other sensor that we currently experiment with is a result of the joint efforts of M. J. Clot, M. Briot, and G. Giralt from Laboratoire d'Automatique et d'Analyse des Systèmes, Toulouse, France. Finally, the work at University of Pennsylvania was supported by NSF grant MCS-81-96176. All the support is gratefully acknowledged.

REFERENCES

[1] Bajcsy, R., Brown D., Wolfeld, J., and Peters, D., What can we learn from one finger experiments? Submitted for publication to *Int. Journal in Robotics Research*, 1982.

[2] Bejczy, A. K., Effect of hand-based sensors on manipulator control performance. In *Mechanism and Machine Theory*, Vol. 12, New York: Pergamon Press, 1977, pp. 547–567.

[3] Bejczy, A. K., Sensor systems for automatic grasping and object handling. *Int. Conference on Telemanipulators for the Physically Handicapped*, Rocquencourt, France, September 4–6, 1978.

[4] Bejczy, A. K., Smart sensors for smart hands. *AIAA/NASA Conference on "Smart Sensors."* Hampton, Virginia, November 14–16, 1978.

[5] Binford, Thomas O., Sensor systems for manipulation. *Remotely Manned Systems Conference Proceedings*, 1972, p. 283.

[6] Binford, Thomas O., Preliminary work on implementing a manipulator force sensing wrist, Personal communication, 1981.

[7] Briot, M., La Stéréognosie en robotique appliation au tri de solides, Doctoral Thesis, L'Universite Paul Sabatier, Toulouse, France, 1977.

[8] Broit, C., Optimal registration of deformed images. Ph. D. Dissertation, University of Pennsylvania, Philadelphia, August 1981.

[9] Clot, J., Rabischong, P., Peruchon, E., and Falipou, J., Principles and applications of the artificial skin. *5th Intern. Symposium on External Control of Human Extremities*, Dubrovnik, August, 1975.

[10] Clot, J., and Falipou, J., Réalisation d'ortheses pneumatiques modulaires: étude d'un detecteur de pressions plantaires, *Publication LAAS* No. 1852, Toulouse, December 1978.

[11] Crossley, F. R., Erskine, and Umholtz, F. G., Design for a three-fingered hand. *Mechanism and Machine Theory, 12*, 85–93, 1977.

[12] Dane, Clayton A., III, An object-centered three-dimensional model builder. Ph. D. Dissertation in the Computer Science Department, University of Pennsylvania, Philadelphia, 1982.

[13] Dario, P., Bardelli, D., de Rossi, L. R., and Wang, P. C., Touch-sensitive polymer skin uses piezoelectric properties to recognize orientation of objects. *Sensor Review*, October, 1982.

[14] Ernst, Heinrich A., MH-1, A computer-operated mechanical hand, In 1962 *Spring Joint Computer Conference, AFIPS Conference Proceedings*, Vol. 21. Natural Press, May 1962, pp. 39–51.
[15] Flatau, C., Design outline for mini-arms based on manipulator technology. MIT Artificial Intelligence Laboratory, Memo No. 300 May 1973.
[16] Gibson, J. J., Observation on active touch, *Psychology Review*, *69*, 477, 1962.
[17] Grimson, W. E. L., *From images to surfaces: a computational study of the human early visual system.* Cambridge, MA: MIT Press, 1981.
[18] Gupta, K. C., and Roth, B., Design considerations for manipulator workspace, *Transaction of the ASME*, September, 1–8, 1981.
[19] Harmon, Leon D., Touch-sensing technology: A review, Society of Manufacturing Engineers, Dearborn, MI, MSR80-03, 1980.
[20] Harmon, Leon D., Automated tactile sensing. *Int. Journal of Robotics Research, 1*, (2), 3–32, 1982.
[21] Harmon, Leon D., Robotic taction for industrial assembly, Working Paper, December 1982.
[22] Hill, J. W., McGovern, D. E., and Sword, A. J., Study to design and develop remote manipulator system. Prepared for *National Aeronautics and Space Administration*, Ames Research Center, Moffett Field, CA, Contract NAS2-7507, 1973.
[23] Hillis, Daniel W., A high-resolution touch sensor. *Int. Journal of Robotics Research, 1*, (2), 1982, pp. 33–44.
[24] Inoue, H., Force feedback in precise assembly tasks. Artificial Intelligence Memo No. 308, MIT Artificial Intelligence Laboratory, 1974.
[25] Ivancevic, N. S., Stereometric pattern recognition by artificial touch. *Pattern Recognition*, Vol. 6, New York: Pergamon Press, 1974, pp. 77–83.
[26] Jacobsen, S. C. et al., Development of the Utah artificial arm, *IEEE Transactions on Biomedical Engineering, BME-29* (4), 1982.
[27] Jaksic, D., Mechanics of the Belgrade hand, *Proc. 3d Int. Symp. on External Control of Human Extremeties*, Dubrovnik, August 1969.
[28] Kinoshita, G., Aida, S., and Mori, M., Pattern recognition by an artificial tactile sense, *Second International Joint Conference on Artificial Intelligence*, 1971.
[29] Kinoshita, G.-I. et al., A pattern classification by dynamic tactile sense information processing. *Pattern Recognition, 7*, 243–251, 1975.
[30] Korein, James U., and Badler, N., Techniques for generating the goal-directed motion of articulated structures. *IEEE Computer Graphics and Applications, 2* (November), 71–81, 1982.
[31] Loomis, Jack M., and Lederman, Susan J., Tactual perception. In the *Handbook of Perception and Human Perception*, to appear in 1983.
[32] Mason, T. Matthew, Compliance and force control for computer controlled manipulators. *IEEE Trans. on Systems, Machine and Cybernetics, SMC-11* (6) 418–432, 1981.
[33] Nerozzi, Andrea, and Vassura, Gabriele, Study and experimentation of a multi-finger gripper. *Proc. 10th ISIR*, Milan 1980.
[34] Okada, T., On a versatile finger system. In *Proc. of 7th Int. Symposium on Industrial Robots*, October 19–21, 1977, pp. 345–352.
[35] Okada, T., Object-handling system for manual industry. *IEEE Trans. on Systems, Man, and Cybernetics, SMC-9* (2), 79–89, 1979.
[36] Okada, T., Computer control of multijoined finger system for precise object-handling. *IEEE Trans. on Systems, Man, and Cybernetics, SMC-12* (3), 289–299, 1982.
[37] Okada, T., and Tsuchiya, S., Object recognition by grasping. *Pattern Recognition, 9*, 111–119, 1977.
[38] Ozaki, H., Waku, S., Mohri, A., and Takata, M., Pattern recognition of a grasped object

by unit-vector distribution, *IEEE Trans. on Systems, Man and Cybernetics, SMC-12* (3), 315–324, 1982.

[39] Paul, Burton, *Kinematics and dynamics of planar machinery*, Englewood Cliffs, NJ: Prentice-Hall, Inc., 1979.

[40] Paul, Richard P., *Robot Manipulators: Mathematics, Prorgramming, and Control*, Cambridge, MA: The MIT Press, 1981.

[41] Purbrick, John A., A force transducer employing conductive silicone rubber. Personal communication, 1980.

[42] Rabishong, P., Stojiljkovic, Z., and Peruchon, E., Automatic control of the grasp with transducer in the finger. *Proc. 3d Int. Symposium on External Control of Human Extremeties*, Dubrovnik, August 1969.

[43] Raibert, Marc H., and Craig, J. J., Hybrid position/force control of manipulators. *ASME Journal of Dynamic Systems, Measurements and Control, 102* (June), 126–133, 1981.

[44] Raibert, Marc H., and Tanner, John E., Design and implementation of a VLSI tactile sensing computer, *Int. Journal of Robotics Research, 1* (3), 3–18, 1982.

[45] Roth, Bernard, Performance evaluation of manipulators from a kinematic viewpoint. NBS special publication on *Performance Evaluation of Programmable Robots and Manipulators*, pp. 39–61, 1975.

[46] Rovetta, Alberto, On specific problems of design of multipurpose mechanical hands in industrial robots. *Proc. 7th ISIR*, Tokyo, 1977.

[47] Rovetta, A., and Casarico, G., On prehension of a robot mechanical hand: Theoretical analysis and experimental tests. *Proc. 8th ISIR*, Stuttgart, 1978.

[48] Salisbury, Kenneth J., Kinematic and force analysis of articulated hands, Ph. D. Dissertation, Mechanical Engineering Department, Stanford University, Stanford, May 1982.

[49] Scheinman, V. D., Design of a computer controlled manipulator. Stanford Artificial Intelligence Project Memo AIM-92, Stanford University, Stanford, California, June 1969.

[50] Shah, J., A force-sensitive wrist for a hand-arm manipulator, 1972.

[51] Shimano, B., The kinematic design and force control of computer controlled manipulators. Stanford Artificial Intelligence Laboratory Memo AIM-313, March 1978.

[52] Silver, D., The little robot system. Artificial Intelligence Memo No. 273, MIT Artificial Intelligence Laboratory, 1973.

[53] Skinner, Frank, Design of a multiple prehension manipulator. *Mechanical Engineering*, September, 30–37, 1975.

[54] Soroka, Barry I., Understanding objects from slices. Ph. D. Dissertation, Department of Computer and Information Science, University of Pennsylvania, Philadelphia, 1979.

[55] Sutro, Louis L., and Kilmer, William L., R-582 assembly of computer to command and control a robot. Presented at the *Spring Joint Computer Conference* Boston, MA, 1969.

[56] Sword, A. J., and Hill, J. W., Control for prosthetic devices with several degrees of freedom. Presented to the *17th Annual Human Factors Society Convention*, Washington, D.C., October 16–18, 1973.

[57] Taylor, Craig, and Schwarz, Bob., The anatomy and mechanics of the human hand, 1955.

[58] Terzopoulos, D., Multi-level reconstruction of visual surfaces, MIT AI Memo No. 671, April 1982.

[59] Tomovic, R., The human hand as a feedback system. *Int. Federation of Automatic Control*, Moscow, 1960, vol. 2, pp. 1119 (translation by Butterworth's Scientific Publications, London).

[60] Tomovic, R. A new model of the Belgrade hand, *Proc. 3d Int. Symp. on External Control of Human Extremeties*, Dubrovnik, August 1969.

[61] Whitney, D. E., Force feedback control of manipulator fine motions. *Journal Dynamic Syst. Measurement Control*, June, pp. 91–97, 1977.

[62] Wolfson, Mark, The calibration of the finger. CSE 400 paper, University of Pennsylvania, Philadelphia, PA, May 1982.

Chapter 7

VEHICULAR LEGGED LOCOMOTION

Robert B. McGhee

1. INTRODUCTION

Up to the present time, nearly all vehicles for rough-terrain locomotion have made use of systems of wheels or tracks for support and propulsion. This is in striking contrast to the locomotion of man and cursorial animals in which this function is accomplished by a set of articulated mechanisms, constituting individual limbs capable of independently powered and flexibly coordinated motion. It has often been suggested that this difference results from an inability of nature to evolve a rotating joint and that animals have therefore been forced to adopt an "inferior" legged locomotion scheme in order to satisfy their mobility needs. A little reflection reveals, however, that the true situation is somewhat more complex. Specifically, while automotive and rail systems are clearly superior to animals for efficient long-distance transportation, this advantage applies only for movement over prepared surfaces [1]. That is, it is the synergistic combination of wheels with rails or roads that

Advances in Automation and Robotics, Volume 1, pages 259–284.

ISBN: 0-89232-399-X

produces an effective system. In fact, for off-road locomotion, the situation is quite the reverse. Whereas off-road vehicle speeds are typically limited to a few miles per hour, and power requirements are in the range of 10 hp per ton, large mammals are able to traverse the same terrain with an order-of-magnitude more speed and at a considerably reduced energy cost [2]. It is this observation which has motivated much of the research to date on the possible application of legged locomotion principles to the design of "walking machines" [3] for use either as off-road vehicles [4] or as rough-terrain robots [5,6].

Another advantage of legged locomotion is to be found in the superior mobility exhibited by animals and humans in comparison to automotive vehicles. One has only to recall personal experiences involving extrication by human beings of vehicles stalled in soft ground or on slippery surfaces to realize that the interaction of wheels or tracks with terrain is in some way fundamentally different from that of legs. The origins of this difference are to be found in soil mechanics and are, in fact, quite simple. Specifically, whereas a wheel or track sinks into a supporting surface and creates a depression which it is continuously trying to climb out of, legs produce only discrete footprints in which slippage, if any, pushes up material behind the foot, thereby improving traction [7]. A consequence of this difference is that while roughly 50% of the land surface of the Earth is inaccessable to conventional wheeled or tracked vehicles [8], nearly all of this terrain can be traversed by man and various animals.

From the above discussion, it is apparent that legged locomotion offers three potential advantages over wheels or tracks for off-road transportation: increased speed, improved fuel economy, and greater mobility. If these points are granted, then one must ask why are there no commercially viable legged vehicles? It is the author's opinion that, almost unnoticed, such machines have in fact already joined the work force and that major improvements in these vehicles are to be expected in the near future. The available machines are sometimes called "climbing hoes" [9] and have evolved gradually and unobtrusively out of conventional earth-moving and construction vehicle technology. Figure 1 shows one such machine. This development has progressed to such an extent that the most advanced models of this class of vehicles now possess up to 15 independently controllable degrees of freedom [10]. The improvements to be expected in the future arise from research in progress on microcomputer coordination of joint motions to increase agility and speed [11,12] and from anticipated innovations in mechanical power distribution leading to better fuel economy [13]. These issues are discussed in the following sections of this chapter, with the major emphasis being placed on the control asspects of limb motion coordination for vehicular legged locomotion.

Figure 1. Menzi Muck Climbing Hoe showing steep-slope operation in timber harvesting applications [73]. Pushing action by boom provides climbing power. Dirt claws on middle legs stabilize body while boom is moved to next "foothold."

2. AN OVERVIEW OF THE LIMB MOTION COORDINATION PROBLEM

The earliest solutions to the problem of limb motion coordination were of course produced by natural systems through evolutionary processes. Thus, today, a remarkable variety of animals making use of legged locomotion can be found, each specialized in such a way as to maximize its chances of survival in its particular environment. One might hope that a study of such animals would yield useful coordination algorithms for vehicular legged locomotion. Unfortunately, it has been found that the complexity of the nervous system of even very simple animals is so great that the principles of neural control of locomotion are only dimly understood. Indeed, the very possibility of a quantitative mathematical theory for control in living systems was never suggested until the appearance of Norbert Weiner's renowned book *Cybernetics*, which first introduced this word into the English language in 1948 [14]. Despite intensive worldwide research since that time, the current understanding of neural control, from both the behavioral and electrophysiological points of view, has remained largely descriptive [15–18]. Specifically, to the author's knowledge, not one complete algorithm for limb motion coordination has as yet been proposed in any publication in the world in the field of biological science. This being the case, engineers and scientists working on the limb motion coordination problem for vehicles or robots can, at present, look to biology for approaches, but not for solutions.

Recognizing the absence of an engineering theory of legged locomotion, and being motivated by the potential advantages of legged vehicles, the General Electric Company, in the time period 1965 to 1968, attempted to bypass the motion coordination problem altogether by incorporating human sensing and neural control into a quadruped vehicle using the *teleoperator* approach [3,4]. Specifically, this vehicle, called the *Quadruped Transporter*, carried with it a human operator provided with four control levers, one attached to each of his four limbs. This arrangement can be seen in Figure 2. Each of the control levers of this vehicle possessed three degrees of freedom (corresponding to the two degrees of freedom at the hip and the single

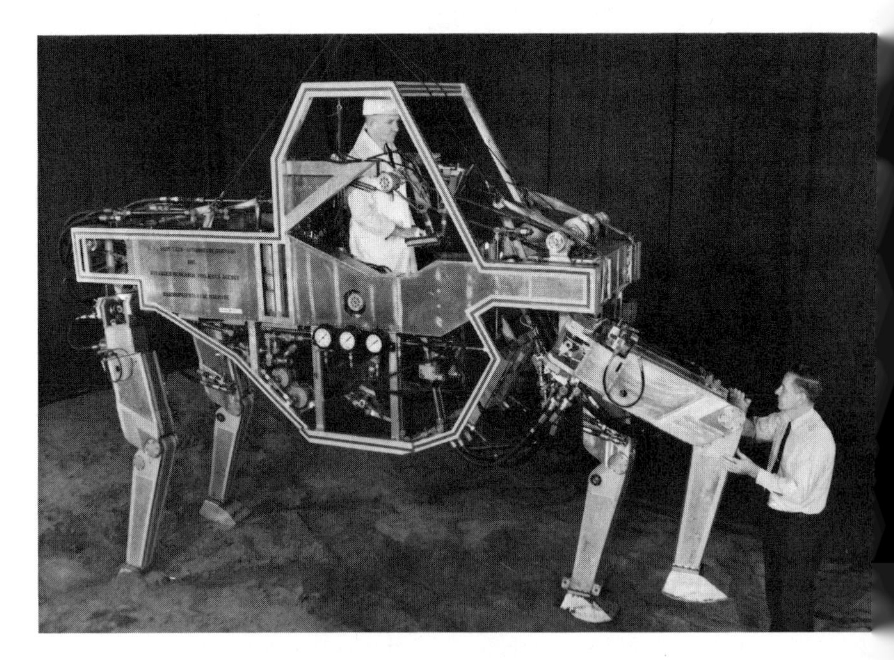

Figure 2. General Electric Quadruped Transporter showing three-axis control levers [4]. Levers for front legs are controlled by operator's hands, those for the rear legs are controlled by his feet.

degree of freedom at the knee of each leg) and amounted to a kind of mechanical computer for coordination of the joints of individual legs. Thus, to cause the vehicle to walk, the operator simply executed the desired motions with his own limbs and the machine produced an appropriately magnified replica of this input together with the necessary force amplification. In order to provide the operator with an elementary sense of ground–foot interactions, the control levers were powered, reflecting approximately 1% of the vehicle joint torques into the operator's limbs. This control scheme, derived from remote manipulator technology, can thus be said to represent a "position-following, force-feedback" strategy, sometimes referred to as *master–slave control* [3]. In the particular instance of the Quadruped Transporter, the implementation of this control was accomplished entirely through hydraulic technology, with no electrical or electronic components of any type being included in the vehicle.

The Quadruped Transporter was a research vehicle and not a machine designed for any particular application. It walked successfully in 1968 and subsequently exhibited an impressive ability to surmount obstacles and to traverse soft soils. Figure 3 illustrates one of the unusual maneuvers possible with this vehicle. As this photograph perhaps suggests, this machine in some ways amounted to a kind of mechanical elephant, at least with respect to its low-speed mobility characteristics. However, despite its obvious success in demonstrating that animal-like mobility can be achieved in a machine, experiments with this system revealed some fundamental difficulties which prevented its further development into a practical vehicle. Among the most important of these was the inefficiency of the hydraulic configuration used to provide power to the joints of each leg. Since fuel economy was not an issue for this project, a single hydraulic power supply was used to provide pressurized fluid for all actuators. This system was operated at near to maximum pressure at all times, with excess pressure at each joint being dropped across flow control valves. This configuration, while quite appropriate to a laboratory experiment, required an engine of approximately 90 hp, operating at nearly full power to achieve a speed of 5 mph which, not surprisingly, resulted in a fuel consumption rate an order of magnitude greater than that of other types of off-road vehicles.

The other major practical difficulty with the Quadruped Transporter related to the method used for limb motion coordination. While it is too soon to say that master–slave control cannot be used in legged vehicles, it does appear from the results of this experiment that such an approach places a heavy cognitive workload on the operator for a machine with 12 degrees of freedom. In fact, the majority of those who tried to control this vehicle were never able to master it. Evidently, the motor coordination skill level needed for this task is very high, much as in the control of a helicopter, except that the quadruped problem is even worse, since a helicopter possesses only 6

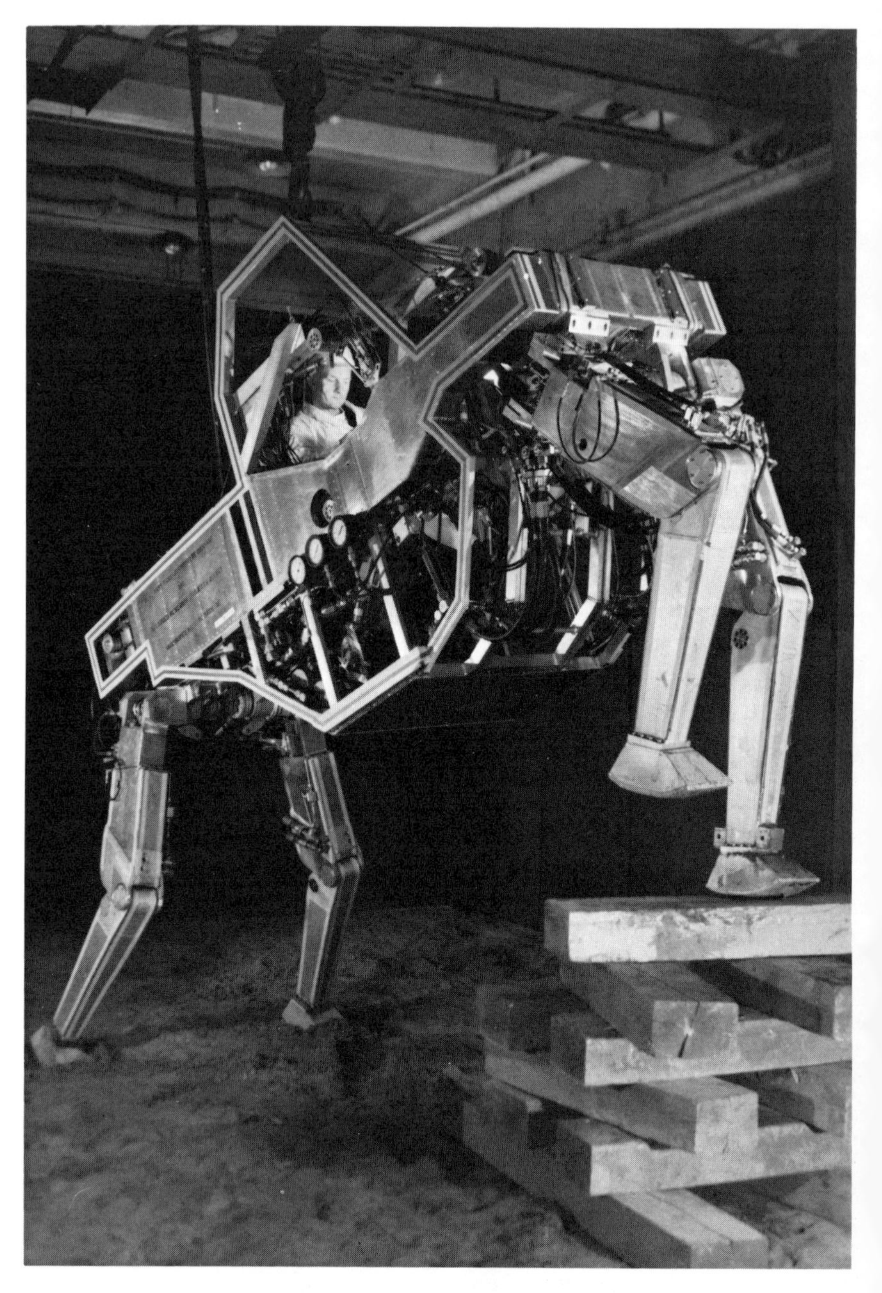

Figure 3. General Electric Quadruped Transporter showing ability to climb large obstacles with manual coordination of limb motions [4].

degrees of freedom rather than 12. Furthermore, even those who could control the Quadruped Transporter found it very demanding and were able to operate it only for a short period of time. Thus, it appears that some type of *supervisory control* [19] might be more appropriate than individual limb control for this class of vehicles. For example, it is conceivable that a human operator could control only body motion (a 6-degree-of freedom problem), leaving the coordination of individual leg motions to a computer [20,21].

Despite the practical difficulties encountered, the Quadruped Transporter project constituted a very basic experiment which opened up the entire field of vehicular legged locomotion. As noted above, work in this area has progressed to the point that there are now practical walking machines utilizing manual coordination with improved hydraulic systems [9,10]. In addition, the supervisory control concept has been developed sufficiently to permit laboratory demonstration of rough-terrain locomotion by a hexapod vehicle with operator control of speed and direction [11,12,22]. Figure 4 shows this machine. Finally, at least two fully autonomous robot vehicles employing

Figure 4. OSU Hexapod Vehicle showing ability to maintain body level while traversing rough ground. Attitude control is achieved using force feedback from feet and vertical gyro for body attitude sensing [22].

legged locomotion principles have been tested on rough terrain in a laboratory environment [6,23]. The remainder of this chapter provides a summary of this work and suggests areas for further research.

3. SUPERVISORY CONTROL

3.1 Control Modes

A fundamental issue in supervisory control concerns the way in which the control task is divided between the human operator and the computer [3,19]. For control of locomotion, this subdivision depends strongly on the nature of the terrain and upon the desired behavior of the vehicle [11,24]. For example, for routine locomotion over relatively smooth terrain, an "automotive" mode of control in which the operator merely steers the vehicle and controls its speed might be appropriate. In this style of control, the computer could use some type of terrain sensor (which could be either of a remote or contact variety) along with inertial sensors to vary each limb cycle so as to maintain the body of the vehicle at a desired attitude and altitude while minimizing shock and vibration induced by terrain roughness. A slight variation of this approach could provide for three-axis control, permitting a body "crab angle" relative to its velocity vector, much as in fixed-wing aircraft control. A suitable name for this type of control could be "cruise mode" [11].

It is possible to imagine many variations on cruise mode control. For example, if the vehicle turning radius were allowed to go to zero, and if arbitrary crab angles were to be permitted, then something more like helicopter control should be possible, with automatic regulation of altitude and attitude if desired. By analogy to some military aircraft control schemes, such a mode of operation could be called "terrain following" [25]. On the other hand, in circumstances where speed is more important than maneuverability or efficiency, it would be desirable to provide a "dash" mode in which both crab angle and minimum turning radius would be restricted sufficiently to permit optimization of limb cycles to maximize foot velocities relative to the body without concern for possible limb collisions.

The control modes discussed thus far are all literal "top-down" modes in the sense that a human operator controls the general behavior of the body of the vehicle while a computer decides what to do with the legs. Modes of this type are appropriate for use in terrain in which every point accessible to a given leg during the swing phase of its motion is suitable for load bearing—that is, terrain for which every "reachable" point is a potential "foothold" [5]. When this is not the case, then a "bottom-up" control strategy in which successive footholds are selected on the basis of both reachability and suitability for load bearing becomes necessary. The bottom-up approach can be formalized by dividing the terrain into discrete cells, each roughly the size of a footprint, and then designating each cell as *permitted* (suitable for weight bearing) or *forbidden* [5]. Various phases of leg motion can likewise be discretized into several states, including at least a supporting state, a swing state, and a "ready" state in which leg motion relative to the body is halted at an appropriate position at the end of swing phase. A leg in this state is available for transition to the support state by a suitable command from either a coordination computer or a human operator. If computer coordination is used, then evidently a coordination algorithm is required. For the degenerate case of terrain in which all cells are permitted, a preset stepping sequence can be used and the corresponding bottom-up coordination algorithm can be quite simple [26]. However, in the case that some terrain cells are forbidden, optimal coordination requires a search over the permitted reachable cells of legs in the ready state to find the best "support point" among the available footholds [5]. Such a search is likely to produce a nonperiodic sequence of leg placing and lifting events, leading to a "free" gait [5,24,27].

With the present state of the art in scene analysis and terrain sensing, it seems unlikely that a computer could reliably classify cells of natural terrain as being permitted or forbidden. Rather, it appears more reasonable to assume that human intelligence must be included in this process in some way. At the lowest level, the human operator could simply place each foot upon a selected terrain cell as in the control of climbing hoes or the G.E. Quadruped. At a higher level, the operator could designate acceptable terrain areas to the computer by using a light pen on a CRT display or by some other means. The computer could then search over the designated cells to find the one most suitable for use as a foothold, based upon considerations such as limb kinematic limits, vehicle stability, and maneuverability [5,24]. However this is accomplished, one appropriate name for such a control mode would be "precision footing."

Research on bottom-up modes has been hampered by both the unavailability of suitable sensors and by the lack of an adequate formalization of the terrain classification problem. These problems are both currently under study at Ohio State University as part of a project aimed at the eventual realization of computer control of hydraulic machines such as the Menzi Muck vehicle shown on Figure 1. As a first step toward this goal, the OSU Hexapod Vehicle shown in Figure 4 has been fitted with two CID television cameras capable of obtaining the range to a given terrain cell by means of triangulation. Figure 5 shows this arrangement. With this type of remote sensing, areas suitable for load bearing are "painted" on the terrain by a human operator using a hand-held helium–neon laser. Since each camera is provided with a narrow-band interference filter, only the laser spot is seen. The control computer's knowledge of the world is thus limited to *permitted* terrain cells. Experiments with this scheme are under way at the time of this writing [28]. Future work will include mounting the laser on a servo-driven

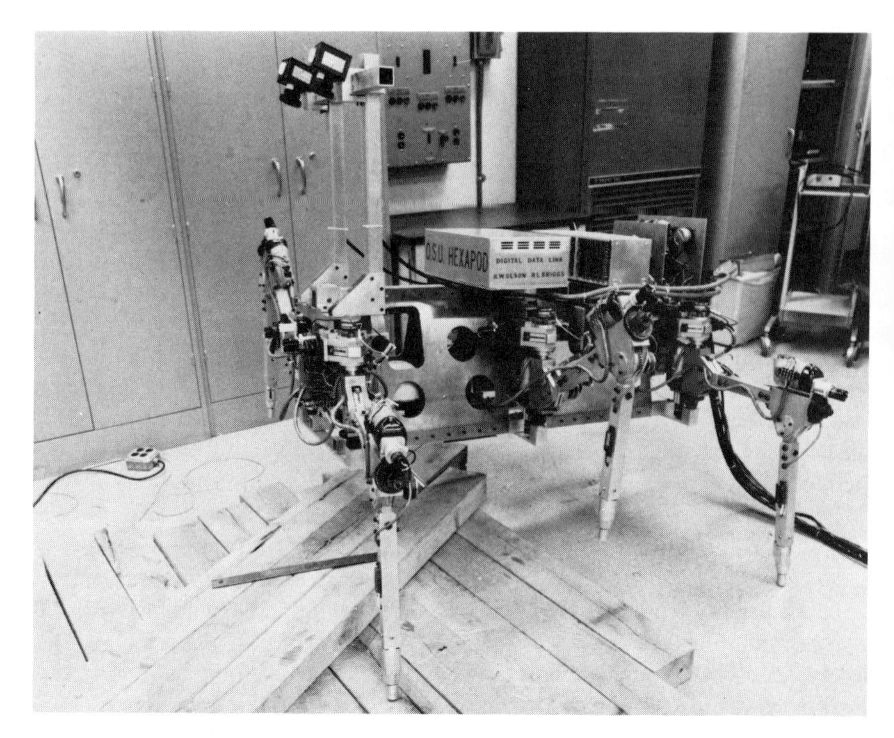

Figure 5. Modified OSU Hexapod showing addition of CID television cameras for remote terrain-sensing using laser illuminator [28].

platform on the vehicle to permit remote designation of permitted terrain cells by the human operator and, eventually, some degree of automatic terrain scanning and classification by the vehicle itself. It is presently anticipated that all such work will involve the use of range information reduced to a local elevation map of the terrain and will not depend on the alternative of two-dimensional scene analysis [29].

3.2 Motion Stability

In the precision footing style of motion coordination, a complex interaction of body and limb motions occurs in such a way as to accomplish a desired maneuver while taking maximum advantage of available footholds. In high-speed motion by cursorial animals, this process involves careful management of both potential and kinetic energy to achieve "dynamic balancing" [30]. Specifically, stabilization of motion in this manner generally involves successive fall and recovery cycles as limbs are placed upon and removed from the supporting surface. Appropriate foot placement during this process produces a type of *limit cycle stability* in which perturbations in body velocity are corrected by small changes in stride [31]. This mechanism has been studied in detail mainly for bipeds [31–34] and for monopods (hopping motion [30]), although one investigation of quadruped lateral motion stabilization by side-step control has also appeared in the literature [35].

Despite the evident success of animals and man in using control based on dynamic balancing and precision footing, the sensory, computing, and actuation problems associated with this approach to vehicular locomotion transcend both the present state of knowledge and the capabilities of current components [36]. Instead, all successful walking machines to date have made use of the principle of *static stability* in which legs are placed and lifted in such a way that the vertical projection of the center of gravity of the machine is always contained within the succession of *support patterns* determined by feet in contact with the ground [37]. Figure 6 shows an example of such a sequence for a quadruped animal or vehicle. Formally, each support pattern can be defined as a two-dimensional point set in a horizontal plane consisting of the *convex hull* of the vertical projection of all points of contact of each supporting foot [37]. Evidently, in Figure 6, the shape of each foot has been

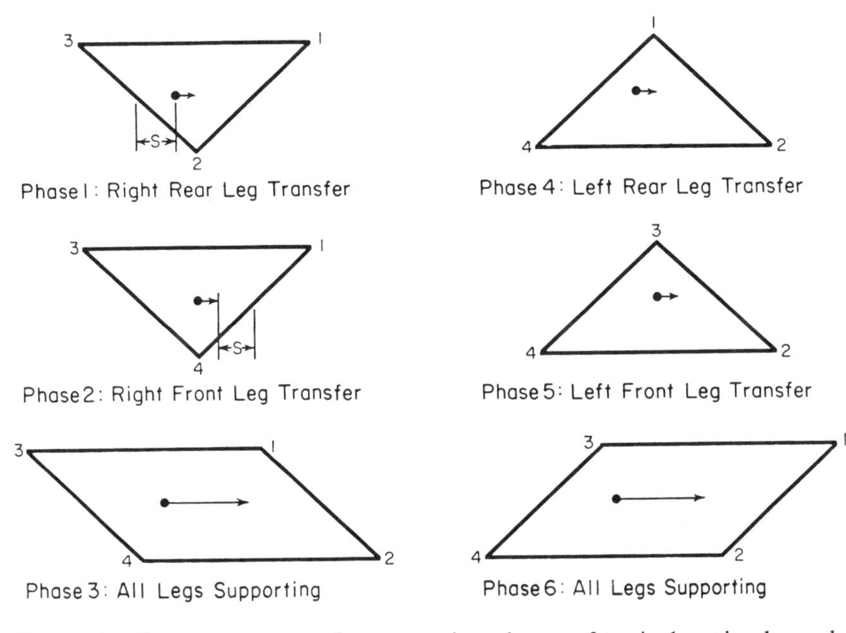

Figure 6. Support patterns for successive phases of typical optimal quadruped crawl gait illustrating static stability. Arrows indicate total motion of vertical projection of center of gravity during each phase. Longitudinal stability margin S is equal to the shortest distance over an entire cycle of gait from the center of gravity to an edge of the support pattern as measured in the direction of travel [37].

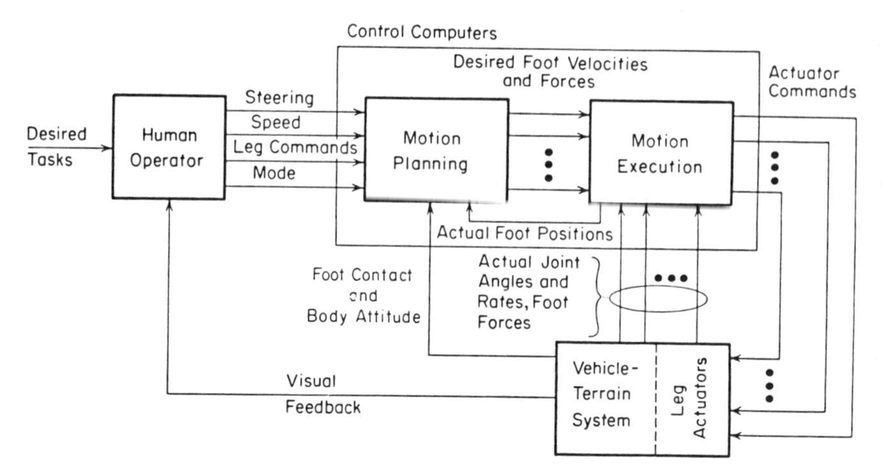

Figure 7. Supervisory control scheme used with OSU Hexapod Vehicle [11].

idealized to a point, or equivalently, to a freely rotating ball-and-socket connection between a leg and its associated foot.

To the author's knowledge, no operating walking machine with supervisory or full robotic control has as yet made use of the precision footing style of control. Rather, in every case, the desired body motion has determined the horizontal movement of legs, and it is only the vertical motion which has been made adaptive to terrain irregularities. That is, no currently operational legged vehicles have any ability to automatically detect and avoid forbidden terrain cells. This being the case, a supervisory control scheme organized as shown in Figure 7 is suitable for such machines. The balance of this section of this chapter is therefore devoted to a discussion of issues relating to this particular partitioning of the limb motion coordination problem.

3.3 The Gait Selection Problem

Referring to Figure 7, it can be seen that in this approach the motion planning block involves software which combines inputs from a human operator with feedback from vehicle sensors to synthesize both a desired velocity relative to the body and a desired terrain interaction force for each foot. Due to the kinematic complexity of walking vehicles, this is a very complicated process, involving both discrete decisions and continuous path control in a manner quite analogous to current practice in the control of industrial robots [38]. The discrete decisions required in motion planning include the problem of choosing a suitable sequence for lifting and placing of the legs, that is, of selecting a particular *gait* [39]. The combinatorial complexity of this problem turns out to be surprisingly great, even for restrictive classes of gaits. Specifically, each complete motion cycle for a given limb obviously involves one lifting event and one placing event. For an n-legged machine or animal, if every limb operates with the same cycle time, then the gait is said to be *periodic* [39]. A periodic gait evidently involves $2n$ discrete events in one complete cycle of locomotion. If the further restriction is imposed that no two of these events ever occur simultaneously, then there are $(2n)!$ event permutations possible. Since any one of these can be chosen as a reference event, the total number of distinct gaits of this type is [39]

$$N(n) = \frac{(2n)!}{2n} = (2n - 1)! \tag{1}$$

This class of gaits is called *nonsingular* [39] in reference to the absence of simultaneous occurrence of any two events in a given locomotion cycle. While animals use nonsingular gaits, they also use *singular* gaits in which certain feet are placed or lifted in synchronism with the lifting or placing of other feet. The total number of gaits including the singular ones is much greater then $N(n)$ and is in fact known only up to $n = 4$. Specifically, if $M(n)$ is the total number of periodic gaits for an n-legged locomotion system, then it can be shown [40] that

$$M(2) = 16 \tag{2}$$

$$M(3) = 704 \tag{3}$$

$$M(4) = 63{,}136 \tag{4}$$

Each of these numbers is considerably larger than that given by Eq. (1), a fact which further complicates the gait selection problem.

The larger number of possible gaits for quadrupeds produced a crisis at one point in the G.E. Quadruped development program. Shortly before initial testing of this machine, the question was raised as to whether or not any of these gaits possessed the property of static stability at all times. To resolve this issue, the notion of a minimax *longitudinal stability margin* was introduced to permit quantitative comparison of the degree of static stability all 63,136 possible quadruped gaits. Figure 6 illustrates the meaning of this criterion, which is formally defined as follows [37]:

Definition 1: The *longitudinal stability margin* S for a periodic gait G is the shortest distance over an entire cycle of locomotion from the vertical projection of the center of gravity to an edge of the support pattern as measured in the direction of travel.

Evidently, from a stability point of view, a desirable gait is one which maximizes S. Further consideration reveals, however, that it is not only the sequential aspects of gait which enter into the determination of S, but that kinematics must also be considered. That is, it matters not only *when* a leg steps, but also *where* it steps. Thus, a more complete mathematical description of a given stepping pattern is required. One such description results from the following definitions relating to periodic gaits for legged locomotion systems [37]:

Definition 2: The *duty factor* β_i is the fraction of a locomotion cycle during which leg i is in contact with the supporting surface.

Definition 3: The *relative leg phase* ϕ_i is the fraction of a locomotion cycle by which the contact of leg i with the supporting surface lags the contact of leg 1.

Definition 4: The *stride length* λ of a gait is the distance by which the center of gravity of the system is translated during one complete locomotion cycle.

Definition 5: The *dimensionless foot position* (x_i, y_i) for leg i of a legged locomotion system is a pair of coordinate values that specifies the position of the contact point of any supporting leg. The origin of the $x-y$ coordinate axes is the center of gravity of the locomotion system. The x coordinate axis is aligned with the direction of motion, with positive x directed forward. The y coordinate axis is normal to x and oriented so that it is positive on the right side of the x axis. The scale of the x and y coordinate axes is chosen so that $\lambda = 1$.

Definition 6: The *dimensionless initial foot position* (γ_i, δ_i) is the value for the pair x_i, y_i that exists at the time leg i first contacts the supporting surface during any locomotion cycle.

Definition 7: A *kinematic gait formula* k for an n-legged locomotion system is the $(4n - 1)$-tuple

$$k = (\beta_1 . \beta_2, \ldots, \beta_n, \gamma_1, \gamma_2, \ldots, \gamma_n, \delta_1, \delta_2, \ldots, \delta_n, \phi_2, \phi_3, \ldots, \phi_n) \tag{5}$$

Definition 8: The *period* τ of a periodic gait is the time required for the completion of one cycle of locomotion.

For constant-speed, straight-line locomotion over a horizontal plane, it is easily shown that a kinematic gait formula completely specifies the position of supporting feet up to a multiplicative factor of λ in space and τ in time. Moreover, if the mass of each leg is considered to be negligible in comparison to the body mass, then the position of the center of gravity is also specified to the same degree [37]. This observation motivates the following definition of a criterion function for gait selection [37]:

Definition 9: The *dimensionless longitudinal stability margin* S^* for a gait G implied by a kinematic gait formula k is defined by

$$S^*(k) = \frac{S(k)}{\lambda} \tag{6}$$

Evidently, maximization of $S^*(k)$ over all possible kinematic gait formulas yields an optimally stable gait under the stated assumptions. However, with the above formalization of the gait selection problem, this leads to a trivial solution in which, for all i,

$$\beta_i = 1 \tag{7}$$

That is, for any legged locomotion system, the most stable gait is one in which all legs are on the ground at all times. In this case, the period must be infinite and motion becomes impossible. Moreover, even if $\beta_i < 1$, for any gait formula such that $S^*(k)$ is positive, it is easily shown that $S^*(k)$ is a non-decreasing function with respect to β_i. To resolve this impasse, in Ref. [37] the following additional constraint is introduced:

$$\beta_i = \beta, \qquad i = 1, \ldots, n, \quad \beta < 1 \tag{8}$$

Such gaits are called *regular gaits* [39] in reference to the equal participation of all legs in supporting and propelling the body. Regular gaits are generally preferred by animals for routine locomotion [41–43]. With the constraint of regularity, and with β specified, optimization of $S^*(k)$ leads to a unique solution for quadrupeds. That is, out of the 63,136 possibilities, there is exactly one gait, called the "quadruped crawl," which optimizes $S^*(k)$. The resulting stability margin is [37]

$$S^*(k) = \beta - \tfrac{3}{4}, \qquad \tfrac{3}{4} < \beta < 1 \tag{9}$$

Figure 6 illustrates this gait for the case $\beta = \frac{11}{12}$. Because of its optimal stability properties, this gait was used during initial testing of the G.E. Quadruped.

The optimal quadruped crawl is an example of a *symmetric* gait [39,41] in which the motion of the legs of any right–left pair is exactly half a cycle out of phase. In 1973, Bessonov and Umnov [44] reported a generalization of the earlier work on quadruped gait optimization to encompass hexapod locomotion. Again, as for quadrupeds, it was discovered that the optimally stable gait is unique for any given duty factor. Moreover, for hexapods, statically stable gaits exist for $\frac{1}{2} < \beta < 1$, a range twice as great as for quadrupeds. The gaits which maximize $S^*(k)$ for hexapods had previously been observed in nature and are known collectively as *wave gaits* [42]. Optimally stable wave gaits involve a progression of stepping events from the back to the front of a hexapod with a phase increment $\Delta\phi$ between successive leg motions on either the right or the left side given by

$$\Delta\phi = 1 - \beta \tag{10}$$

In addition, the motion of the entire left-hand side is exactly out of phase with the right-hand side, so the optimally stable hexapod wave gaits are members of the *regular symmetric* class of gaits. Unlike quadrupeds, however, rather than a single gait, the optimal hexapod gaits exhibit six different stepping patterns, depending on the value of β. Table 1 lists each of these gaits [45].

In 1974, McGhee and Sun [46] determined the stability margin for all regular symmetric wave gaits for any even number of legs, using the phase increment given by Eq. (10). The results of this calculation are shown in Figure 8. It should be noted that the stability margin on this figure is normalized with respect to body length rather than stride length. This comes about because the optimization carried out in [37] goes beyond maximation of $S^*(k)$ and also optimizes stride length. For geometries in which leg overlapping is not possible, it turns out that, for quadrupeds, optimal stepping involves placing each foot directly behind the supporting foot just ahead of it at the time feet are exchanged. This action is evident in Figure 6 and can be carried over to the hexapod case [44,46,47], although it is not known if this is optimal. Figure 8 therefore relates λ to β as follows [37]:

$$\lambda = \frac{d}{\beta} \tag{11}$$

where d is the spacing in the fore–aft direction between the points of attachment of successive pairs of legs to the body.

Table 1. Relationship Between Duty Factor and Number of Supporting Legs for Optimally Stable Hexapod Wave Gaits [45]

Duty Factor Range	*Number of Supporting Legs*
$\beta = 1/2$	3 (tripod gait)
$1/2 < \beta < 2/3$	3 or 4
$\beta = 2/3$	4 (parallelogram gait)
$2/3 < \beta < 5/6$	4 or 5
$\beta = 5/6$	5
$5/6 < \beta < 1$	5 or 6

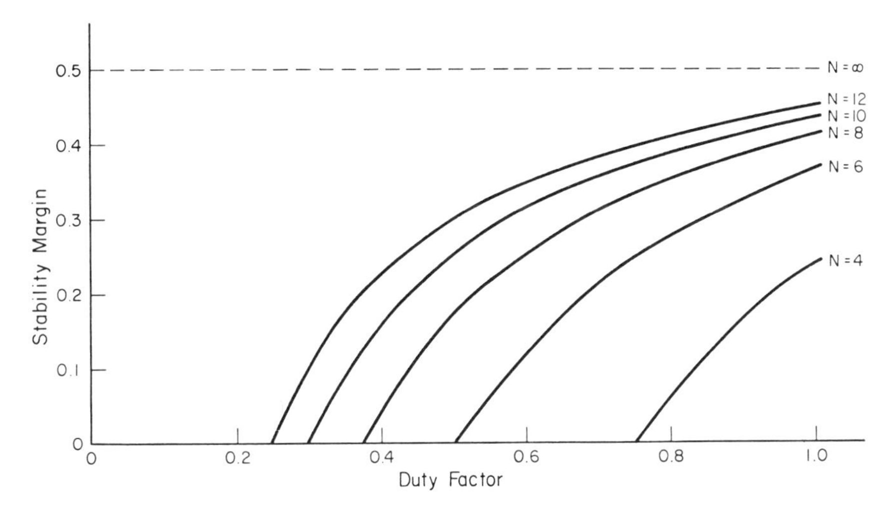

Figure 8. Optimal wave-gait stability margin as fraction of body length vs. duty factor for *n*-legged locomotion systems [26]. Legs are assumed to be arranged in evenly spaced pairs along longitudinal axis of body. Body length is considered to be distance between front and rear leg attachment points.

From Figure 8, it is evident that increasing the number of legs of a walking machine from four to six provides a great improvement both in static stability and in the allowable range of duty factors. While adding more legs produces further improvements in this regard, the change is not so dramatic, especially in view of the added cost of each additional pair of legs. For this reason, since 1974, much of the research on computer coordinated walking machines has been concentrated on hexapod vehicles.

3.4 Motion Planning

While gait optimization studies prior to 1974 involved many interesting theoretical questions, from a practical point of view legged locomotion offers little advantage to vehicles for motion over smooth, level ground. In recognition of this fact, about 1972 work began in the Soviet Union on the problem of *terrain-adaptive* gait synthesis [48,49] in which the motion of individual limbs is altered to accommodate terrain irregularities and to avoid forbidden terrain cells. Because of the known optimality of wave gaits for straight-line locomotion over flat terrain, the first approach taken in this work was to modify the spatial aspects of wave gaits to accomplish ditch crossing, obstacle avoidance, turning, and so forth. This approach was further encouraged by the then prevalent belief among biologists that insects also use this strategy [42]. Research based on the same assumptions began in the United States at about the same time, and by 1976 several detailed computer simulation studies dealing with hexapod locomotion over uneven terrain had been reported in the literature [49–51]. One quadruped simulation study of this nature was also conducted during this time period [52].

In 1977 two hexapod vehicles were completed for the purpose of laboratory investigation of the terrain-adaptive coordination algorithms developed in earlier simulation studies. The first of these, the OSU Hexapod shown in Figures 4 and 5, took its first steps in January 1977 [53,54]. The second, the Moscow University Hexapod (MGU), illustrated in Ref. [6], walked a few months later. Initially, these machines were capable of only straight-line locomotion on level ground. However, each vehicle subsequently demonstrated an ability to accomplish complex maneuvers over rough terrain under supervisory control by a human operator [6,11,55]. Research in both of these projects is now aimed principally at the incorporation of optically sensed terrain data into the motion planning aspect of supervisory control [28,56].

While details are lacking, the MGU Hexapod is controlled by a hybrid analog–digital computer. Foot trajectories relative to the body are synthesized by the digital computer as a series of straight-line segments defining six distinct phases of leg motion [6]. These trajectories are then converted to joint angle commands by analog computer "resolvers" making use of inverse Jacobian matrices to transform foot position errors into corrective joint rates. That is, the resolvers accomplish implicit function generation [57] by using the vector equation

$$\dot{\theta}_c = KJ^{-1}(x_c - x_A) \tag{12}$$

where x_c is the commanded position for any given foot relative to the body; x_A is the actual foot position; $\dot{\theta}_c$ is the vector of commanded joint rates; K is a diagonal gain matrix; and J is the Jacobian matrix given by

$$J = \left[\frac{\partial x_i}{\partial \theta_j}\right] \tag{13}$$

In this approach, the desired joint angles θ_c are obtained by analog integration of Eq. (12), thereby avoiding the difficult *reverse kinematics* problem [58] in which θ_c must be determined explicitly or numerically from x_c.

If only straight-line locomotion over level terrain were desired, then x_c could be precomputed using optimal wave gait kinematics, and the above scheme would then yield joint angles as a function of time. However, for maneuvering, it is necessary to alter x_c in response to commands from the human operator or an automatic navigation system. In the MGU machine, the underlying approach to this problem has been to "steer" the limb cycles relative to the body as if they were wheels, while retaining the basic wave-gait stepping sequence. Step length and step cycle time are also dynamically altered by the operator, or by higher-level software, in order to control vehicle speed and to avoid limb interference. For locomotion over rough terrain, limb cycles are in addition modified in the vertical direction using both body attitude and contact sensor information [6,56]. In all cases, desired foot positions relative to the body are converted to joint angle commands which provide inputs to conventional position servos controlling each joint independently.

In contrast to the MGU vehicle, the OSU Hexapod is entirely digitally controlled. The motion-planning software makes explicit use of a terrain data base acquired by optical or tactile sensing together with body attitude information, proprioceptive (joint angle) information, and operator inputs to synthesize a motion plan. Moreover, unlike the MGU machine, the OSU Hexapod motion planning is expressed in terms of desired foot velocities rather than positions. While experiments to date have used preset stepping sequences, these are not confined to wave gaits, but have also involved other periodic sequences, including gaits optimized for obstacle negotiation and ditch crossing [28]. The latter studies have been aided by recent biological research findings which show that insects also abandon wave gaits in such circumstances [59]. Finally, rather than explicitly altering limb cycles for control of speed and turning rate, the central idea in the OSU Hexapod software is the creation of an "electronic linkage" in which both the gait phase rate and foot placement at the end of swing phase are variable and are derived from operator commands regarding desired body motion [11]. Since the output of such calculations represents commanded foot velocities in body coordinates, the task of determining the corresponding joint behavior is relegated to the subsequent motion execution software block.

When legs are in contact with the ground, force feedback can be used to modify either position or velocity commands to regulate "fine motions" [60] of the feet to prevent antagonistic foot forces from developing and to equalize limb loading for weight bearing. That is, whenever a walking machine has two or more legs in contact with the ground, closed kinematic chains are established through the supporting surface. Such chains can sustain forces which cause the machine to expend energy through "isometric exercise" without contributing to the desired motion. More mathematically stated, with n supporting legs, a total of $3n$ ground reaction force components exist. On the other hand, the body of a walking machine has only six degrees of freedom, leading to six dynamical equations, each linear in the ground reaction forces [20,51]. Since at least three legs are required to obtain a stable support pattern, the problem of calculating the leg forces needed for a prescribed body motion is always underdetermined.

Two different formalisms have been proposed for dealing with the indeterminancy of the leg force problem. The first, motivated by earlier work relating to statically unstable quadruped gaits [52], assumes that the energy required to move a joint is proportional to the output work of the joint with one efficiency factor for positive work, and another for negative work [51]. With this assumption, and with limitations on the maximum torque available at any joint, it is possible to formulate the force optimization problem as a linear programming problem. This approach can be elaborated to constrain vector foot forces to the interior of the friction cone associated with a given type of surface, thereby avoiding foot slippage [51]. Unfortunately, desirable as such optimization might be for a practical vehicle, the dimensionality of this problem is very large for a hexapod and real-time solution does not seem to be feasible with computers of acceptable size and cost [61]. Instead, the current OSU Hexapod makes use of a suboptimal pseudo-inverse solution for the $3n$ force components which minimizes a weighted sum of squares of the commanded foot forces while satisfying the six equations of motion for the body [22]. While no quantitative results are as yet available for the additional energy cost of this approach, qualitative comparisons indicate that the pseudo-inverse method is entirely satisfactory and may in fact be preferable to linear programming since it does not require modeling of either actuator or terrain characteristics. Further research on this question is in progress and could eventually lead to other force allocation strategies.

3.5 Motion Execution

The lowest level of any supervisory control scheme must at last involve execution of the planned motions. This level is sometimes called the "servomechanism" level and necessarily involves the full dynamic complexity of the vehicle–terrain system. Modeling errors at higher levels finally show up as execution errors at this level. For walking machines, the inevitable coupling between the actions of individual legs requires either very accurate anticipation of such effects at the motion-planning level or some scheme for accommodating relatively larger errors during motion execution. Roughly speaking, the MGU Hexapod makes use of the first approach while the OSU Hexapod uses the latter. More precisely, the MGU control system utilizes measured forces in its motion planning software to "fine-tune" joint angle commands which are then executed by precision servomechanisms. In contrast, the OSU Hexapod software allows more local autonomy at the motion execution level through implicit control of joint angles in which force and velocity commands are treated as recommendations rather than requirements. The approach used, sometimes called "Jacobian" control, incorporates both force and position errors for any leg into the single equation [62]

$$\dot{\theta}_c = J^{-1}[\dot{x}_c + K_p(x_c - x_A) + K_f(f_c - f_A)] \tag{14}$$

where x_c is the commanded foot position derived by integration of the commanded rate [11]; x_A is the actual foot position; f_c is the commanded foot force; and f_A is the actual foot force. The values of the elements of the gain matrices K_p and K_f determine the "accommodation" [60] of force errors by position errors and, for diagonal matrices, lead to the concept of "active compliance" [62]. That is, provided commanded joint rates are accurately reproduced by precision rate servomechanisms, it can be shown [62] that for any component x of foot motion as measured in body coordinates, it follows that

$$f_{xA} - f_{xc} = -\alpha(\dot{x}_A - \dot{x}_c) - k(x_A - x_c) \tag{15}$$

where

$$k = \frac{k_p}{k_f} \tag{16}$$

and

$$\alpha = \frac{1}{k_f} \tag{17}$$

This predicted damped spring behavior has been experimentally verified both for single legs [62] and for coupled legs [22]. In the latter case, experimental validation included rough-terrain locomotion.

Current research at Ohio State University relative to motion execution is concerned with further improvements in control algorithms and sensing so as to permit local modification of swing-phase foot trajectories [63]. The approach taken in this work involves changing the motion planning software for swing-phase control so that an anticipated touchdown point is specified for each foot along with recommended velocities to attain this point with a prescribed ground clearance. The motion execution software then treat these inputs as "vectoring" commands and uses a proximity sensor on each foot both to maintain the desired midcourse altitude and to achieve a "soft landing" of the foot in the presence of motion planning errors. Successful realization of this scheme will serve to further decentralize control of individual legs and leads naturally to a multiprocessor implementation of supervisory control [12].

4. COMPONENTS

The preceding section of this chapter has dealt with issues relating mainly to control algorithms and computer software. While this is the intended focus of this contribution, it is certainly true that algorithms are of little value without hardware appropriate to their implementation. In the case of walking machines, such hardware involves not only a suitable computer, but a complex mechanical system as well. Moreover, since useful applications of such vehicles appear to be confined to extreme terrain conditions, the need for both direct and remote sensing devices is much greater than for conventional automotive systems.

With regard to computers, work on supervisory control of walking machines has generally made use of minicomputers connected to the vehicle by an umbilical cord. While this is very convenient for laboratory experimentation, it is evident inappropriate for outdoor operation of a practical vehicle. The recent emergence of 16-bit microprocessors has solved this problem, and the first self-contained walking machine with an on-board computer for supervisory control was successfully tested in 1982 [30]. A much larger machine, called an *adaptive suspension vehicle* (ASV), currently under construction at Ohio State University [63], will make use of a multiprocessor control computer similar in architecture to that described in [12], but utilizing 13 single-board computers, each including a hardware floating-point arithmetic unit. A breadboard version of this computer has been constructed, and it presently appears that a fully satisfactory on-board computer designed along these lines can be completed in time for field testing of the ASV machine in 1985.

Mechanical system design for walking machines presents many challenges, especially in the areas of limb kinematics and efficient distribution of power. Figures 1 through 5 illustrate three different solutions to these problems, each specialized to suit a particular operational environment. Based on rather limited experience, it appears that electric power is suitable for walking machines with a total weight of a few hundred pounds and with top speeds of the order of 1 mph or less. Widely spread legs similar to those of insects can be used to achieve a high degree of lateral stability in machines of this class. For walking machines exceeding about 1000 lb in weight, hydraulic power distribution [30,64,65] seems to be more advantageous than electric power. This is especially true if speeds in excess of 1 mph are desired. Finally, scaling effects seem to favor "mammalian" leg configurations for machines as large as the G.E. Quadruped, since excessive bending moments result if an attempt is made to design legs with an insect-like geometry for heavy vehicles. The ASV machine currently under construction at Ohio State University is roughly the same size as the G.E. Quadruped and will probably use legs of similar geometry, but with pantograph coordination of joint motions [23,66] in order to decouple actuator loads, thereby permitting design of specialized hydraulic circuits for drive, lift, and lateral motion of legs [30,64,65]. This approach ought to improve vehicle fuel economy and, if successful, should increase the potential range of applications for walking machines.

The terrain-sensing needs of robotic or vehicular walking machines resemble those of modern aircraft, except that the shorter distances involved seem to favor optical ranging devices rather than radar [67–70]. Current technology appears to be capable of producing the needed systems. An important feature of the ASV experiment at Ohio State University will be the evaluation of a time-of-flight optical ranging system which will scan the terrain ahead of the vehicle at about 2-Hz rate. A simpler, nonscanning proximity detector type of ranging device will also be needed for each foot if local control of swing-phase leg motion is to be achieved. Both optical and acoustic sensors are currently being studied for this purpose [63].

5. SUMMARY AND CONCLUSIONS

Serious work on terrain-adaptive vehicular legged locomotion began a little more than 15 years ago. This work initially consisted of two almost independent streams of research, one dealing with limb motion coordination and the other with mechanical design and actuation. These two lines of research have now been joined and have lead to a number of successful laboratory experiments demonstrating both fully robotic and supervisory control of walking machines operating over rough terrain. In parallel with this research, simpler manually coordinated vehicles making use of legged locomotion concepts have been developed, and several thousand of these are now in service. Research currently in progress at Ohio State University and elsewhere should soon lead to major improvements in these machines and eventually ought to extend their range of applications to such diverse areas as off-road transportation, space assembly [24], nuclear reactor inspector and servicing [72], and forestry [73]. It is hoped that this chapter will prove to be useful as a guide to the literature for those interested in such possibilities.

ACKNOWLEDGMENT

Research on vehicular legged locomotion is sponsored at Ohio State University by the Defense Advanced Research Projects Agency under Contract MDA903-82-K-0058.

REFERENCES

[1] Gould, S. J., Kingdoms without wheels. *Natural History, 90* (3), 42–48, 1981.
[2] Gabrielli, G., and Von Karmen, T. H., What price speed? *Mechanical Engineering, 72* (10), 775–781, 1950.
[3] Corliss, W. R., and Johnsen, E. G., *Teleoperator controls* NASA Report SP-5070, Washington, D.C., December, 1968.
[4] Mosher, R. S., Exploring the potential of a quadruped. SAE Paper No. 690191, International Automotive Engineering Conference, Detroit, MI, January, 1969.
[5] McGhee, R. B., and Iswandhi, G. I., Adaptive locomotion of a multilegged robot over rough terrain. *IEEE Trans. on Systems, Man, and Cybernetics, SMC-9* (4), 176–182, 1979.
[6] Gurfinkel, V. S. et al., Walking robot with supervisory control. *Mechanism and Machine Theory, 16*, 31–36, 1981.
[7] Bekker, M. G., *Introduction to Terrain-Vehicle Systems.* Ann Arbor, Mi: University of Michigan Press, 1969.
[8] Anon., Logistical Vehicle Off-Road Mobility. Project TCCO 62-5, U.S. Army Transportation Combat Developments Agency, Fort Eustis, Va., February, 1967.

[9] Anon., *Menzi Muck Climbing Hoe.* Fayettville, GA: Climbing Hoe of America, Ltd., 1981.

[10] Anon., *Kaiser Spyder Model X5M*, Industrial and Municipal Engineering Corp., Galva, I11., 1982.

[11] Orin, D. E., Supervisory control of a multilegged robot. *International Journal of Robotics Research, 1* (1), 79–91, 1982.

[12] Klein, C. A., and Wahawisan, W., Use of a multiprocessor for control of a robotic system. *International Journal of Robotics Research, 1* (2), 45–59, 1982.

[13] Waldron, K. J., and Kinzel, G. L., The relationship between actuator geometry and mechanical efficiency in robots. *Proceedings of Fourth CISM-IFToMM Symposium on Theory and Practice of Robots and Manipulators*, Warsaw, Poland, September, 8–12, 1981, pp. 366–374.

[14] Weiner, N., *Cybernetics.* New York: John Wiley & Sons, Inc., 1948.

[15] Herman, R. M. et al., *Neural Control of Locomotion.* New York: Plenum Publishing Corp., 1976.

[16] Pearson, K. G., The control of walking. *Scientific American, 235* (6), 72–86, 1976.

[17] McMahon, T. A., *Muscles, Reflexes, and Locomotion.* Cambridge, MA Harvard University, Division of Applied Sciences, 1983.

[18] Franklin, R., Bell, W. J., and Jander, R. Rotational locomotion by the cockroach blattella germanica. *Jour. of Insect Physiology, 27* (4), 249–255, 1981.

[19] Ferrell, W. R., and Sheridan, T. B., Supervisory control of remote manipulation. *IEEE Spectrum, 4* (10), 81–88, 1967.

[20] Frank, A. A., and McGhee, R. B., Some considerations relating to the design of autopilots for legged vehicles. *Journal of Terramechanics, 6* (1), 23–35, 1969.

[21] McGhee, R. B., and Pai, A. L., An approach to computer control for legged vehicles. *Journal of Terramechanics, 11* (1), 9–27, 1974.

[22] Pugh, D. R., *An Autopilot for a Terrain-Adaptive Hexapod Vehicle.* M. S. thesis, The Ohio State University, Columbus, Ohio, September, 1982.

[23] Hirose, S., and Umetani, Y., The basic motion regulation system for a quadruped walking machine. ASME Paper No. 80-DET-34, *Design Engineering Technical Conference*, Los Angeles, CA, September, 1980.

[24] Klein, C. A., and Patterson, M. R., Computer coordination of limb motion for locomotion of a multipled-armed robot for space assembly. *IEEE Trans. on Systems, Man, and Cybernetics, SMC-12* (6), 913–919, 1982.

[25] Kellington, C. M., An optical radar system for obstacle avoidance and terrain following. Report AGARD-CP-148, U.S. Army Avionics Laboratory, May, 1975.

[26] McGhee, R. B., Robot locomotion. In R. M. Herman et al. (eds.), *Neural Control of Locomotion.* New York: Plenum Publishing Corp., 1976, pp. 237–264.

[27] Kugushev, E. I., and Jaroshevskij, V. S., Problems of selecting a gait for an integrated locomotion robot. *Proc. of Fourth International Conference on Artificial Intelligence*, Tbilisi, Georgian SSR, USSR, September, 1975.

[28] Tsai, S. J., *An Experimental Study of a Binocular Vision System for Rough-Terrain Locomotion of a Hexapod Walking Robot.* Ph. D. dissertation, The Ohio State University, Columbus, OH, March, 1983.

[29] Moravec, H. P., Obstacle avoidance and navigation in the real world by a seeing robot rover. Report CMU-RI-TR3, Robotics Institute, Carnegie-Mellon University, Pittsburgh, PA, 1983.

[30] Raibert, M. H., and Sutherland, I. E., Machines that walk. *Scientific American, 248* (2), 44–53, 1983.

[31] Gubina, F., Hemami, H., and McGhee, R. B., On the dynamic stability of biped locomotion. *IEEE Trans. on Biomedical Engineering, BME-21* (2), 102–108, 1974.

[32] Vukobratovic, M., Frank, A. A., and Juricic, D., On the stability of biped locomotion. *IEEE Trans. on Biomedical Engineering, 17* (1), 25–36, 1970.
[33] McGhee, R. B., Computer simulation of human movements. In A. Morecki (ed.), *Biomechanics of Motion.* New York: Springer-Verlag, 1980, pp. 41–78.
[34] Formalskii, A. M., *Motion of Anthropomorphic Mechanisms.* Moscow: Nauka Press, 1982 (in Russian).
[35] Hemami, H., and Lee, Y. H., Stabilization of a quadruped locomotion system by side step control. *Journal of Terramechanics, 12* (2), 1975.
[36] McGhee, R. B., Robot locomotion with active terrain accommodation. *Proc. of NSF Robotics Workshop*, University of Rhode Island, Newport, Rhode Island, April, 1980.
[37] McGhee, R. B., and Frank, A. A., On the stability of quadruped creeping gaits. *Mathematical Biosciences, 3* (3), 331–351, 1968.
[38] McGhee, R. B., Future prospects for sensor-based robots. In G. G. Dodd and L. Rossal (eds.), *Computer Vision and Sensor-Based Robots.* New York: Plenum Publishing Corp., 1979, pp. 323–334.
[39] McGhee, R. B., Some finite state aspects of legged locomotion. *Mathematical Biosciences, 2* (1), 67–84, 1968.
[40] Koozekanani, S. H., and McGhee, R. B., Occupancy problems with pairwise exclusion constrains—An aspect of gait enumeration. *Journal of Cybernetics, 2* (4), 14–26, 1972.
[41] Hildebrand, M., Analysis of the symmetrical gaits of tetrapods. *Folia Biotheoretica, 4*, 9–22, 1966.
[42] Wilson, D. M., Insect walking. *Annual Review of Entomology, 11*, 103–121, 1966.
[43] Wilson, D. M., Stepping patterns in tarantula spiders. *Journal of Experimental Biology, 47*, 138–151, 1967.
[44] Bessonov, A. P., and Umnov, N. V., The analysis of gaits in six-legged vehicles according to their static stability. *Proc. of CISM-IFToMM Symposium on Theory and Practice of Robots and Manipulators*, Udine, Italy, September, 1973.
[45] McGhee, R. B., Klein, C. A., and Chao, C. S., Interactive computer control of an adaptive walking machine. *Proc. of 1979 Midcon Conference*, Chicago, November, 1979.
[46] McGhee, R. B., and Sun, S. S., On the problem of selecting a gait for a legged vehicle. *Proc. of VI IFAC Symposium on Automatic Control in Space*, Armenian SSR, USSR, August, 1974.
[47] Sun, S. S., *A Theoretical Study of Gaits for Legged Locomotion Systems.* Ph. D dissertation, The Ohio State University, Columbus, OH, 1974.
[48] Okhotsimski, D. E., and Platonov, A. K., Control algorithm of the walker climbing over obstacles. *Proc. of the Third International Joint Conference on Artificial Intelligence*, Stanford, Calif., August, 1973.
[49] Okhotsimski, D. E., and Platonov, A. K., Walker's motion control. *Proc. of Second CISM-IFToMM Symposium on Theory and Practice of Robots and Manipulators*, Polish Scientific Publishers, Warsaw, 1976, pp. 221–230.
[50] Orin, D. E., *Interactive Control of a Six-Legged Vehicle with Optimization of Both Stability and Energy.* Ph. D. dissertation, The Ohio State University, Columbus, Ohio, 1976.
[51] McGhee, R. B., and Orin, D. E., A mathematical programming approach to control of joint positions and torques in legged locomotion. *Second CISM-IFToMM Symposium on Theory and Practice of Robots and Manipulators*, Polish Scientific Publishers, Warsaw, 1976, pp. 225–232.
[52] Park, W. T., and Fegley, K. A., Control of a multi-legged vehicle. *Proc. of V IFAC Symposium on Automatic Control in Space*, Genoa, Italy, June, 1973.
[53] McGhee, R. B., Control of legged locomotion systems. *Proc. of Eighteenth Joint Automatic Control Conference*, San Francisco, Calif., June, 1977, pp. 205–215.

[54] Jaswa, V. C., *An Experimental Study of Real-Time Computer Control of a Hexapod Vehicle.* Ph. D. dissertation, The Ohio State University, Columbus, Ohio, June, 1978.

[55] Okhotsimski, D. E., Devyanin, E. A., and Gurfinkel, V. C., *Model Six-Legged Walking Apparatus with Supervisory Control.* Moscow: Institute of Mechanics, Moscow University Press, 1978 (in Russian).

[56] Devjanin, E. A. et al., The six-legged walking robot capable of terrain adaptation. *Proc. of Fourth CISM-IFToMM Symposium on Theory and Practice of Robots and Manipulators,* Warsaw, Poland, Sept. 8–12, 1981, pp. 375–384.

[57] Levine, L., *Methods for Solving Engineering Problems Using Analog Computers.* New York: McGraw-Hill, 1964.

[58] Orin, D. E., and Oh, S. Y., Automated motion planning for articulated mechanisms. *Proc. of National Electronics Conference,* Chicago, IL, October 16–18, 1978, pp. 174–179.

[59] Pearson, K. G., *Cinematographic Analysis of Animal Walking.* Contract RF714250-01, University of Alberta, Edmonton, Alberta, Canada, September 30, 1982.

[60] Whitney, D. E., Force feedback control of manipulator fine motions. *Proc. of 1976 Joint Automatic Control Conference,* Purdue University, West Lafayette, IN, July, 1976, pp. 687–693.

[61] Chao, C. S., *A Software System for On-Line Control of a Hexapod Vehicle Utilizing a Multiprocessor Computing Structure.* M. S. thesis, The Ohio State University, Columbus, Ohio, 43210, August, 1977.

[62] Klein, C. A., and Briggs, R. L., Use of active compliance in the control of legged vehicles. *IEEE Trans. on Systems, Man, and Cybernetics, SMC-10* 7, 393–400, 1980.

[63] Broerman, K. R., *Development of a Proximity Sensor System for Control of Foot Altitude During Locomotion of a Hexapod Robot.* M. S. thesis, The Ohio State University, Columbus, Ohio, 43210, June, 1983.

[64] Waldron, K. J., Frank, A. A., and Srinivasan, K., The use of mechanical energy storage in an unconventional, rough-terrain vehicle. *Proc. of 17th Intersociety Energy Conversion Engineering Conference,* Los Angeles, August 8–13, 1982.

[65] Srinivasan, K. et al., The design and evaluation of a hydraulic actuation system for a legged rough-terrain vehicle. ASME Winter Annual Meeting, Phoenix, Arizona, November 14–16, 1982.

[66] Kessis, J. J., Rambaut, J. P., and Penne, J., Walking robot multi-level architecture and implementation. *Proc. of Fourth CISM-IFToMM Symposium on Theory and Practice of Robots and Manipulators,* Warsaw, Poland, Sept. 8–12, 1981, pp. 374–355.

[67] Albus, J. S., *Brains, Behavior, and Robotics.* New York: McGraw-Hill, 1981.

[68] Shen, C. N., and Kim, C. S., A laser rangefinder path selection system for martian rover using logarithmic scanning scheme. *Proc. of IFAC Symposium on Automatic Control in Space,* Oxford, U. K., July 29, 1979.

[69] Paine, G., The automation of remote vehicle control. *Proc. of 1977 Joint Automatic Control Conference,* San Francisco, June 22–24, 1977, pp. 216–224.

[70] Nitzan, D. A., Brain, A. E., and Duda, D. O., The measurement and use of registered reflectance and range data in scene analysis. *Proc. of IEEE, 65* (February), 206–220, 1977.

[71] Ozaki, N. et al., Inspection and maintenance of fission reactors. *Proc. of 1982 Annual Meeting of American Nuclear Society,* Los Angeles, June 6–10, 1982, p. 637.

[72] Anon., *Komatsu Seabed Robot* Tokyo, Japan: Komatsu, Ltd., 1982.

[73] Arola, R. A. et al., *Felling and bunching small timber on steep slopes.* Research Paper NC-203, North Central Forest Experiment Station, U.S. Dept. of Agriculture, Houghton, Michigan, 1981.

SUBJECT INDEX

Adaptive controls, 70, 91–112
  using autoregressive model, 95–97
Adaptive perturbation control, 97–106
Adaptive suspension vehicle, 280
Advances in Automation and Robotics, ix
Approach vector of hand, 41
Approximation algorithms, 8
Arm matrix, 40–48
Artificial intelligence, 4–6
Automation, ix–x
Autoregressive model, adaptive control using, 95–97

Base coordinates, 36
Belgrade hand, 252
Binary vision methods, 142–144, 147–175
  decisions based on region features, 171–175
  high-contrast images, 148–155
  region analysis, 160–171
  threshold methods, 155–160
Bone structures of human hand, 242–243
Boundary features, 166–167
Bounding rectangle, 169–170

Cartesian arm, 22
Cartesian motion, 122–123
"Chain product" rule, 44
Cognitive systems, 4
Color discrimination, 152
Compliance
  passive, 142
  specifying, 133–134
Computation, real-time, 117–118
Computed torque technique, 72–74
Computer-controlled manipulators, 10–17
Contact configurations, 249
Containment tree, 165
Control(s)
  adaptive, 92–112
  intelligent, 2–3, 4
  robot arm (*see* Robot arm control)
Control modes, locomotion, 266–268
Convolution value, 176
  zero crossings in, 182
Coordinate frames, 36
Coordinate representation,

homogeneous, 29–33
Coordinate systems, 136–137
Coordination level, 6
task, 14–16
Cylindrical arm, 22

d'Alembert equations of motion, generalized, 56–60
Data structures, transform equation, 127–131
Decision schema, linguistic, 7–10
Decision tree method, 171–172
Degrees of freedom, 246, 248–252
Denavit-Hartenberg representation of linkages, 25, 36–40
Directional illumination, 153–155
Dynamic coefficients of manipulators, 53
Dynamics, robot arm, 49–60, 66–69

Edge detectors, 175–184
model-driven, 184–187
Energy calculation
kinetic, 51–52
potential, 52
Euler-Lagrange equations, 13

Force sensing, 211
Forgetting factor, 96
Form perception, three-dimensional, 210–211
French finger, 220–231

Gait formula, kinematic, 272
Gait selection problem, 270–275
Gaussian function, 178
Geometry of triangulation, 192–193
Global scaling factor, 30
Graph matching, 172–175
Gray-level methods, 144–145, 175–191
Grippers, 209

Hand coordinate system, 82–85
Hand designs with different applications, 251–253
Hand-eye system, 120–122
Hand movements, 246
Hand patterns, resting, 244, 245
Hand scenario, 210, 242–255
Hand subsystem, 13
Hands
human, anatomy of, 242–246
mechanical, design considerations for, 246–251
pattern recognition using, 253, 255
prosthetic, 251
Hardware control level, 6–7
High-contrast images, 148–155
High-level programming languages, 123–135
Histograms, 155–160
Hysteresis curve, 224

Illumination
directional, 153–155
laser, 151
multiple-stripe, 200–202
multiple-wavelength, 203–207
scattered, 188
secondary, 199–200
structured, 197–202
ultraviolet, 152–153
Image intensity, surface position and, 187–191
Image processing, 132
Industrial automation, ix–x
Industrial robots (*see* Robots, industrial)
Input language, 8
Integration time, 194
Intelligence
artificial, 4–6
increasing, with decreasing precision, principle of, 12, 17
machine, 2–17

Intelligent control theory,
  hierarchically, 5–10
Intelligent controls, 2–3, 4
Inverse kinematics solution, 44–49

"Jacobian" control, 278–279
Joint axis, 34
Joint-link pairs, 34
Joint motion controls, 69
Joint parameters, 35
Joint variable, 37

Kinematic equations for manipulators, 40–44
Kinematic gait formula, 272
Kinematics
  inverse solution, 44–49
  robot arm, 24–49
Kinetic energy calculation, 51–52

Lagrange-Euler equations of motion, 52–53
Lagrange-Euler formulation, 51–53, 67–68
Laser illuminators, 151
Legged locomotion, vehicular, 259–281
Length of link, 35
Limb motion coordination, 261–266
Line contact, 237–239
Linguistic decision schema, 7–10
Link coordinate system assignment, 38–39
Link coordinate systems, 34–40
Link paameters, 35
Links, 35
Locomotion, vehicular legged, 259–281
Longitudinal stability margin, 271, 273

Machine intelligence, 2–17
Manipulative systems, 3
Manipulator control, 65–113
Manipulator trajectory, 121
Manipulators, xi, 65–66
  computer-controlled, 10–17
  dynamic coefficients of, 53
  kinematic equations for, 40–44
  programming, 117–138
  remote, 251–253
Mask function, 176–178
Master-slave control, 263
Menzi Muck Climbing Hoe, 261
M.I.P.2 Gripper, 254
Model-driven edge detection, 184–187
Model referenced adaptive control, 93–95
Moscow University Hexapod, 276–278
Motion, functionally defined, 134–135
Motion-defining categories, 23
Motion execution, locomotion, 278–279
Motion planning, locomotion, 275–278
Motion stability, locomotion, 268–270
Motor control processor, 216–217, 218–219
Motor interface unit, 216
Multiple-stripe illumination, 200–202
Multiple-wavelength illumination, 203–207
Muscles of human hand, 244

Near-minimum-time control, 74–76
Newton-Euler formulation, 54–56
Nonlinear feedback control, 78–81
Nonsingular gaits, 271
Normal vector of hand, 41
Numerical control machines, x

Okada's finger system, 252–253
One-finger scenario, 210, 213–231

Organization level, 6
task, 16–17
Orthogonal transformation, 27
OSU Hexapod Vehicle, 265, 268, 270, 276–278
Output language, 8

Pad sensors, 232–233
Parallax angle, 193
Pattern recognition using hands, 253, 255
Performance index, 13–14
Pixel neighborhood, 176
Pixels, 147
Plane contact, 237–239
Point contact, 237–239
Position vector, 30
of hand, 41
Potential energy calculation, 52
Precision, decreasing, principle of increasing intelligence with, 12, 17
Prehension patterns, 244–246
Programming, manipulator, 117–138
Programming languages, high-level, 123–135
Prosthetic hands, 251
PUMA robot arm control strategy, 70–72

Quadruped Transporter, 262–265

Range sensor, 203–204
Real-time computation, 117–118
Region analysis, 160–171
Region features, 164–171
decisions based on, 171–175
Regular gaits, 273
Remote manipulators, 251–253
Resolved motion, 81–82
Resolved motion accelerated control, 87–89
Resolved motion adaptive control, 106–112
Resolved motion control, 70, 81–92
Resolved motion force control, 90–92
Resolved motion rate control, 85–87
Robot arm categories, 21–24
Robot arm control, 69–92
resolved motion control, 81–92
Robot arm dynamics, 49–60, 66–69
Robot arm kinematics, 24–49
Robotic manipulators (*see* Manipulators)
Robotic vision, 119, 141–207 (*see also* Vision)
Robotics, intelligent control for, 1–17
Robots, industrial, x–xi, 118–119
defined, x, 23
sensor-controlled, 119, 120
Rotation matrices, 25–29
homogeneous, 31–33
Rotational motion, 58

Scale factor, 30
Scattered illumination, 188
Scheinman wrist, 211
Screw, defined, 238
Secondary illumination, 199–200
Segmentation methods, 189–191
Sensor-controlled robots, 119, 120
Sensors, 211–213
homogeneity among, 220
pad, 232–233
range, 203–204
repeatability of, 233–234
time response of, 221–223
touch, 213
Servo mode, 133
Shape
analysis of, 235–240
determining, from touch, 209–256
Silhouette generation, 148–151
Singular gaits, 271
Sliding vector of hand, 41

Software organization, 131–133
Spherical arm, 22
Stanford/JPL hand, 254
State space representation, 68–69
Static stability, 269
Stride length, 272
Structured illumination, 197–202
Supervisory control, locomotion, 266–279
Surface description, 241–242
Surface information acquisition, 240
Surface position, image intensity and, 187–191
Surface roughness, 228–231
Symmetric gaits, 273–274

Tactile image analysis, static, 233–235
Tactile interface unit, 216
Tactile sensing processor, 216–217, 218
Task description, 124–126
Tendons of human hand, 244
Term record, 127
Terrain-adaptive gaits, 275–278
Texture detection, 225–226, 227
Three-dimensional form perception, 210–211
Three-dimensional vision, 145–147, 191–207
Threshold methods, 155–160
Tomek method of linear segmentation, 171
Touch, determining shape from, 209–256
Touch sensors, 213
Transform equation data structure, 127–131
Transformation matrix
  D-H, 38
  homogeneous, 30–34
Transforms, 126–127
Translation schema, 8–10, 17
Translational motion, 58
Triangulation geometry, 192–193
Triangulation implementation, 193–197
Triangulation principle, 193
Twist angle of link, 34
Twist motion, 238
Two-dimensional array, 199
Two-finger scenario, 210, 232–242

Ultraviolet illumination, 152–153
Unimation PUMA 600 arm, 10–17

Variable structure control, 76–78
Vectors, 126, 136
Vehicular legged locomotion, 259–281
Vision
  approaches to, 142–147
  binary methods (*see* Binary vision methods)
  gray-level methods, 144–145, 175–191
  high-contrast, 148–155
  robotic, 119, 141–207
  three-dimensional, 145–147, 191–207
Vocabulary optimal decision schema, 8–9

Walking machines, 279–281
Wave gaits, 273–274
WAVE system, 122
Wavelength discrimination, 151–153
Work volume, 22
Wrench axis, 238
Wrist subsystem, 13

X-Y-X arm, 22

Zero crossings in convolution value, 182